Well Met!
Friends and Travelling Companions of Rev. Thomas Bowles

Well Met!

Friends and Travelling Companions of Rev. Thomas Bowles

Journals of Travels in Egypt, Petra and the Near East, 1854

David Kennedy

ARCHAEOPRESS

Archaeopress Publishing Ltd
Summertown Pavilion
18-24 Middle Way
Summertown
Oxford OX2 7LG
www.archaeopress.com

ISBN 978-1-80327-483-6
ISBN 978-1-80327-484-3 (e-Pdf)

© David Kennedy and Archaeopress 2023

All rights reserved. No part of this book may be reproduced, or transmitted, in any form or by any means, electronic, mechanical, photocopying or otherwise, without the prior written permission of the copyright owners.

This book is available direct from Archaeopress or from our website www.archaeopress.com

Contents

List of Figures iv

List of Tables vi

Preface and Acknowledgements vii

Introduction ix

Part 1 The Grand Tour

Chapter 1. Rev. Thomas Bowles: Biography 3
   Early Life and Education of Thomas Bowles 6
   The Rev. Thomas Bowles 12
   Nothing known - perhaps the following item? 13
   Retirement and Death 16
   Religious Views of Thomas Bowles 20

Chapter 2. Thomas Bowles's Grand Tour 21
   Bowles's Travelling Companions from Plymouth to Australia and Beyond 22
      Rev. Henry Francis Stobart (Chester-le-Street, 26 April 1824 – Funchal, Madeira, 30 December 1895) 22
      Lord Henry Scott (5 November 1832 - 4 November 1905) 23
      Lord Schomberg Kerr (2 December 1833 – 17 January 1900) 24
      Servants 24
   The Journey 24

Chapter 3. Sailing from Galle via Aden to Suez and Cairo 28
   Introduction 28
   S. S. Bombay 29
   Aden 30
   Suez to Cairo 31
   Shepheard's Hotel 35

Chapter 4. Travelling Companions and Encounters from Galle to Cairo 37
      Rev. William George Tupper 37
      Stephen Lushington 39
      Dr Patrick Gammie 40
      Walter Scott Campbell Seton-Karr 44
      Rev. George Percy Badger 45
      Rev. Charles Tombs 49
      (Later) Major-General Sir Henry Tombs, VC, KCB 49

Chapter 5. Thomas Bowles in Egypt 51
Western visitors in Egypt and the Near East in the 19th century............................. 51
Motivations 51
Tourism.. 51
Health 52
Bowles Tourism 52
Companions in Cairo 55
    George Strachan 55
    William Palmer.. 56
    Alfred Henry Pierpont Edwards 57
    Mr and Mrs Samuel Briggs 59
    Rev. Pierce Butler 60
    Capt. Henry Thomas Butler... 60
    Antonio Schranz 64

Chapter 6. Petra and Visiting Petra in the 19th century: 66
Petra.. 66
Handbooks 70
Logistics of Travelling to Petra 71
The Journey 74
Conditions of Travel 78

Chapter 7. Companions on the Long Desert Route: 1 80
Travelling Companions 82
    Henry Hyndman Kennedy 82
    Samuel Greame Fenton 85
    Rev. George Metcalfe Fenton 85
    Mr John Leverett Rogers... 87
    Mrs Virginia Beverley Rogers (née Wood) 87
    Henry Veazey Ward 92
    Henry Clay Yeatman... 96

Chapter 8. Companions on the Long Desert Route: 2 102
    Rev. James Henry Eames 102
    Jane Anthony Eames 102
    'Mr F[ish] from Alabama'... 106
    Charles Rodewald = Carl Reinhard Conrad Rodewald 107
    Robert Ross-of-Bladensburg 110
    Dr William Bryce 114
    John Edward Wakefield 119
    'Mr Freeman' 120

Chapter 9. Thomas Bowles at Petra 123
The Alawin 124
The Inhabitants of the Ruins.. 126
Thomas Bowles inside Petra.. 130

Chapter 10. Jerusalem and Associated Trips 141
Palestine 141
James Finn 144
Elizabeth Finn (née McCaul) 144
Colonel and Mrs West 147
Edward Heneage 149
Renee Elizabeth Levina (née Hoare) Heneage 149
James Graham 151
Caroline Cooper 152
Matilda Creasey 152
Rev. John Wheeler Hayward 155

Chapter 11. Palestine, Syria and Lebanon and Return Home 158
Lucius Bentinck Cary 159
Amelia FitzClarence, Lady Falkland 159
General the Honourable Sir St. George Gerald Foley, K.C.B. 162
Nelson Rycroft 164
Arthur Lannay Coussmaker 166

Part 2 The Journals

Chapter 12. The Journals: Character, writing, composition, survival 171

Chapter 13. Abbreviations and Symbols; Editing 174

Chapter 14. Transcription of the Travel Journals of Rev. Thomas Bowles 176
Travel Journal 2 176
Travel Journal 3 189

Appendices 291
Appendix 1: The Anglo-American Group at Petra 26-29 March 1854. ... 291
Appendix 2: Timeline of Bowles' Journey in Egypt and the Levant. 292

Bibliography 295
Unpublished Sources 295
Published Sources 295

List of Figures

Figure 1.1:	A photograph of Rev. Thomas Bowles displayed in the Vestry of the Church of St Augustine of Canterbury at East Hendred where he was Rector for fifteen years.	3
Figure 1.2:	Family tree for the Rev. Thomas Bowles (Drawn: Julie Kennedy-Pugh).	4
Figure 1.3:	Milton Hill House. Although significantly remodelled not least to its current condition as a hotel, the family home of Thomas Bowles is largely the white building. It is set still in a large park and includes several magnificent cedar trees (Photo: David Kennedy).	5
Figure 1.4:	Church of St Blaise, Milton (Photo: David Kennedy).	5
Figure 1.5:	Monument to members of the Bowles family in the churchyard of St Blaise, Milton. Although extensively inscribed with family names few are readable without cleaning (Photo: David Kennedy).	6
Figure 1.6:	Monument and detail of the inscription to Thomas Bowles in the churchyard of St Andrew's, Grafham, Surrey.	15
Figure 1.7:	Church of St Augustine of Canterbury at East Hendred (Photo: David Kennedy).	17
Figure 1.8:	The Rectory at East Hendred where Thomas Bowles lived from 1875-1890 (Photo: David Kennedy).	17
Figure 1.9:	Church of St Barnabas, Jericho, Oxford (Photo: David Kennedy).	18
Figure 1.10:	TBs Oxford home at 24 Leckford Road, Oxford (Photo: David Kennedy)	19
Figure 1.11:	Grave monument of Rev. Thomas Bowles in the Old Cemetery at Abingdon. The end panel records that it is also the resting place of his nephew Francis Wildman Selwode (sic) Bowles in 1940 (Photo: David Kennedy).	19
Figure 3.1:	A stereophoto of 1862 of the S. S. Bombay in Mort's dry-dock in Balmain, Sydney (Mitchell Library, State Library of New South Wales) (Public Domain).	29
Figure 3.2:	Port of Aden, Arabia in 1891 (From F. E. Chadwick et al. 1892: 272) (Public Domain).	31
Figure 3.3:	'Desert Van' on the Suez-Cairo route (from Barber 1845: 34) (Public Domain).	32
Figure 4.1:	Indian Mutiny Medal of Dr Patrick Gammie (Courtesy David Galt, Mowbray Collectables www.mowbraycollectables.com).	42
Figure 4.2:	Grave of Patrick and Mary Gammie, Brompton Cemetery, London (Photo: David Kennedy).	43
Figure 4.3:	George Badger - back row, right, with long white beard, with the Frere Mission to Zanzibar while in Cairo on 22nd December 1873 (Coupland 1939: facing p. 186).	48
Figure 4.4:	Photo of Henry Tombs in 1857 (Public Domain); Epitaph on his tombstone at Newport.	49
Figure 5.1:	Portrait of William Palmer.	57
Figure 5.2:	Rev. Pierce Butler in Crimea in 1855 (Roger Fenton Collection, Library of Congress. PH - Fenton (R.), no. 252) ('No known restrictions on publication').	62
Figure 6.1:	The Isle of Graia. Coloured lithograph by Louis Haghe after David Roberts who was there in 1839. (Library of Congress) (Public Domain).	67
Figure 6.2:	Lithograph of Johann Ludwig Burckhardt, c. 1820 (© The Trustees of the British Museum)	68
Figure 6.3:	The Long Desert Route to Petra (Drawn: Travis Hearn).	72
Figure 6.4:	Convent of St. Catherine, with Mount Horeb. Coloured lithograph by Louis Haghe after David Roberts who was there in 1839 (Cleveland Museum of Art Collection) (Public Domain).	75
Figure 7.1:	The fort on the shore at Aqaba, residence of the Ottoman governor. The dusty and remote place of 1854 is today a port city and bustling tourist centre of hotels and private beaches (APAAME_20141020_DLK-0040C).	89
Figure 7.2:	Tombstone of Caroline Reynolds, first wife of Henry Veazey Ward, who seems to have died in childbirth.	94
Figure 7.3:	Copy of the official Pardon and Amnesty issued to Henry Clay Yeatman shortly after the end of the 'rebellion'.	100

Figure 8.1:	The graffito in charcoal of '[Rode]wald 1854' in the Khazneh at Petra (Courtesy of Prof. Fawzi Abudanah). 107
Figure 9.1:	Engraving from a photograph taken in Cairo in 1855 (Prime W. 1855: 96). 126
Figure 9.2:	Elji as recorded on a photograph by Charles Hornstein from a visit in 1895 (Courtesy Palestine Exploration Fund). 128
Figure 9.3:	Muqabil Abu Zaitoun (Sheick Yomgebel Abouseeton) (Formby 1843: Title page). 130
Figure 9.4:	Jabal Harun: the plateau with the Byzantine Monastery and the Muslim shrine on the peak above (APAAME_20051002_DLK-0086). 131
Figure 9.5:	Khazneh from above (APAAME_19980520_DLK-0200C). 133
Figure 9.6:	The paved road in the Siq. When Thomas Bowles was there, the Siq could be a veritable river in winter, deep in boulders and gravel and thick with oleander bushes, hiding the roadway (APAAMEG_20070409_DLK-0024). 134
Figure 9.7:	The Obelisk Tomb in the Upper Wadi Musa (APAAMEG_20070409_DLK-0150). 134
Figure 9.8:	The Royal Tombs: from the left, the Palace, Corinthian, Silk and Urn Tombs (APAAME_20160918_DLK-0367). 135
Figure 9.9:	The Theatre (APAAME_20151005_DLK-0429). 136
Figure 9.10:	Aerial view of the Colonnaded Street, Great Temple, Temenos Arch and Qasr el-Bint al-Pharoun (APAAME_20051002_DLK-0102). 136
Figure 9.11:	Tomb of of the Roman governor of Arabia Titus Sextius Florentinus. 137
Figure 9.12:	The Deir – note human figures as scales (APAAMEG_20070409_DLK-0126) 138
Figure 9.13:	Tomb of the Soldier (VHE_20171001_IMG_7730). 139
Figure 9.14:	Triclinium Tomb – interior (Brünnow and von Domaszewski 1904: 1, 159, Fig. 181) 140
Figure 10.1:	Pools of Solomon as seen in a photograph of 1890 - they would have been at least as isolated when seen by TB. Today they are in a built-up suburb of west Bethlehem (Library of Congress) (Public Domain). 142
Figure 10.2:	Photograph of James Finn (Public Domain via Palestine Exploration Fund). 145
Figure 10.3:	Photograph of Elizabeth Anne Finn (Public Domain via Palestine Exploration Fund). 146
Figure 10.4:	Tombstone of James and Elizabeth Finn (St Mary's Churchyard, Wimbledon) (Photo Courtesy: Tina Schofield). 147
Figure 10.5:	Young women at work in Miss Cooper's Jewesses Institute as seen in 1854 (Cubley 1860: Plate V). 154
Figure 10.6:	Mar Saba from the air (Copyright: Andrew Shiva). 156
Figure 11.1:	The earliest inscriptions as reproduced by the French artist Cassas in 1799; a photograph of 1922 of the inscription of Caracalla recorded by TB. 160
Figure 12.1:	Travel Journal 3 of Thomas Bowles: (Top) Cover; (Below) 'Title Page'. 172
Figure 13.1:	Page 21 of Travel Journal 3 illustrating the header, handwriting and some of the commonest abbreviations. 175
Figure 13.2:	Common abbreviations and frequent words as written in the Journal. 175
Figure 14.1:	Sketch of Tomb of Mehmet Ali in Cairo. 191
Figure 14.2:	Sketch of features of a tomb at Petra. 235
Figure 14.3:	Greek Text. 250
Figure 14.4:	Sketch of the Convent at Nazareth locality. 265
Figure 14.5:	Pattern on the plates on the door of the Great Mosque in Damascus. 274

List of Tables

Table 1.1:	Key events in the lives of the Bowles siblings.	8
Table 1.2:	Census of 6 June 1841 for Wandsworth Road, Vauxhall, Borough of Lambeth.	10
Table 1.3:	The Bowles brothers at school and Oxford and a few contemporaries mentioned in the text.	11
Table 1.4:	Outline of the life and career of Thomas Bowles before and after his Grand Tour.	13
Table 2.1:	Main stages in the Grand Tour of Thomas Bowles	26
Table 3.1:	Timeline of TB's journey from Ceylon to Cairo	28
Table 5.1:	Tourism of Thomas Bowles while at Shepheard's Hotel, Cairo and Hotel de l'Europe, Alexandria	53
Table 6.1:	Early western travellers to reach Petra	69
Table 6.2:	Western visitors to Petra between Laborde (1828) and Bowles who published extensive accounts of their journeys	71
Table 7.1:	Timeline of the Long Desert Route Cairo to Jerusalem	81
Table 7.2:	The parties making up the Anglo-American Group leaving Aqaba for Petra on 23rd March 1854.	82
Table 7.3:	Summary of the four Yeatman travel journals in the Tennessee State Library.	101
Table 10.1:	Tourism around Jerusalem and associated trips	142
Table 11.1:	The travels of Thomas Bowles in the Levant from 17th April till 8th May 1854.	159
Table 11.2:	Travels of Thomas Bowles from Beirut to Home, 1854	164
Table 12.1:	Structure and coverage of the three Bowles Journals	171
Appendix 1:	The Anglo-American Group at Petra 26-29 March 1854.	291
Appendix 2:	Timeline of Bowles' Journey in Egypt and the Levant.	292

Preface and Acknowledgements

Almost fifty years ago, as preparation for a first season of fieldwork in Jordan and its subsequent publication, I undertook a search for accounts by scholars, visitors and travellers who had preceded me. Almost all of those had been published mostly in the well-known accounts by the great early western travellers – Burckhardt, Seetzen, Buckingham and a score of others referred to elsewhere in this book. For researchers, fifty years ago is not just half a century but a different world. The wholesale digitization of entire libraries and archives has been transformational. Many hundreds of pre-1914 published books and articles have emerged which were quite unknown to me in the 1970s and 80s but are now commonly freely available for instant download. Many 'published' works now digitized were privately printed 'for family and friends' and in very small numbers; those too have now been digitized. Then there are the archival deposits including unpublished journals/diaries and letters which are revealed through online searches and can often be accessed easily by purchasing digital copies.

Amongst the last of these is the journal of the Rev. Thomas Bowles. Three years ago, I was not even familiar with his name as someone who had followed the Long Desert Route from Cairo to Petra and on to Hebron and Jerusalem. At that point, my friend Andrew Oliver, specializing in American travellers in Egypt (Oliver 2014), and a wonderful source of information on those of them who went on to Petra and the Levant, drew my attention to an item in the online catalogue of a New York antiquarian bookseller. The selling price was beyond my own resources, but the Friends of the Library at the University of Western Australia (UWA) stepped in and purchased this 'Diary' for the library collection. Although I have access to the journal - as Bowles correctly calls it, an early task was the prepare a high-quality digital copy. It is from the latter – which can be 'read' on-screen enlarged as required, that the transcription was done. I am immensely grateful to Drew Oliver in Washington, Deanne Barrett, Travis Hearne, Rebecca Repper, Kirsty Nicholson and Alana Colbert all of UWA. Especially important has been the last of these – acting as a Research Assistant, Alana Colbert undertook the gruelling task over several weeks of 'reading' and transcribing the handwriting of Thomas Bowles. From my own many hours subsequently trying to identify place and personal names which were especially problematic and trying to read words and phrases that had defied her, I know how challenging the work was.

The next group of names are of those many people who provided significant information about the Bowles family, some of whom are descendants (albeit not of the unmarried Thomas). Four stand out - Caroline Cannon-Brookes, Tom Laporte, Lucinda Lewis-Crosby and Lisa Parkinson. In different ways each has been of immense help and to that I can add a special thanks to Caroline Cannon-Brookes who kindly hosted a visit to her home near Oxford and then arranged for a joint visit to several of the places associated with Thomas Bowles. Lisa Parkinson was equally generous with hospitality in her home and sharing family material about Henry Stobart. Associated with these 'relatives', is Crispin Powell, archivist of the Duke of Buccleuch's papers who has been immensely helpful in providing information about the writings of the four men who set sail from Southampton for Australia in 1852 and arranged a very informative visit to Boughton House in 2022.

Thanks to the internet, it has often been possible to contact institutions and individuals far from my home in Western Australia. It has been immensely touching how swiftly and helpfully total strangers have responded. Most notable are those who trekked to a neighbouring church to photograph a wall plaque or search out a gravestone. More anonymous have been many others who – through the Find-My-Grave website, have supplied photos taken at my request. Finally, a few names deserve special mention: Gary Collins at Rugby School, Rev. Camilla White at Grafham, Terence Madigan at Camberley, David Galt in New Zealand, and Henry Warriner in Wiltshire.

Next are a swathe of friends who helped along the way: David Treloar, Julie Kennedy, Julie Kennedy-Pugh, Dawn Christenson, Norah Cooper, and Zbigniew Fiema

Lastly my wife. Veronica has been the classic supportive wife. By definition academics are more than a little obsessive even over matters of small interest to almost everyone else. She has been there to listen while I spouted excitedly details newly discovered of people of whom she had till that moment been entirely – and probably happily, ignorant. And she has been understanding when I found the need to work too many evenings and weekends and even during the night when insomnia provided further opportunity while everything was quiet. And of course she understands it is not 'work' – I have loved every minute of it and am well into my next book.

In selecting illustrations, I have tried to pick at least some of the roughly contemporary sketches, paintings or photographs of places and structures as they would have been seen by Thomas Bowles. (I wonder if he ever bought one of the sets of photographs of the Nile, or Petra or the Holy Land which were soon widely available in western countries). For Petra itself, I have a large selection of my own ground and aerial photographs to choose from. Having visited Petra many times and been flown into it to view the site and structures from a helicopter at a few hundred feet, I remain convinced it is the single most remarkable archaeological site I have ever seen. I can only suppose that when Thomas Bowles and his companions visited it, without the thousands of tourists and souvenir shops and hotels, the impression will have been greater still.

Introduction

Twenty years ago, the late Norman Lewis shared with anyone interested a typescript list of western visitors to Petra in the 19th century,[1] the basis of several articles/chapters he then published (Lewis 2003; 2004; 2007). In the years since then, hundreds of additional sources have emerged as discussed above. Although the list is just that, Lewis was well aware that the account published by – for example, Lord Castlereagh, of his visit actually involved several people besides his lordship That is true of almost every visitor who wrote a surviving account whether published or a manuscript in an archive. The result is that the crude impression that several dozens or one or two hundred westerners visited Petra between Burckhardt in 1812 and the outbreak of the First World War in 1914 is a significant undercount. My current best guesstimate is that the true figure was probably nearer 2000, not counting the larger parties of tourists taken there by Messrs Cooke and Lunn. The much higher number matters for several reasons, most importantly for the impact so many westerners had on the societies they encountered and employed and on their economies. The regular requests to the Ottoman authorities and frequent complaints about the beduin lodged by so many westerners inevitably drew the former's attention to this lawless part of their empire and ultimately played a role in their assertion of control. Finally, the reports published by visitors of wonderful archaeological remains at Petra and elsewhere, stimulated exploration of the wider region of what is now Central and Southern Jordan and of publications devoted wholly to these remains (most magisterially Brünnow and von Domaszewski (1904-09)).

Each new discovery in some obscure and now defunct 19th century periodical of an account of a visit to Petra or of a manuscript – diary/journal or letters, or a collection of sketches or photographs, adds to our ability to analyse and interpret this phenomenon and its impact. Most of the people undertaking this journey were at least modestly well-to-do; some were very wealthy. All were rich by the standards of the native inhabitants they encountered. Most were people who otherwise made little impact at home whether in North America or Europe. Many are known only from their publication. Today, thanks to a plethora of family history tools, it is possible to flesh-out the writers and those they met along the way but who are not remembered for any publication or archival document.

Thomas Bowles is one of these latter. Even now with a far longer list of known visitors to Petra, his unpublished journal represents a welcome addition to the corpus. The process of getting to Petra and the experience of western travellers on this Long Desert Route and at Petra itself, was a constantly changing one. Bowles undertook the journey in March 1854, just 42 years after Burckhardt's re-discovery, and several decades before the Ottoman grasp on security and construction of the Hejaz Railway wrought a transformation on conditions for travellers.

The Bowles Journal is significant, too, for who he met along the way. Previously, most accounts provided names of fellow-travellers but, unless they were famous, could add little more. At first sight that is true of this Journal. The writer met numerous people and often spent a great deal of time in their company. He seldom identifies them beyond a surname. Family history

---

[1] My copy is annotated 'Updated 2004'.

tools have transformed our ability to identify and provide at least a simple biography. What emerges is a remarkable inventory of people of all kinds and nationalities from at least the travelling class of the mid-19th century: merchants and soldiers, tourists, etc. A wonderful cross-section of the professional and leisured classes on the move – business, pilgrimage, tourism, adventure. Hence the title of this book. As far as we know, this was by far the most important tour overseas Bowles ever undertook until an apparent return visit to at least Jerusalem in March 1868. Even today it would be regarded as a major undertaking – in 1854 it was a serious adventure and one involving considerable dangers for Bowles as for almost everyone he met along the way this would have remained a major landmark in their lives. Some of those who did publish accounts of their visit explicitly say they had been urged to do so by family and friends simply because of its remarkable novelty. It is unlikely Bowles had much subsequent contact with most of his travelling companions and others encountered. He may well have noted references to some of them in the press or other publications; it would be interesting to know if he ever purchased any of the accounts published soon after his own visit, not least that of Jane Eames who was his companion for more than a month. In a world in which obtaining information was so much harder than today, he may well have wondered what became of X or Y or Z: did they marry, have children, their careers and achievements. I like to think he would have been glad to read this book with its biographical notes on some many of those he met: gladdened in many cases but saddened elsewhere at how many died relatively young.

Included here is a wonderful random selection of people encountered in and along the periphery of the British Empire. A middle class and aristocratic selection by and large but there are unexpected exceptions such as George Badger (Ch. 4).

Identifying the People Bowles Met

In his two surviving Journals, only twice (Palmer and Coussmaker) does Bowles give their full name to any of the many people he met, even when they were together for lengthy periods. Indeed, despite having been friends since their shared time as students at Oxford, both Bowles and Stobart refer to one another in their writings by their surnames alone. A common practise at the time.

In the case of men, he usually omits formal title after the first mention: e.g. 'Mr Strachan' at first usually becomes just 'Strachan'. If the title is a professional one - Dr, Major, Lord etc, he is more likely to retain it: e.g. Captain Hill he meets on the *S. S. Bombay* is almost invariably 'Captain Hill', spelled out in full. Women are always given their title - Mrs, Lady, the Duchess etc: 'Mrs Rogers' or 'Lady Falkland'. Occasionally, usually with someone he does not much like, he may give them their title but no name at all: e.g. the 'French Bishop' who joined the *S. S. Bombay* at Aden or the 'Irish priest' met in Jerusalem.

'British' is never used as a nationality - his fellow-countrymen are all called 'English' even when he will have known some were Irish or Scottish. For other nationalities he uses just a broad term - e.g. 'American'. For none does he say anything about where any of his companions originate or now live in their home country. Although he was surely conscious of significant differences between the accent, modes of speech and cultural views of Yeatman from Alabama and the Eameses from Massachusetts and New Hampshire, there is no hint of it in his journal.

Finally, he does not note that Rodewald, though American now, was a German immigrant, having seemingly shifted country as an adult and quite probably still distinguished by a German accent. We owe that piece of information to Jane Eames.

Although it is highly likely Bowles will have learned in the course of conversations over days, weeks or even months, a great deal about many of his companions, he evidently felt little need to record it. His lack of interest in the full names and antecedents of companions and acquaintances carries through in most cases to any *recorded* interest in what they were. Clergymen are (usually) identified as such and he notes that some of his companions are 'Dissenters' while the French Bishop is explicitly an 'R. C. Bishop'. Bryce is identified as a medical doctor but only in passing when he is called upon to treat an injury. Palmer is a famous writer so identified more fully as 'Mr Palmer - William Palmer of Magd(alen) College'. Although soldiers are identified by their military ranks he only once tells us their regiment: Captain Keane was an officer in the Ceylon Rifles. Most astonishing of all is 'Mr Kennedy', his close travelling companion for almost three months who is never given his full name; he characterises him as 'English' although Kennedy surely regarded himself as a Scot. We get only a chance hint of Kennedy's occupation and what he had been doing before they met on the ship from Galle (below).

As noted above, the *S. S. Bombay* probably carried about 45-50 passengers and Bowles mention 16 and provides a few details for some. Oddly, although Strachan was on the ship, he is never mentioned till they all reach Cairo. There was the potential to become well-acquainted with some and – in the case of Kennedy, to find him agreeable enough to subsequently spent two further months in his company. However, the reverse was also true and Bowles explicitly mentions trying to evade certain fellow passengers but how – in the confined space of a steamer and with - in this case, 14 days at sea, fellow-passengers could become tiresome. On 8th February 1854, his 10th day since Galle:

> *I walked sometime alone, shirking Miller, who altered his pace & tried to fall in with me, till at last I thought he would see I was avoiding him & so gave up my <u>solitary</u> & joined him & we walked together sometime. He went below & I was rejoicing in a solitary walk again, when it was again broken by Wood joining himself to me. And he staid walking with me till I went down to bed. It does not do to be selfish on board ship. One gets into the way of suiting oneself to any one & every one. It used to be torment to me to be with <u>some</u> people - there seemed to be a natural antipathy - but it wears off of necessity in travelling.*

Although it is exasperating, we must remember that his Journals were being written for himself and without the need to explain or illuminate what would have been included in letters sent home (above). It need not surprise us that Bowles can be exasperating in his vagueness about people he meets and travels with; it would be useful to have some of his letters home for comparison. However, while his journal for this period frequently refers to letters received and his pleasure in them, he seldom mentions sending any: e.g. on Saturday 25 February 1854 he notes: *'wrote journal & finished my letter to Alice, besides writing a few lines to Tupper.'*

And two days later *'Since then I have been finishing letters, …'.* In one case (Tupper's) it was not even a letter home but to someone he had met in Ceylon (below).

Once in Egypt Bowles could be more selective of his companions than on a crowded ship. Most notably almost all of the group of sixteen who arrived in Petra together had travelled with or near one another for weeks.

Thomas Bowles confided to his journal almost nothing about the people he met. And what people they were! Mostly British and American but mainly drawn from particular categories of the middle and upper classes – politicians, administrators of empire, soldiers, and businessmen; clergymen visiting the scenes of the Bible they had been steeped in and inspired by; people of all professions or 'living on their own resources', but turning their attention to serious amateur adventures. All will have been familiar with the Bible in a way not common today and many will have studied the Graeco-Roman Classics. Most had been privately educated and many had university degrees. Badger and Finn are unusual is their modest background.

Despite their haggling over prices and occasional exasperation with demands for 'baksheesh', they were prosperous enough to undertake this travel and by the standards of almost everyone they met, were rich ... and would return home to comfort and plenty after their adventure. Despite the privations of camping or poor hotels, these travellers will have seemed almost unbelievably wealthy to beduin or peasants looking at tents, camp furniture, clothing and lavish provisions of food and drink. Understandably their dragomen sometimes sought to win advantage by claiming their charges were great aristocrats.

Bowles surely learned a great deal of the back-stories of the people with whom he conversed for hours over many days at dining tables, while walking around the deck of a steamer or trudging across the desert. Sadly almost all of that is lost but our research tools today can unearth some of what Bowles may have known, more that he probably did not know and some of what became of these people in later years. A good – but rare example, is Bowles reference to Dr Gammie's account of his shipwreck some years before but which we can flesh out considerably then follow his career and life after Bowles met him.

In later years, Bowles would have encountered in the Victorian media, some of those he had met – aristocrats, clergymen, politicians and soldiers in particular, but he probably never met them again in person. In many cases – especially the non-British or the British who lived overseas, he may never have even heard of them again. We are more fortunate and can construct at least brief biographies.

Some originally went to Egypt for their health, but even the Nile Valley could be deadly and the rigours of travel in the wider region surely impacted at least temporarily on the health of many. A few certainly died of within a few years (Tupper and Wakefield); others were killed or injured or their lives transformed in the Crimean War (Captain Butler), Indian Mutiny (Dr Gammie) and – even more convulsively, the American Civil War (Mr Yeatman).

Several married soon after the end of their travels; some married but had no children; and some – like Bowles himself, never married. Many of those who had children have descendants today and some – the landed gentry in particular, are traceable. Some were already notable people (the Falklands); others achieved renown of some kind in the years ahead (William Palmer); and others were the parents of people of prominence in later generations (Bryce);

some were the parents or grandparents of people who went to Petra in future decades (Bryce); and, half a century after the event, I discovered that one of the most influential and fondly remembered of my teachers as a student at the University of Manchester was the grandson of the German-American-British Charles/ Carl Rodewald who travelled to Petra with TB and wrote his name in charcoal on the wall of the Khazneh.

The circumstances in which his people ended their lives was very varied. Tupper, already seriously unwell for weeks, died at sea off Malta; Captain Butler was killed in the Crimea; Matilda Creasy was murdered four years after Bowles met her; and James Eames – a terrible sailor, died on shipboard entering harbour in Bermuda; Yeatman was killed by a train in 1910; and Bowles himself seems to have died soon after surgery.

Only eight of the 45 biographical sketches are of women and just two of the sixteen members of the Anglo-American Group with whom TB visited Petra were women. TB has little to say about the women he encountered but happily we have Lady Falkland's chapter on her visit to Palestine and – far more extensive and full of interest, are the published letters of Jane Eames. I suspect writings – if any, of Mrs Rogers, a 27 year old woman crossing the desert with her 45 year old husband whom she outlived by over 30 years, would have made interesting reading.

Part 1
The Grand Tour

Chapter 1

Rev. Thomas Bowles: Biography

(Born at Milton, Berks (now Oxon.) on 5th January 1822 - died Marylebone, London on 12th January 1899)[1]

The Bowles family emerge into relatively modern history with the grandfather of the diarist - John Bowles (1725-1796). In 1777 this man, a solicitor and Town Clerk at Abingdon,[2] bought Heath Farm, c. 2.5km/1.5 miles south of the village of Milton and began building what came to be Milton Hill (House). It was enlarged and developed further by Thomas Bowles (1790-1837), John's illegitimate son by his housekeeper.[3] Evidently the latter made good use of over £50,000 (plus landed property) bequeathed him (Cannon-Brooke 2008: 28) to create the delightful Georgian mansion one sees today, set in a park of c. 20ha/50 acres (Figure 1.3). The house then passed to this Thomas's son, John Samuel Bowles (1815-1884) - brother of the diarist,[4] and finally sold in the next generation by John's eldest son - our diarist's nephew, Col. Thomas John Bowles (1852-1933) in 1905.

Milton Hill House was soon owned by Sir Adam Mortimer Singer, the Anglo-American son of the founder of the Singer Sewing Machine company, who provided it as a hospital during the First World War. The house is today a hotel and venue for conferences and weddings.[5] The family remained prominent, of course, owners also of Abbey House in Abingdon and of Westridge Manor near Streatley where they were Lords of the Manor, and which became the principal residence of the main line of the family after the sale of Milton Hill House. Descendants of the diarist's wider family - still live in the area and as far afield as Canada.

Figure 1.1: A photograph of Rev. Thomas Bowles displayed in the Vestry of the Church of St Augustine of Canterbury at East Hendred where he was Rector for fifteen years.

---

[1] In completing this chapter, I have benefitted enormously from correspondence with several people who generously shared their knowledge and data about Thomas Bowles, both research and unpublished documents in their possession: Caroline Cannon-Brookes, Tom Laporte, Lucinda Lewis-Crosby and Lisa Parkinson. None is responsible for how I have made use of that information.
[2] Now, through boundary changes, in Oxfordshire but then in north Berkshire.
[3] Will of 'John Bowles of Abingdon' (TNA Prob 11/1284) 'my natural or reputed son Thomas Bowles who now also liveth with me and is the child of the said [housekeeper] Martha Morrell'
[4] Bowles's oldest brother, John Samuel Bowles (1815-1884), Eton and Exeter College, Oxford, J. P., D. L., inherited Milton Hill House, was inevitably on the board of various local associations, was a local magistrate and went on to be Sheriff of Berkshire in 1852.
[5] https://www.miltonoxfordshire.co.uk/hotels-abingdon

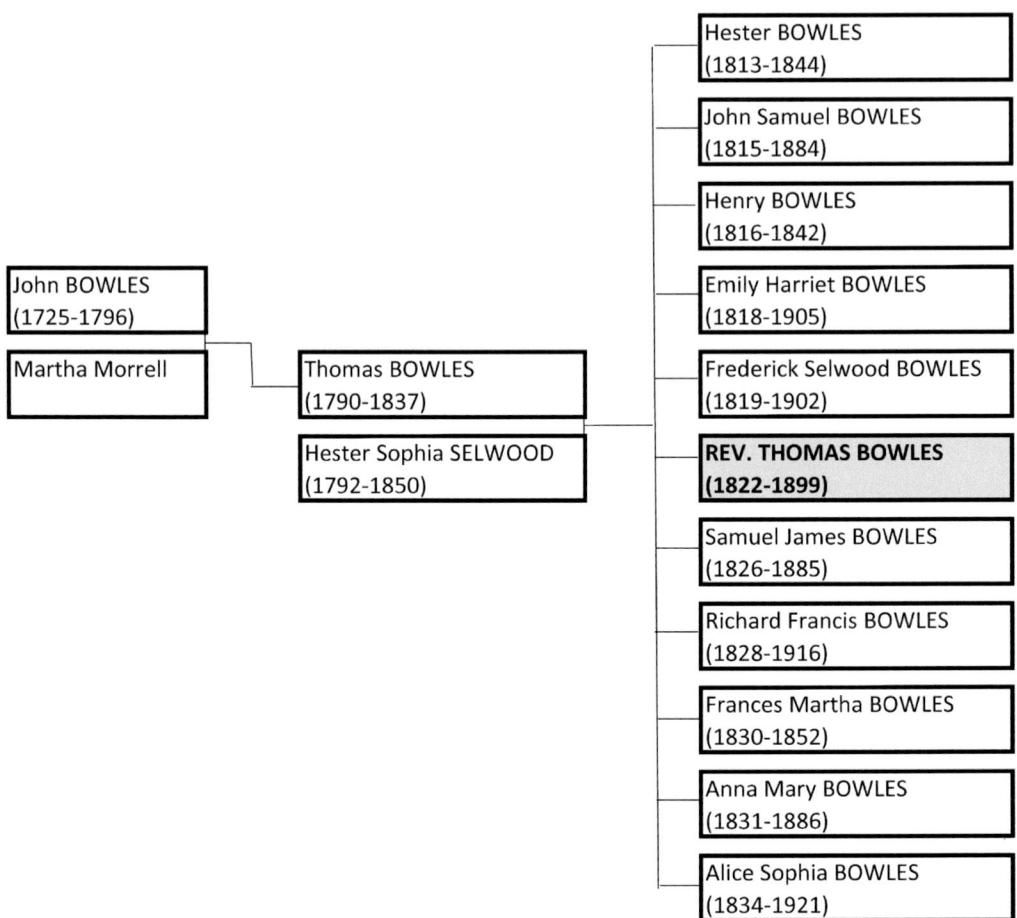

Figure 1.2: Family tree for the Rev. Thomas Bowles (Drawn: Julie Kennedy-Pugh).

Milton was - and still is, a modest village, by-passed by the main current highway and the railway and preserving much of its 18th and 19th century character. It has a notable church, dedicated to St. Blaise (Figure 1.4) - rebuilt extensively in 1851 by the church architect, Henry Woodyer, brother-in-law of the Rev. Thomas Bowles - henceforward 'our diarist' or 'TB'. It has many associations with the Bowles family: TB's parents are commemorated in two memorial windows and his sister, Frances Martha Woodyer, is buried in the crypt. An impressive monument in the churchyard commemorates several members of the family (Figure 1.5).

The landscape to northeast and north is dominated by the River Thames. The nearest town is Abingdon, c. 8km/5 miles to the north, with Oxford 10km/6 miles beyond that, also on the Thames. The delightful village of Sutton Courtney is just 6km/4 miles to the northeast; Dorchester-on-Thames (near but not actually *on* the Thames despite the name) with its old inns and thatched houses is 10km/6 miles to the east.

The Cotswold Hills begin just beyond Oxford to the northwest and the Chiltern Hills lie to the southeast. In short it lies is something of a broad bowl, with some of the most beautiful landscape in England (Vincent 1906).

Figure 1.3: Milton Hill House. Although significantly remodelled not least to its current condition as a hotel, the family home of Thomas Bowles is largely the white building. It is set still in a large park and includes several magnificent cedar trees (Photo: David Kennedy).

Figure 1.4: Church of St Blaise, Milton (Photo: David Kennedy).

Figure 1.5: Monument to members of the Bowles family in the churchyard of St Blaise, Milton. Although extensively inscribed with family names few are readable without cleaning (Photo: David Kennedy).

The geographical and environmental context is worth stressing not just as that of the diarist's childhood, but because he subsequently spent much of his working adult life nearby, retired to Oxford then to Abingdon and is buried in the latter. Much of what would have been familiar in his lifetime remains.

As so often throughout Britain, people would have been conscious of a landscape moulded by previous settlers (cf. Hughes 1859) whether it was the prehistoric remains along the Ridgeway just *c.* 6km/4 miles to the south of Milton Hill, the Roman town and its Saxon successor at Dorchester or the Norman origins of so many of the churches in villages in the region - including an abbey in Abingdon (where the Bowles family inherited Abbey House through TB's mother). About 1820, a dozen years before our diarist's birth, the 84km/52-mile-long Wiltshire and Berkshire Canal was cut just west of Milton Hill, passing through Steventon on its way to join the Thames at Abingdon. In the mid-1830s railway lines from London opened up the area, one passing on the east through Didcot and Radley (for Abingdon) and a second forking at Didcot to pass between Milton and Milton Hill. Towards the end of his life, TB would have seen road improvements to meet the needs of the growing number of cars but died four years too soon to hear of the first powered flight in 1903.

Most of the airfields that blossomed in the region during the Second World War have merged back into the landscape. The later 20th century disfigurement of the rolling landscape of this part of Oxfordshire-Berkshire by the immense cooling towers of the Didcot Power Station, has been reversed by their recent demolition but the village of Milton is now dwarfed by the large Milton Industrial Park on its south side. From the elevation of the prehistoric Ridgeway one can see East Hendred where TB held his last rectorship, but also the University of Oxford's science and technology hub at Harwell just 3.5km/2 miles from Milton Hill House. TB would have been baffled by such a place and the puzzling names of its streets - Fermi, Rutherford, Curie, Faraday ..., and equally by the immense fields of solar panels just to the west of Milton Hill.

Early Life and Education of Thomas Bowles

Our diarist was the sixth child and fourth son[6] of Thomas (1790-1837) and Hester Sophia Bowles (née Selwood) (1792-1850) who had married in 1812. It was a large family: five girls

---

[6] Not '3s' as commonly printed in published records such as *Oxford Alumni*. The early death of his brother Henry in 1842, a year before TB himself matriculated at Oxford, is probably the cause of the confusion.

and six boys (Figure 1.2).[7] Rather remarkably for the time, all survived into adulthood; indeed, four lived into their 80s. On the other hand, the familiar 'seasonality of death' in the 19th - even for a well-to-do family, is on display: seven died in either December or January, the eldest three all in the last week of a December (Table 1.1).

The home life and very early education of Thomas Bowles are preserved to some extent in the pages of a diary kept by his sister Emily (1818-1905) which continues till 1844 when Tom - as he was known in the family, was *c*. 22.[8] Emily's diary paints a picture of a homely existence summarised recently by a descendant of the family:[9]

> *Thomas and Hester Bowles provided an ordered childhood for the family at Milton Hill. They were encouraged to create and invent their own occupations, and they all learnt to ride. They were given a thorough educational start with introduction to literature 'with serious reading' after tea. Music was important, they played instruments and their father played the violin.[10] Hester painted.[11] The older girls had a governess, the younger boys had a nursery governess to start with. Their grandmother was also an influence on the family and a friend of Miss Tombs whose brother was in the service of the Indian Company. Later Fred was to take Henry Tombs (afterwards Sir Henry) to fish near Abingdon.[12]*

Where he began his education can be inferred: under 12th August 1833 - when TB was aged eleven and half, Emily's diary records 'Fred and Tom going back to school at Rose Hill' and a reference in Mildred Bowles' diary to a prep school - 'Slatters at Rose Hill', suggests that at least those two children were sent to the small school run by the Rev. William Slatter at Rose Hill[13] on the south-eastern outskirts of Oxford. Presumably they were there as boarders. Possibly not a happy episode as Emily records a year later on 6th August 1834: *'Poor Tom went back to school.'* Presumably he continued there until moving to Rugby in August 1836.

TB's eldest brother, John, had been sent to Eton but the next four brothers - Henry, Tom, Samuel and Richard, went to Rugby after preparatory school.[14] Tom is recorded as entering in August 1836, aged 14, by which time his elder brother Henry had already left and had completed a year at Oxford. On the other hand, his younger brother Samuel was enrolled

---

[7] Most are known only from their baptismal records; whether there were other children who died before baptism is (currently) unknown.
[8] Emily's diary - a typescript of 537 pages, is in the hands of a family descendant who is preparing it for publication. A further diary kept by TB's niece, (Constance Louisa) Mildred Bowles (1867-1962), one of 12 children of John Samuel Bowles, adds further details of the lives of her father and his siblings and is held by the Birmingham Oratory Archives.
[9] I am grateful to Caroline Cannon-Brookes for this summary.
[10] Emily claimed Frederick was the best musician in the family.
[11] Many sketches and watercolours attributed to the eldest brother John Samuel Bowles have recently been auctioned (https://www.sulisfineart.com/catalogsearch/result/?q=John+Samuel+Bowles). TB himself frequently mentions sketching during his expedition from Suez to Beirut and included several thumbnail sketches of buildings.
[12] Two members of the Tombs family of Abingdon are 'met' in TBs journal (below Ch. 3). The burgeoning career of Henry Tombs he heard about from a travelling companion on the ship from Galle to Suez; his brother Charles had died in Aden and TB visited his grave there during a re-coaling on this same voyage (below Ch. 3).
[13] This William Slatter is listed as a Freeman of Oxford in 1834. His son, Rev. John Slatter appears in the record of Oxford Alumni: *Slatter, John, 1s William, of Iffley, Oxon, cler. Lincoln Coll., matric 19 March, 1835, aged 18; exhibitioner 1835-42, B.A. 1838, M.A. 1841, perp. curate Sandford-on-Thames 1852-61, vicar of Streatley, Berks, 1861-80, hon. canon Christ Church, Oxford, 1876, rector of Whitchurch, Berks, 1880.* Is it coincidence that he became vicar of Streatley - location of a Bowles Manor, in 1861-80; he would have been almost the same age as John Samuel Bowles who was by then resident at Streatley. The Slatter family are the subject of a web site:- https://sites.rootsweb.com/~haslatter/history/slatters/OxfordSlatters/connection.htm
[14] Frederick seems not to have been sent to any public school.

Table 1.1: Key events in the lives of the Bowles siblings.

	Name	DOB	Place	DOD	Aged	Place	Education	Occupation	Buried
1	Hester	26 Sept. 1813 (bapt.)	St Blaise, Milton	29 Dec. 1844	31				
2	John Samuel	22 July 1815 (Bapt.)	St Blaise, Milton	25 Dec 1884	70	Milton Hill, Berks	Eton; Exeter, Oxford	Landowner, JP (Berks.), High Sheriff (Berks. 1852)	
3	Henry	Sept. 1816 12 Sept. 1816 (bapt.)	St Blaise, Milton	27 Dec 1841	c. 26		Rugby; Oriel, Oxford	Anglican Priest	
4	Emily Harriet	28 Jan 1818	St Blaise, Milton	4 August 1905	87	Southampton		Catholic Nun (after 1843); Author; Translator	
5	Frederick Selwood	15 Aug. 1819	St Blaise, Milton	1 Dec. 1902	83	Hope Cottage, Harrow-on-the-Hill, London	Slaters; Exeter, Oxford	Anglican Priest then Converted (1845). Chaplain to Dominican Convent, Harrow.	
6	Thomas	5 Jan. 1822	St Blaise, Milton	12 Jan. 1899	77		Rugby; Queen's, Oxford	Anglican Priest	Old Cemetery, Abingdon
7	Samuel James	22 Nov 1826	St Blaise, Milton	10 August 1885	c. 59	Beaconsfield, Bucks	Rugby; Magdalen, Oxford	Anglican Priest	Beaconsfield, Bucks
8	Richard Francis	22 June 1828 (bapt.)	St Blaise, Milton	16 August, 1916	88	South Moreton, Wallingford	Rugby; Exeter, Oxford	Lawyer, East India Merchant, Stationer; Landowner	
9	Frances Martha	18 April 1830 (bapt.)	St Blaise, Milton	21 June 1852	22	Grafham, Surrey		Wife of Henry Woodyer and mother of Hester Fanny Lake	St Blaise crypt, Milton
10	Anna Mary	15 Jan 1832 (bapt.)	St Blaise, Milton	Dec. 1886	c. 54				
11	Alice Sophia	23 Feb 1834 (bapt.)	St Blaise, Milton	13 Jan. 1921	c. 87			Converted (after 1892). Living with Sisters of Mercy	

three years later, aged just 12, and the two would have overlapped by one or two years. By the time Richard enrolled in 1841, aged 13, Samuel would still have been there but Thomas would have left.

Rugby School was already an old school - founded 1567, but, until the late 1820s, it had long been obscure. It would have been a brutal place for small boys as noted about one of its peers, Eton (Briggs 1954: 145):

> *The living was sometimes so austere that writers could complain that the inmates of a workhouse or a jail were better fed and lodged than the scholars of Eton.*

During the time the Bowles brothers were at Rugby in the 1830s, conditions were changing sharply for the better. A new Headmaster appointed in 1828 and in charge till 1841, was the famous Thomas Arnold, whose improvements in the treatment and education of the boys influenced most of the other public (i.e. private) schools and whose stewardship established Rugby as one of the seven most prestigious in England.

An exact contemporary of Thomas Bowles at Rugby was Thomas Hughes (1822-1896), author of *Tom Brown's Schooldays* (1857), a book that had sold 28,000 copies by 1862 and gone through 53 editions by 1892. It represents a fictionalised image of life at the school from the rife and savage bullying of before Arnold to the changing attitudes under his influence and reforms. How much of the former TB experienced can only be guessed at and many basic features continued. The boys slept in communal beds of up to six together - though parents could pay to have double or single beds. One pre-Arnold master at Rugby remarked that that 'the boys were the excrescences of pond life' (quoted in Briggs 1954: 142) and Arnold is reported to have been shocked by their behaviour (Briggs 1954: 143):

> *Here in the nakedness of boy-nature, one is quite able to understand how there could not be found so many as even ten righteous in a whole city.*

As Eton seems to have been resistant to the reforms Arnold introduced as Rugby, this may be why TBs parents decided to send most of their sons to the latter. For many of the boys the impact of Arnold was profound. Hughes happily sent his own son there a generation later and attempted to found an idealized community in far-away Tennessee which he called Rugby. Arthur Stanley ('Dean Stanley'), another contemporary of the Bowles brothers at Rugby, was to write an admiring and affectionate biography of Arnold (Stanley 1844).[15]

A 'Dot Bowles' appears in the pages of *Tom Brown's Schooldays* but is apparently enrolled at under aged 10. TB was 14 when he entered Rugby and Hughes himself had been 11. There may not be any intention to evoke any of the Bowles brothers beyond use of the name.

We may guess that TB left Rugby in c. 1840, aged 18, profoundly and positively influenced by the experience. Briggs has summarised the complementary opinions of Hughes and Stanley (1954: 154):

---

[15] Arthur Stanley and Theodore Walrond had both been at Rugby, the former as an approximate contemporary of Henry Bowles; the latter as an exact contemporary of TB. In March 1853, a year before TB and, of course, long before Stanley published his book (1856) (below), these two visited Petra.

[Stanley] was one of the small group of Rugby boys who went on to Oxford University stamped with a gravity and deep seriousness of purpose in life which were not to be found in most other young public school boys. To their enemies Arnold's serious 'disciples' were 'prigs'; to themselves they were dedicated men, bound to their headmaster with what Stanley described as 'idolatrous affection.' Stanley's picture of Arnold was colored by his own reverence, and, powerful though it is, it only becomes complete when supplemented by Hughes's more homely description. If Stanley saw what happened in Arnold s mind, Hughes saw what happened in many of his pupils' minds, in the minds of all the Tom Browns who made up 'the masses' of the school.

By chance, one of TB's fellow passengers from Galle to Suez was a contemporary from Rugby and they reminisced about the school and what had become of other boys they could remember (below, Ch. 3, Seton-Karr). There were many other boys in the school: the Year Book for 1836, the year TB enrolled, there were about 280 pupils and subjects were grouped into three categories - Classics, Mathematics and Modern Languages. The second of these would be important for the next few years of TBs life.[16]

All six Bowles brothers went on to Oxford, two matriculating aged 17 and three aged 18 (Table 1.3). The remaining one - Thomas, is striking as he did not matriculate till 1843, presumably some 3 or 4 years after he left Rugby and now 21. A clue is provided in the diary of his sister Emily under 1 February 1841: *'Tom left us for London (lodging with Mr Cattermole).'* Under 23rd October 1842 she wrote:

Of all those who were left, the one who caused me the keenest heartache was Tom he did not like his destined profession the law, London made him really ill, the loss of Abingdon and its tranquil hours on the river was a real disaster and the loss of my mother's presence and soothing voice was one that to him could never be replaced. After John's marriage she had grown more and more accustomed to consult him and confide in him so that next to Hester [= TB's sister] I think he was her greatest earthly joy.

TB can be traced in London in the Census of 6 June 1841 (Table 1.2) confirming him as a law student lodging at an unnamed address in Wandsworth Road, Vauxhall, London. It is likely records for one of the Inns of Court will define his dates of enrolment. It may be that he dropped out soon after this diary entry in 1842, a year of family tragedy and crisis (below).

As it happens Richard Cattermole (1795-1858) is well-known as is his more famous brother, George (1800 -1868), as artists,

Table 1.2: Census of 6 June 1841 for Wandsworth Road, Vauxhall, Borough of Lambeth.

Name	Age	Profession, Trade, Employment or of Independent Means
Richd Cattermole	45	Clergyman
Ellen Cattermole	30	
Charles Do	9	
Richard Do	7	
Ellen Do	2	
Walter Do	1	
Jane Millington	25	F(emale) S(ervant)
Elizabeth Mason	30	Do
Martha Smith	25	Do
Thomas Bowles	18	Law Student

---

[16] I am grateful to Gary Collins, the Archivist at Rugby School, for this information.

Table 1.3: *The Bowles brothers at school and Oxford and - in italics, a few contemporaries mentioned in the text. The dates in the 'Rugby' column are their DOBs; the Rugby records for Thomas have him listed as entering in August 1836.*

Child	Rugby	Entry Year	Matric at Oxford
John (1815-1884)	Eton	1829?	Exeter, 15 May, 1833, aged 17
[Arthur Penrhyn Stanley (1815-1881)	*Aged 13, Dec. 13*	*1829*	*Balliol, 30 Nov. 1833, aged 17*
Henry (1816-1842)	Aged 15, Aug. 15	1831	Oriel, 15 May 1835, aged 18
Frederick (1819-1902)	???	???	Exeter, 9 Nov. 1837, aged 17
[Theodore Walrond (1824-1887)	*Aged 10, Feb. 12*	*1834*	*Balliol, 10 March, 1842, aged 18*
[Thomas Hughes (1822-1896)	*Aged 11, Oct. 19*	*1834*	*Oriel, 1841, aged 19*
[Walter Scott Campbell Seton-Karr (1822-1910)	*Aged 12, Jan. 23*	*1834*	*East India College/ Haileybury*
Thomas (1822-1899)	**Aged 14, Jan. 5**	**1836**	**Queen's 2 Nov. 1843, aged 21**
Samuel (1826-1885)	Aged 12, Nov. 22	1839	Christ Church, 15 May 1845, aged 18
Richard (1828-1916)	Aged 13, May 19	1841	Exeter, 4 June 1846, aged 18

authors and architects (Hawes 2004; Sperling 2004). Richard continued his artistic pursuits even after he entered the Anglican priesthood where he was vicar of the South Lambeth Chapel (now St Anne's and All Saints at the intersection of Miles Road and South Lambeth Road, just a few hundred metres from where he lived).

Evidently TB was allowed to abandon his legal studies and in 1843 he matriculated at Queen's College, Oxford, aged 21. The official record of Oxford Alumni records:

> **Bowles**, *Thomas, 3s. Thomas, of Milton, Berks, arm.* QUEEN'S COLL., *matric. 2 Nov., 1843, aged 21; B.A. 1847, M.A. 1850, rector of East Hendred, Berks, 1875.*

Elsewhere he is said to have been awarded Honours in Mathematics in 1847 (Foster 1893: 183).

It was while he was at Oxford that he met in the same college, a near exact contemporary, Henry Stobart, with whom he was to undertake a major part of his Grand Tour (below, Ch. 2).

By the time TB graduated in 1847 he had a mixed education behind him of Rugby, probably 2-3 years of training in the law, then a university degree specialising in Mathematics. All that was about to change as he followed a growing family preoccupation with religion. His brother Henry had graduated from Oriel College, Oxford in 1840, where he had been a keen follower of one of its Fellows, John Henry Newman (below) and might well have converted to Catholicism had he not died young in 1842, aged 26. The following year (1843) TB's sister Emily converted to Catholicism while in Rome. Then came Frederick, who graduated from Exeter College, Oxford in 1842, then studied at Chichester Theological College, before he moved to Littlemore just outside Oxford to join the group of young converts gathered around Newman. He became a Roman Catholic in 1845, joined the Birmingham Oratory, 1848-60, and was later Chaplain to the Dominican Convent at Harrow, Middlesex, for twenty years, where he died. Thomas was the next sibling but there were to be no more converts for half a century until the youngest sister, Alice in 1892. Thomas did enter the church but ordained as a deacon (1848) then an Anglican priest (1849) which he remained despite the probable 'pull' of sibling converts and his own preferences (below).

The impetus to closer participation in religion may have been increased by family events. TB experienced two family tragedies in swift succession aged 20 and 22. On 27th December 1842 his brother Henry, a recently ordained Anglican priest, died aged *c*. 26. Earlier that same year, the health of his eldest sister, Hester, led their mother to take her and the younger daughter, Emily, to Italy. Thomas and his older brother John travelled with the ladies as far as Paris and in 1844 he again crossed the Channel to meet Hester's party in Switzerland on their way home.[17] They were all returned home in July 1844 but in October Hester had a will drawn up and died aged 31 on 29th December. A further death followed eight years later. On 5th August 1851, his younger sister Frances had married the up-and-coming church architect, Henry Woodyer (Quiney 1995), who had been an exact contemporary of TB's brother John at Eton and Oxford (1829-35 and 1835-38). Tragically, she died in childbirth the following June, aged 22. The child – Hester Fanny Woodyer, survived and was to be a legatee in TB's will.[18]

It was against that background of deaths, especially that of Frances, that Thomas set off a few months later on 22nd October 1852 on his Grand Tour (Ch. 2). He was not new to foreign travel as noted above, but this new venture was to be of quite a different order and did not end till he arrived back at Milton Hill on 27th May 1854, a year and seven months later.

Before turning to the tour in the next chapter, we can sketch the life of Thomas Bowles before and after that trip. The details can be tabulated (Table 1.4).

The Rev. Thomas Bowles

After 2-3 years training as a lawyer, TB matriculated at Queen's College Oxford in 1843 where an exact contemporary was Henry Francis Stobart who was to become a travelling companion and life-long friend (below).[19] Having graduated B.A. in 1847 (and M.A. in 1850), he was swiftly ordained first as a deacon (1848) then as an Anglican priest (1849). Initially he was one of three Curates of a group of three churches in Wiltshire centred on Westbury. A recollection about one of the three - A. H. Mackonochie - by a visitor, presents a picture of a demanding and ascetic life for the curates:[20]

---

[17] I am grateful to Caroline Cannon-Brookes for these details culled from the diary of Emily Bowles which in turn had drawn on those of her siblings.
[18] She married, aged 39, and died and was buried along with her husband, Lt. Gen. Sir Percy Lake, in Victoria, British Columbia, to which they had emigrated.
[19] Their closeness from University days is underscored in a letter dated El-Arish 1846, in which Stobart told his mother he had had a letter from Bowles.
[20] Bowles and/ or the parishes in which he was then serving, appear in a number of documents collected as part of the online *Project Canterbury* which is the source of the quotations presented below unless otherwise cited. http://anglicanhistory.org/england/ahmackonochie/towle1890/03.html

Table 1.4: *Outline of the life and career of Thomas Bowles before and after his Grand Tour.*

Dates	Activity	Place	Note
5 January 1822	Born	Milton Hill House, Milton, (then) Berks (now) Oxon.	Sixth child and third son of Thomas and Hester Bowles.
c. 1832-36		Rev. W. Slatter Prep School, Rose Hill, Oxford	Diaries of Emily and Mildred Bowles and by inference.
1836	Rugby School	Rugby, Warwicks.	'... son of T. Bowles, Esq. Milton Hill, near Abingdon, aged 14 Jan. '5.'
c. 1840-1842?	Training as a lawyer	London	Diary of Emily Bowles
Mid-1842	Travel on continent with mother, sisters Emily and Martha and brother John	To Paris	
Mid-1844	Travel on continent with mother and sisters Emily and Martha returning from Italy	To Switzerland	
1843-1847	Queen's College	Oxford	'3s. (sic) Thomas, of Milton, Berks, arm. QUEEN'S COLL., matric. 2 Nov, 1843, aged 21; B.A. 1847, M.A. 1850'. Honours: Mathematics.
1848	Ordained as Deacon		'Deac. 1848 by B(isho)p of G(loucester) and B(ristol) for Sarum.'
1849	Ordained as Anglican priest.	Westbury, Wilts.	'Pr. 1849 by Bp of Salis.'
1848-1850	Curate	Westbury + Old Dilton and Bratton. Wilts.	Curate shared with two other curates of 3 churches: All Saints; St Mary's; St James the Great
1850-1852			Nothing known - perhaps the following item?

Dates	Activity	Place	Note
'Summer' + 1852	'Curate to [Rev. William John] Butler'	Wantage	St Peter and St Paul
22 October 1852 - 27 May 1854	Travelling	By sea to Australasia, then Asia, Middle East, France	Initially with Rev. Henry Stobart and Lords Henry Scott and Schomberg Kerr
1854-1858	Perpetual Curate	St Andrew, Grafham, Surrey	Temporary church at Grange House, the home of his brother-in-law, the church architect, Henry Woodyer
1858-1864	Curate	Wantage	St Peter and St Paul
1868	Curate	Garsington, Oxon	St Mary
1868-1871	Curate	Clifton, Bristol	All Saints
1872-1873	Rector	Wallingford	St. Mary and St. Leonard
1875-1890	Rector	East Hendred, Berks	St Augustine of Canterbury
1890-1899	Retirement		
1891; 1895 [Census; Kelley's Directory Oxfords]	Retirement	24 Leckworth Road, Oxford	Occasionally officiated at St Barnabas
1899 [Probate]	Retirement	4 Spring Terrace Abingdon	
12 January 1899	Death	St Luke's Hostel, London while resident at 16 Nottingham Place, Marylebone, London	Died during surgery
21 January 1899	Burial	Abingdon, Oxon	Old Cemetery, Abingdon

Mr. Bowles, who furnishes us with some recollections of this time, writes: It was certainly hard work for three curates, all young. We had three churches. One of them (Bratton) was more than three miles off; another (Ditton) one and three-quarter mile. Bratton had its two services, Ditton two also, and there were three at the parish church

…

The curates lived together, and 'I do not think,' writes one who knew them intimately, 'that they ever had a shade of disagreement. The only fault was that they did not know the limits of human strength, fasted too much and worked too hard. Mr. Mackonochie's powers of work were wonderful. At one time, when single-handed in the parish, he had four daily services in different parts of it, in addition to his other work.

By 'Summer' 1852 TB was a Curate of the Rev. William John Butler at Wantage - drawn, perhaps, by the death of his sister Frances in June that year and the desirability of being nearer his mother: Milton Hill is just 10km/6 miles to the east. In October 1852 he set off on the 19 months of his Grand Tour. When he returned, his widower brother-in-law, Henry Woodyer (above), appointed him 'Perpetual Curate' of the temporary church Woodyer established in his home at Grange House near Grafham, Surrey and a monument in the churchyard of St Andrew's - the church Woodyer later built, commemorated Bowles as their 'First Incumbent' (Figure 1.6):

> ALSO IN MEMORY
> OF THE REVD THOMAS BOWLES, MA
> FIRST INCUMBENT OF THIS PARISH
> DIED JANUARY 12 1899 AGED 77 YEARS
> BURIED AT ABINGDON
> IN THE COUNTY OF BERKS

After four years - and before Woodyer's church itself had been built, TB had returned to Wantage for six years. A document sheds light on the vicar, the parish and its duties:[21]

Figure 1.6: (a) Monument and (b) detail of the inscription to Thomas Bowles in the churchyard of St Andrew's, Grafham, Surrey.

---

[21] http://anglicanhistory.org/england/ahmackonochie/towle1890/03.html

Already, in 1852, Wantage was celebrated for its elaborate parochial machinery and efficient organisation. Under the energetic rule of the Rev. W. Butler, the present Dean of Lincoln, it was especially fitted to be a training school for the younger clergy.[22] There was no overwhelming population with which to contend, no crushing sense of multiplied duties some of which must needs be left undone; neither was there the isolation and monotony of a single-handed ministry in a country village. There was a sufficient staff of clergy with varied duties, in the town itself, in the large Church schools in the outlying hamlets, without speaking of the different works of mercy carried on by the Sisters, who co-operated with them. Men were not slow to avail themselves of the advantages offered to them of associated labour and systematic training, and it is a curious fact, which has frequently been commented upon, that so many who afterwards occupied important posts, or were in other ways especially distinguished, had been, at one time or another, curates of Wantage.

He may have spent the next two years as a Curate at Garsington near Oxford then a final curacy at Clifton, near Bristol for 3 years. He is next found at Wallingford but now as 'Rector' and just for a single year. Finally in 1875 he began the rectorship that was to last him for 15 years through to retirement in 1890. Although neither Wantage, nor Wallingford, nor even Garsington, are very far from his family roots, East Hendred brought TB back to just 2.5km/1.5 miles from the family home at Milton Hill as Rector of the Church of St Augustine of Canterbury. His sister Emily - a convert to Catholicism, lived in this same rather unspoilt village. The parish was the subject of a monograph a century ago (Humphrey 1923) while Kelly's Directory of 1883 (63) reports:

The living is a rectory, yearly value £600 with the residence in the gift of the Bishop of Oxford and held since 1875 by the Rev. Thomas Bowles M.A. of Queen's College, Oxford.

The church itself was one of those (many) at which TBs brother-in-law, Henry Woodyer, had undertaken additions, restorations and fittings (in 1860-61) (Quiney 1995: 217) (Figure 1.7).

The census of 1881 lists the 59 year old Thomas Bowles as resident in the Rectory (Figure 1.8) with three unmarried servants: a 62 year old housekeeper, 25 year old housemaid and a 24 year old groom and gardener. A brass plate in the Chancel records his rectorship and there is a photograph of him still today in the Vestry (Figure 1.1).

Retirement and Death

When TB retired, he moved to Oxford where he was licensed to preach and was associated with the Church of St Barnabas in Jericho, consecrated in 1869 by a couple dedicated to the Anglo-Catholic tradition in Anglicanism. It is a fine, rather striking building contrasting with typical English churches in the region: a 'Romanesque basilica, possibly modelled on San Clemente in Rome ... with a distinctive square tower, in the form of an Italian campanile.' The church is often named in literature from Thomas Hardy to more recent writers; John Betjeman wrote a poem about it (Figure 1.9).

---

[22] William John Butler (1818–1894): Vicar of Church of St Peter and St Paul, Wantage (1847-1881), where he founded the Community of Mary the Virgin. The parish is Church of England but in the Anglo-Catholic tradition. Butler was Dean of Lincoln from 1885-1894.

Figure 1.7: Church of St Augustine of Canterbury at East Hendred (Photo: David Kennedy).

Figure 1.8: The Rectory at East Hendred where Thomas Bowles lived from 1875-1890 (Photo: David Kennedy).

Figure 1.9: Church of St Barnabas, Jericho, Oxford (Photo: David Kennedy).

TB himself lived nearby: the 1891 census places him a kilometre away at 24 Leckford Road with two servants: Joseph and Eve Johnson, respectively a Gardener and Housekeeper (Census 1891) (Figure 1.10). How long TB remained in Oxford is unclear - he was still in Leckford Road in 1895 (Kelly's Directory 1895: 280; 317).

By *c*. 1898 he had moved to Abingdon and apparently living at 4 Spring Terrace. He died the following year, but in London where his address is given as 16 Nottingham Place, Paddington. The East Hendred guide book explains the circumstances (Humphreys 1923: 211):

> He officiated at St Barnabas, Abingdon (sic),[23] on Christmas Day, 1898, and shortly afterwards died under an operation at St Luke's Hostel, London, on 12 Jan. 1899, aged seventy-seven, and is buried in the cemetery at Abingdon.

A week after officiating at St Barnabas,[24] he officiated at St. Michael and All Angels in Abingdon in whose Chancel a brass plaque records:

---

[23] Repeated in several publications. An error: churches dedicated to Barnabas are rare, there is none in Abingdon but there is the well-known one in Jericho, Oxford, not far from where TB lived, and which would have appealed to his Anglo-Catholicism.
[24] It is just possibly a slip for the Church of St Nicholas in Abingdon which has a Bowles family connection and is hard by their family property of Abbey House. TBs sister Hester had dedicated a stained-glass window in the church: *'Given by Hester Bowles for the Glory of God and in memory of Frances Sellwood her grandmother who died in the faith, March 28th 1842'*. Another family member, the Reverend Richard Bowles had endowed a fellowship providing for the chaplain of Trinity College on Oxford to deliver sermons which continue to this day (Cannon-Brookes 2020: 11).

In memory of Thomas Bowles, Priest, who (died) 12th January 1899. He offered the Holy Sacrifice for the last time at this altar on the preceding New Year's Day. May he rest in Peace.

He was buried in the Old Cemetery at Abingdon, near where he had lived and the two churches in which he had officiated most recently. An impressive but moss-covered monument marks the spot. The epitaph reads (Figure 1.11):

IN LOVING MEMORY OF REV^D THOMAS BOWLES OF MILTON HILL BORN JAN 5 1822 DIED JAN 12 1899. BY LOVE SERVE ONE ANOTHER. ALSO, HIS NEPHEW FRANCIS WILDMAN SELWOOD BOWLES. DIED DECEMBER 1940. GRANT HIM LORD ETERNAL REST.

Thomas Bowles never married. His life was spent as a curate or rector, looked after by (female) housekeepers,

Figure 1.10: TBs Oxford home at 24 Leckford Road, Oxford (Photo: David Kennedy).

Figure 1.11: Grave monument of Rev. Thomas Bowles in the Old Cemetery at Abingdon. The end panel records that it is also the resting place of his nephew Francis Wildman Selwode (sic) Bowles in 1940 (Photo: David Kennedy).

housemaids and other servants. Although he was one of eleven children, all of whom lived until at least 22, only four married.

We know he officiated at the wedding of his sister Frances to Henry Woodyer in 1851 (Goodwin 1996: 1). A chance survival in a contemporary newspaper shows TB not only maintained contact with his friend of Oxford and the Grant Tour, Henry Stobbart, but in 1891 when the latter's daughter Alice married, newspapers for 28th October recorded, *'the officiating clergy were the father of the bridegroom, assisted by the Rev. Canon Bramley, ... uncle of the bridegroom, and the Rev. Thomas Bowles, godfather of the bride, late rector of East Hendred, Berks'*.

He may well have been godfather to more of Stobart's eleven children and to his own nephews and nieces. And of course, he officiated alone or with others at the funerals of siblings, six of whom pre-deceased him.

His will records assets as valued at £543-11-11 net with £195 in securities gifted to first his sister Emily then, on her death, to his niece, Hester Fanny Lake (above). His executor was his nephew Francis Wildman Selwood Bowles.

An obituary described TB as having *'a kindly disposition (who) made friends wherever he went'* (*Reading Mercury*, 21st January 1899).

Religious Views of Thomas Bowles

The mid-19th century was a period of ferment in Anglicanism leading to a considerable number of prominent people converting to Roman Catholicism.[25] Others stayed with the Anglican Church but embraced what came to be called Anglo-Catholicism. As noted above, TB's brother Frederick converted in 1845 and went on to practice as a Catholic priest till his death. Two of his sisters also converted: Emily in 1843 and Alice in 1892. Henry was a keen follower of the future Cardinal Newman but died aged 26 in 1842. TB's mother was evidently unhappy with these conversions, not least Emily who was the first when just 25 and evidently whilst in Rome with her mother and ailing sister, Hester. TB himself remained with the Anglican Church but is then associated with the Anglo-Catholic part. He frequently notes in his travel journal religious discussions he has and refers to Puseyites, Tractarians etc. Interestingly, in 1851, the mother of Lord Schomberg Kerr (Ch. 2) was one of the quite numerous converts to Catholicism from the great aristocratic families.

---

[25] Blain, M. (2019) *The Canterbury [New Zealand] Association (1848-1852): A Study of Its Members' Connections*, updated online edition of a book published in 2000: 8 records that by 1910, '84 graduates of Christ Church Oxford, 41 from Oriel College Oxford, 52 from Exeter College Oxford, and 102 (outstandingly the largest number) from Trinity College Cambridge, and 41 from S John's College Cambridge had converted. Similarly, the secondary school backgrounds show close parallels with the backgrounds of the Canterbury Association members. From Eton 93 converts, Charterhouse 15, Harrow 39, Westminster 21, 33 Winchester.'

Chapter 2

Thomas Bowles's Grand Tour

It is likely that Thomas Bowles explained in the lost first volume of his journal the circumstances in which he joined the Grand Tour being organised by his friend and fellow clergyman, Henry Stobart, and his intended role accompanying the latter and his two young aristocratic charges, Lords Henry Scott and Schomberg Kerr (below). As it is, we can get details and an outline of the first few months from the lengthy letters written by Stobart to his mother and a 'journal' he kept in parallel or sometimes instead. Although much of what Stobart saw and experienced would apply equally to TB during the many months they were together, for the present book, commencing just after they had parted company in Sri Lanka, it is largely irrelevant and only mined below for what it may tell us in general of TB's character.

As noted in the previous chapter, it was against a background of several family tragedies in 1852 - and perhaps spurred especially by that on 21 June 1852, of TB's 22-year-old younger sister, Martha Frances Woodyer, dying in giving birth to her first child, that he set off four months later on 25 October 1852 on his Grand Tour. He was not to return for 19 months.

There may have been hesitation by TB till quite late as a letter from Stobart to his mother, dated 'London. Thursday night. [21] Oct. 1852' says (underlining in original):

> I cannot make out what has happened to Bowles. I have not <u>heard anything</u> from him. I find he wrote to Mr. Green - but I fear now I can have little hope of him going. I am <u>very sorry</u>.
>
> I called by request on Lady Lothian today - who puts her son [Lord Schomberg Kerr (below)] in a measure under my care and charge.

On Tuesday 26th October 1852, writing now from the Royal Hotel in Plymouth, Stobart says:

> The last has been a very exciting week - parting with you all on the Monday - busy day and night in London making necessary preparations - and kept in a constant suspense as to whether or not my friend Bowles would accompany us. William will have told you all about it, up to between 1 and 2 o'clock yesterday when I parted from him. About 3 p.m. we got Bowles' outfit from Silvers and went down to Steventon by the Train at 6.50. We went thence to Milton Hill, Bowles' home, and found him prepared for a start - and we travelled down here by the night mail and were in Plymouth about 7 this morning. Nothing could be done with the young man who had taken the half of the cabin - he was not willing to resign it, tho' I offered him a considerable bribe. Nothing remains but for Bowles and me to take my Cabin and the half of the other between us, and I dare say we shall manage to make ourselves comfortable.

In short, the intention had been to have a cabin each but instead they had to share and to have the use of half of a second cabin with a stranger. Presumably the young aristocrats did have a cabin each while their servants shared. The weather in Plymouth was bad and Stobart anticipated a delay of two or three days; in the event they were held there for 12 days, only finally sailing on 9th November. Time for yet more *'last letters before sailing'* and further news (28 October):

> *I am sure you will be glad for my sake that my friend Bowles is going to accompany us. I should have entered, I fear, on the long sea voyage with a baddish heart otherwise. He will be with us the whole Tour, if spared.*

On 8th November Stobart is writing yet another letter from Plymouth and acknowledging that *'Bowles warned me not to say "positively the last time"'*. In fact, he sent another brief letter from the ship itself the following morning on the back of which he wrote: *'Left Plymouth Breakwater 12.10 pm Tuesday.'* They were seen off after their several false starts, by Lord Henry's parents, the Duke and Duchess of Buccleuch; there is no mention of Lord Schomberg's parents, and it seems the initiative for the trip lay with the former. The high standing of the party is implied by their being transported to their ship on *'the Admiral's Barge'*.

Bowles's Travelling Companions from Plymouth to Australia and Beyond[1]

Rev. Henry Francis Stobart (Chester-le-Street, 26 April 1824 – Funchal, Madeira, 30 December 1895)

Stobart's family were colliery owners in Co. Durham. Henry, the second son, was sent to the short-lived Grange School in Sunderland. Stobart and Bowles were then overlapping contemporaries at Oxford - both enrolled in The Queen's College, though the latter was over two years older because of his prior years training for the law (Chapter 1). Stobart graduated B.A. in 1847[2] and was ordained a deacon in 1849 and as a priest in 1852 leading to a succession of curacies. Between Oxford in the mid-1840s and ordination, however, he had travelled extensively. Earlier letters home to his mother show him at Rome and Naples, Athens, Constantinople, Beirut, Damascus, Jerusalem, cruising on the Nile then home via Malta. Overall, they cover from at least January/February 1846 until after May 1847 - probably 18 months or more.[3] The journey in the Near East was certainly adventurous enough for a 22 year-old. It included a visit to the notoriously difficult of access Roman ruins at Jarash in Jordan; he was to visit those ruins again in 1855 - one of the few westerners to do so in the 19th century, much less twice. That may not have been the total of his travels as he seems to imply in a letter written while sailing from Plymouth to Cape Town in late 1852 and sailing far to the west well towards South America, that he had previously been in Rio. It was surely because of his status as a clergyman and his previous lengthy and adventurous travels that - aged 28, he was employed at a salary of £200 p.a. to accompany Lords Henry and Schomberg to Australia then home via India and perhaps Egypt and the Holy Land.

Presumably Stobart had recommended Bowles join the Grand Tour - at his own expense, and a second clergyman would have seemed a useful addition. Telling his mother about the two

---

[1] Although all three companions – especially Stobart, are deserving of lengthy biographies, none was with TB during any part of the journey treated in this book.
[2] Foster's *Alumni Oxoniensis* gives 1847 but he had matriculated in May 1842.
[3] Copies of the Grand Tour letters, and many transcriptions of them, are held in the National Library of Australia in Canberra and in the Greenwich Museum in London. The originals and copies of many of his other letters are in the keeping of descendants some of whom I have been able to identify and contact. They are a wonderful treasure-trove.

young aristocrats, Stobart noted that *'My friend Bowles, too, they both like very much and feel a great addition to our party.'*

Stobart and Lord Henry parted from TB in Ceylon and, after touring in India in 1854, returned to Europe via Egypt. They seem to have set off for the East again immediately: they spent December 1855 - April 1856 in Egypt and did the same again several more winters, a remarkable series of visits which included the Long Desert Route to Petra in early 1857 (below).

Stobart was a keen collector of antiquities and even published an account of some he took home (Stobart 1855). His letter of January-April 1856 is explicit in mentioning his purchase of papyri and packing of boxes with items he intends to sell. Some he sold but were often then gifted by the new owners to museums; others were later gifted by descendants to museums. The collections - now in the museums in Liverpool, Bristol and Brighton, and in the British Museum and Griffiths Institute in Oxford, include a number of important artefacts and papyri.

On 9th December 1856, St Clair Kelburn Mulholland, a Belfast linen manufacture, and his 26-year-old daughter Annie Mulholland (1831-1926) and her sister, left Shepheard's Hotel in Cairo to embark on a Nile cruise; Stobart and Lord Henry set off from the hotel that same day also to start a Nile cruise. Stobart, Lord Henry and the Mulhollands then travelled across Sinai to Aqaba and Petra together. A few months later on 15 September 1857, Stobart and Annie were married in Lisburn Cathedral. Thomas Bowles was best man and Lord Henry Scott was also present. They were to have nine daughters and two sons. The elder son - as was common, was named for his maternal grandfather: St. Clair Kelburn Mulholland Stobart (1861-1908). He died relatively young and is perhaps better-known as the husband of the redoubtable Mabel Anne Stobart (née Boulton), feminist, author, adventurer who led her own privately raised female nursing units to the front during the First Balkan War in 1912 then - via narrowly escaping being shot by the Germans as a spy in Belgium, to Serbia in the First World War. Naming their younger son Henry John Scott Stobart (1867-1931) - rather than William for Stobart's own father, as we might have expected, was evidently in honour of Lord Henry Scott.

Stobart held various curacies and rectorships but in 1865 the Duke of Buccleuch appointed him Rector of Warkton, Northamptonshire, where he remained until 1881 when he seems to have retired. He was evidently wealthy - he is recorded in 1875 as having land-holdings in Co. Clare in Ireland of 4249 acres/1600ha, so probably acquired from Annie. He died in 1895 at Funchal, Madeira, leaving over £14,000 net to his heirs.

He remained in contact with Bowles throughout his life, the latter being godfather to at least one of his children (Chapter 1).

Lord Henry Scott (5 November 1832 - 4 November 1905)

Henry John Montagu Douglas Scott Montagu was the second son of a great aristocrat, the 5th Duke of Buccleuch. As a younger son of a Duke, for most of his life he was styled 'Lord Henry Scott'. He was educated at Eton but, suffering from asthma, was advised to spend the cold and damp British winters in warmer, healthier climates. The long tour with Stobart in 1852-54 when he was aged 20-22, was part of that and was followed by several successive winters in Egypt with Stobart (above). In 1861 he was elected as a conservative M.P. for Selkirkshire then for South Hampshire which he retained until moving to the House of Lords in 1885, with the newly created title of 1st Baron Montagu of Beaulieu.

On 4 August 1865, aged 33 he married the 30-year old Honourable Cecily Susan Montagu-Stuart-Wortley (1835-1915), a daughter of the 2nd Baron Wharncliffe. Lord Henry had met the latter in Egypt and the Levant in 1855 but this younger daughter had not been with him.

Lord Schomberg Kerr (2 December 1833 – 17 January 1900)

Schomberg Henry Kerr (1833-1900) was also a second son, and also of another great Scottish aristocrat, the 7th Marquess of Lothian. He was educated at Trinity College at Glenalmond in Perthshire, Eton and New College, Oxford. After his return from the Grand Tour in 1854 he embarked on a career in the Diplomatic Service until the early death of his elder brother and his own unexpected succession to the title as the 9th Marquess of Lothian in 1870. Now seated in the House of Lords, he held numerous government posts including Secretary of State for Scotland (1887-1892).

On 23 February 1865, aged 33, he married the 21-year old Lady Victoria Alexandria Montagu-Douglas-Scott (1844-1938), sister of his Grand Tour travelling companion, Lord Henry Scott. In 1889 their daughter, Lady Cecil Kerr, married Lord Henry's son, the 2nd Baron Montagu, her cousin.

Servants

Largely invisible is that the young aristocrats each had a personal servant. On the Grand Tour we get a single reference to 'Hutchenson'. Presumably TB and Stobart had at least some support from the servants while all together, but TB would have travelled without any servant when he parted with Stobart and the two young aristocrats.

The Journey

In the absence of TB's first journal we can turn to Stobart's letters, written on board the '*Ship Resolute*', the first, compiled under various dates along the way then posted at Capetown which they reached on 30th January 1853. They initially experienced awful weather and rough seas causing damage and distress on board even amongst the more privileged passengers such as themselves. They dreaded that the misery of the first fortnight might possibly endure for the entire three or four months to Sydney. Happily, the weather improved, and they settled into a routine: early exercise, morning prayers, reading alone or to the two young aristocrats, chess, conversation and '*evening service about 10 p.m.*' and even a service on deck assisted by Bowles.

They ate well. In the absence of refrigeration, the ship carried live animals to be butchered along the way. The cow had to be killed early on, leaving them with only '*preserved*' milk, '*a poor and greasy substitute*'. However, they were well supplied with fresh meat: '*Our livestock consisted when we started of some 30 Southdowns [sheep], 20 pigs, 26 dozen chickens, and a large number of geese and ducks.*' On the other hand, meals on a rolling ship were tiresome with table items drifting away or towards you including your neighbours' plates, only mitigated a little by a network of cloth-covered ropes and slim sandbags across the table onto which diners could hold.

Stobart gives a few details of their fellow-passengers noting in particular a large family of seven or eight, one of who was a sick young woman, and that Lord Schomberg generously gave up his (presumably single-occupancy) spacious cabin to them. Despite the claustrophobic nature of such a long voyage, Stobart reports that *'they all got on well in the Cuddy'*.

They never saw land for the first month, passing at night even Madeira - where Stobart was to die 43 years later. They arrived at Cape Town on the night of 29th January 1853 and went ashore the next day, after 82 days at sea. They left Cape Town after just two days on 1st February 1853 finally arriving in Sydney on 22nd March 1853, after a further 50 days at sea.

Stobart continued writing letters home and now kept a journal, too. Most were sent to his mother but he also mentions writing to 'the Duchess' (of Buccleuch), presumably his reports on the progress of her son. Lord Henry had been sent on this tour because of his poor health and Stobart refers to it several times including - approvingly, that the young lord was 'stouter and stronger' having put on a stone in weight. In a letter to his mother, he notes he was keeping *'very exact accounts of expenses.'*[4]

Presumably Bowles was assiduously writing his own account in the lost Journal 1. After Cape Town, Stobart's next letter is dated 29 March 1853 from Sydney and a succession of letters some of which substituted as a 'journal' followed thereafter. Initially Bowles was with Stobart and the young lords during their two months in the capital of the New South Wales Colony. However, the latter planned to spend the winter months in the warmer climate around Moreton Bay (Brisbane) and Bowles seized the opportunity to strike out on his own: in late May he set off for New Zealand and was gone for over three months. In the absence of his own Journal 1 and – of course, Stobart's letters, for this expedition and it is only after the latter returned to Sydney from Moreton Bay on 19th September that we hear that Bowles had arrived back two weeks earlier.[5]

On 8th October 1853 they set off again as a group on the *'Early Bird'*, bound for Hong Kong but via trading stops in New Caledonia (Isle of Pines) and Vanuatu (Tanna) where they were to spend nearly three weeks much of it ashore exploring and astonished by the appearance and customs (including cannibalism) of the inhabitants. (From 28th October onwards we now have TB's Journal 2). A further 39 days followed at sea before dropping anchor in Hong Kong on 16th December. They spent a couple of nights ashore, then - undaunted by so much sea travel and much more to come, undertook a brief trip up the Pearl River to Canton (now Guangzhou) before returning to Hong Kong for Christmas and Boxing Day and to board the *'S. S. Singapore'* on 27th December. The journey took them past the coast of Indochina to Singapore on 2nd January 1854. The next day they were off again and after a brief halt in Penang reached Galle in Ceylon (Sri Lanka) on 11th January 1854. On Ceylon the party broke up again with Bowles staying in a hotel while the other three were accommodated in a government building. They also toured separately.

On 23rd January Stobart and the two young aristocrats set off by steamer for Calcutta and several weeks touring in India where Lord Henry encountered two of his brothers. They

---

[4] It seems that some at least of these reports and expenses survive in the archives of the Dukes of Buccleuch.
[5] Stobart says Bowles sailed back to Australia on the 'Old Hashemy', one of the two ships on which the first cargoes of convicts for New Zealand had arrived from England in 1849, just five years earlier.

subsequently returned to Ceylon and sailed to Egypt. At that point Schomberg continued home while Stobart and Lord Henry embarked on a Nile cruise, then the Short Desert Route through El Arish to Palestine. At Jerusalem they undertook tourism in and around the city and to the Jordan and Dead Sea. More adventurously, they joined a party totalling 12 westerners, led by the British Consul in Jerusalem, James Finn (Chapter 10), on an expedition across the Jordan to Jarash. This was Stobart's second visit (above). Back in Palestine, they continued touring the famous places of the Bible, before following in TB's footsteps a few months earlier, to Damascus, Baalbek, and Beirut. They sailed from the latter to Alexandria where, after a brief stop, they sailed via Malta to France.

Meantime, back on Ceylon, Bowles was preparing for the long journey that would take him back to Europe which is treated in detail in the next chapters. In summary, he set off on the *'S. S. Bombay'* on 28th January 1854. Ten days later they stopped in Aden for a few hours to take in coal then on to Suez which they reached on 12th February and two days later he was in Cairo. He had two weeks in Cairo, mainly in the company of two men he had met on the voyage from Galle. On 28th February, he joined several other westerners on the Long Desert Route across Sinai to St Catherine's Monastery, Aqaba, Petra then on to Jerusalem arriving 3rd April. As Stobart was to do the following year, he undertook two weeks of tourism in the city and to the Dead Sea finally departing on 17th April for tourism in northern Palestine before arriving in Damascus. He had just a few days there before continuing via Baalbek to Beirut from which he sailed on 8th May for Alexandria. Then two days in the latter city before boarding the *'S. S. Mentor'* to Malta and on to Marseilles which he reached on 20th May. Exactly a week later, 27th May, having hastened across France, he was home at Milton Hill. He had been away from home for one year and seven months (Table 2.1).

Table 2.1: Main stages in the Grand Tour of Thomas Bowles.
Roman font = Journal 1; **Bold font = Journal 2**; *Italic font = Journal 3.*

Date	Place
Monday 25 October 1852	Departed Milton Hill with Stobart
Tuesday 9 November	Depart Plymouth
Saturday/ Sunday 29/30 January 1853	Arrive Cape Town
Tuesday 1 February	Depart Cape Town
Tuesday 22 March	Arrive Sydney
Tuesday 26 May	At Sydney and says Bowles is about to leave for New Zealand while he and Lords H and S go to Moreton May
Monday 19 September	Says Bowles had arrived back in Sydney on board the 'Old Hashemy' from New Zealand a fortnight earlier
Saturday 8 October	All four sailed from Sydney to Hong Kong on 'Early Bird' via New Caledonia and Vanuatu
Tuesday 18 October	Landed on Isle of Pines (New Caledonia)

Thomas Bowles's Grand Tour

Sunday 23 October	Left Isle of Pines
Wednesday 26 October	Arrived Tanna (Vanuatu)
Friday 28 October	Bowles at Tanna
Monday 7 November	Left Tanna
7 November - 16 December	At sea for 39 days
Friday 16 December	Dropped anchor at Hong Kong
Saturday 17 December	Stayed ashore at the Club House in Hong Kong
Tuesday 20 December	Steamer up the Pearl River from Hong Kong to Canton (Guangzhou)
Saturday 24 December	Left Canton to return to Hong Kong
Sunday 25 December	Christmas and Boxing Days in Hong Kong
Tuesday 27 December	Sailed from Hong Kong on P&O 'S. S. Singapore' for Singapore
Monday 2 January 1854	Arrived Singapore
Tuesday 3 January	Left Singapore via brief stop in Penang on 5th January
Wednesday 11 January	Arrive Galle early in morning after 14 days from HK
	Bowles touring in Ceylon often separately from Stobart
23 January	Stobart, H and S leave on steamer to Calcutta
Saturday 28 January	*Departs Galle for Aden and Suez aboard 'S. S. Bombay'*
Monday 6 February	*Aden. Day ashore*
Sunday 12 February	*Arrival in Suez*
Tuesday 14 February	*Arrive Cairo*
Tuesday 28 February	*Depart Cairo on the Long Desert Route.*
Sunday 26 March	*Arrive Petra*
Wednesday 29 March	*Depart Petra*
Monday 3 April	*Arrive Jerusalem*
Monday 17 April	*Depart Jerusalem*
Wednesday 26 April	*Arrive Damascus*
Saturday 29 April	*Depart Damascus*
Wednesday 3 May	*Arrive Beirut*
Monday 8 May	*Depart Beirut by sea for Alexandria*
Friday 12 May	*Leave Alexandria onboard Steamer 'Mentor' to Marseilles via Malta*
Saturday 20 May	*Arrive Marseilles*
Wednesday 24 May	*Arrive Paris. Depart for Rouen*
Thursday 25 May	*Crossed from Dieppe to Southampton*
Saturday 27 May	*(Home with family at Mill Hill)*

Chapter 3

Sailing from Galle via Aden to Suez and Cairo

Introduction

Bowles arrived in Galle, then the principal port of Ceylon, on 11th January 1854. He was to spend 17 days on the island including excursions to Colombo and Candy which he found very demanding. For much of that time he was joined regularly by one or more of his long-time companions to and in Australasia (above): Rev. Henry Stobart, Lord Henry Scott and Lord Schomberg Kerr. On 23rd January, however, these three boarded the S. S. Bengal to sail to Calcutta. Bowles did not again meet with them though he notes in his journal (11th May) when he was again in Alexandria more than three months later, that Stobart and the two young aristocrats had passed through a fortnight before.

Bowles embarked on the *S. S. Bombay* at Galle on his own westward voyage on 28th January 1854 (Table 3.1). The ship sailed direct from Galle to Aden - eight days, where it re-coaled, some passengers went ashore and another came on board. Then it sailed direct from Aden through the length of the Red Sea to the Gulf of Suez and final disembarkation at the port of Suez on 12th February – a further five days. Accomodation in Suez was limited and poor and there was little to attract western travellers. Bowles, like most, transferred almost immediately to one of the 'vans' for the crossing of the *c.* 150 km/95 miles of desert to Cairo. At Cairo the passengers in transit and keen to get 'home' travelled upsteam on the Nile to Alexandria and a connecting steamer to Europe. Like others with more time, Bowles sought accomodation in one of Cairo's hotels for westerners. While there he made the important decision to undertake the immense adventure of following the Long Desert Route to Jerusalem via Mt Sinai, Aqaba and Petra (Chs 7 and 8), which then led on to extensive touring in Palestine, Syria and Lebanon (Chs 9 and 10).

Table 3.1: Timeline of TB's journey from Ceylon to Cairo (cf. Appendix 2).

Date (all 1854)	Place	Journal Page(s)
Saturday 28 January	Galle. Departs on the steamer *S. S. Bombay* in the evening.	411-414
Sunday 29 January	On board steamer *S. S. Bombay*	414-415
Monday 6 February	Aden. Day ashore	432-439
Tuesday 7 February	Depart Aden overnight. Red Sea: on board steamer *S. S. Bombay*	439-450
Sunday 12 February	Arrival in Suez	450-458
Monday 13 February	'Van' from Suez to Cairo through Desert	458-463 and 1-2

Figure 3.1: A stereophoto of 1862 of the S. S. Bombay in Mort's dry-dock in Balmain, Sydney (Mitchell Library, State Library of New South Wales) (Public Domain).

S. S. Bombay

Several ships bore that name in the 19th century and Bowles says little about the ship beyond remarking on the vibration of the screw and that it was evidently a coal-fired steamer. It can be identified as the one built at Partick, Glasgow in 1852 for P and O. The P and O Heritage Fact Sheet ('Bombay 1852') on it record that in June 1853 it was transferred to the Calcutta-Suez route. In January 1862 it was moved to the Bombay-Galle-Sydney service and that same year it was recorded in an early stereophotograph in dock in Balmain (Sydney) (Figure 3.1).

Although he names 16 passengers and unnamed family members, he only indirectly suggests the overall number. Fortunately, a passenger on the same ship some 15 months later as it sailed from Galle to Singapore left a record, presumably obtained from a crew member (Macintyre 1854):[1]

The Steam ship the "Bombay", on which I embarked on 9th April [1855] in the harbour of Point de Galle for Singapore [arrived 15th April], is of 1200 tons [1186], 270 horse power, of a speed of ten to eleven miles an hour, consumption of coals in the 24 hours (is) 25 to 31 tons. She is propelled by means of the screw worked on the indirect action of the steam engine, at a speed of 60 to 75 revolutions a minute. The direct motion of the piston of the cylinder and other parts of the engine is at 20 to 25 strokes a minute, which a cog-wheel attached to the shaft of the screw multiplies three times and gives the necessary motion to the screw. .... The vessel is nearly new and is built of iron, and the engine and every part of the machinery are made on the most improved principle and are of the best workmanship. The condensation from the steam for distilled water may be estimated at 400 to 500 gallons of pure water in the 24 hours. The water is carried into tanks and pumped up on the lower deck, as wanted, and there is no restriction put on the use of it, within reasonable boundaries.

---

[1] James Macintyre was on a round-the-world trip for which he kept a detailed multi-volumed diary now also held by the National Library of Australia. He interrupts his account of the voyage to Singapore to give a detailed report on the ship, its crew and operations. Some is not relevant for Bowles' journey away from Asia but it is worth consulting for details of meals etc.

The number of persons on board was as under:

Officers and Sailors of all kinds ------- *about 50 men*
Engineers, firemen and Coalmen ----- *about 50 -*
Stewards and Cuddy Servants -------- *about 20 -*
Passengers --------------------------- *45 men and women*
                             165 Persons

We can add from the P&O Heritage Fact sheet that she was 71.30m long x 9.57m.

The number of people mentioned by TB is quite small and we know he certainly had more companions than he names. For example, he never mentions Strachan although he was certainly there. Likewise, when they reached Suez and transferred to 'vans' for the journey to Cairo, he was pleased he was in the first batch of four vans, each carrying six people. Kennedy and Strachan seem to have been in a later batch. The implication is the passengers numbered up to about 48, which accords with Macintyre's figure of 45 (above).

Macintyre contrasted the more Asiatic passenger party on the leg from Galle to Singapore and certainly the same man's report on the passengers he had previously travelled with from Suez to Galle on a different ship was more in harmony with Bowles' experience in the opposite direction on the *S. S. Bombay*. For Bowles's journey, the majority of those named were British military officers, administrators or merchants. Most were coming from India but at least two - Kennedy and Strachan, were coming from China.

Just two weeks before Bowles made the journey, an anonymous traveller, starting and later returning to Sydney, published an account in *The Sydney Morning Herald* of the same Galle to Suez and Cairo section on *The Oriental* (Anonymous 1854). That was a larger ship and older (*c.* 1840), carrying *c.* 100 passengers but he was very pleased with his cabin, the meals and the service (despite no more stewards than the *S. S. Bombay* had for half the number). Like Bowles, he found many passengers were *'a variety of civil and military officers, such as majors, captains, lieutenants, & c.'*. There were several eminent passengers: *'a commodore in H.M. navy, the French governor of Pondicherry and family, a Bombay Catholic bishop'*. Unlike Bowles, who names several fellow passengers and sometimes adds details, this traveller named just one – (the soon-to-be Admiral Lambert).

Aden

Bowles went ashore in Aden (Figure 3.2) while the ship took on coal. He explains that though P&O had a coal depot in Suez, it was so much more expensive there that it was only ever used in emergencies (cf. Barak 2015). With coal needed for the trip to Suez then back to Aden, the decks were *'crowded and besides a great many sheep & stores'*. A useful reminder of how cramped the decks were on ships which carried meat on the hoof and especially now those requiring coal. As Macintyre noted, the crew of the *S. S. Bombay* included about 50 engineers, firemen and Coalmen. Bowles undertook some tourism in Aden but gives most of his entry to his meeting with the Chaplain, Rev. George Percy Badger, and visiting the grave of a friend, Rev. Charles Tombs (Ch. 4).

Figure 3.2: Port of Aden, Arabia in 1891 (From F. E. Chadwick et al. 1892: 272) (Public Domain).

Suez to Cairo

The Pasha's hotel at Suez could accommodate several hundred and served respectable food but few passengers were tempted to stay beyond the time constraint of waiting for their van.[2] The 'vans' were two-wheeled carriages (Figure 3.3), which were drawn by a relay of horses and mules westwards across the *c.* 150km/95 miles of desert to Cairo.

That could normally be accomplished in a day of gruelling travel which could be bitterly cold at night early in the year. It seems batches of 4 set off at 5-hourly intervals. Once again, the account by an anonymous traveller in *The Sydney Morning Herald* a few months later of his experience on 31 January 1854, just two weeks before Bowles, of landing in Suez then of the desert crossing in a van, provides more detail of what Bowles describes more briefly (12th and 13th February) and is worth quoting at length (Anonymous 1854):

> *Before the ship arrives at Suez, all the passengers have formed parties of six in number for crossing the Desert, when one of each draw lots for the 1st, 2nd, 3rd, and 4th batch, and so on. This arises from the fact of the tempting comforts and curiosities of Cairo being so much greater than the miserable and wretched, dusty, dried up city of Suez. Frequently, when there are a large number of passengers, some have to remain 24 and even 36 hours at Suez, where there is nothing, to be seen except a few*

---

[2] The Cairo-Suez railway line was constructed between 1856 and 1858. The Suez Canal was not completed till 1869.

Figure 3.3: 'Desert Van' on the Suez-Cairo route (from Barber 1845: 34) (Public Domain).

filthy, dilapidated, houses, bazaars, &c., whilst there is so much at Cairo, and then they only just arrive at the latter in time to reach the Nile boat, which leaves the moment the last van is in; hence great value is laid upon those who are fortunate enough to secure some of the early vans. The first of our party left at 7 p.m. in 4 vans, containing six in each, and drawn by two mules in the shafts and two Arab horses for leaders. In that position they work best together, the mules being more steady than the horses, which possess the speed but cannot be depended upon for safety. The vans are extremely comfortable, being fitted up with arms for each person and well padded with horsehair and leather, you would hardly know you were not reclining in an easy chair at home, were it not for a few odd stones and holes, which are apt to bring the head and other parts of the body in contact with the roof or side of the van occasionally, just to convince you that you are travelling, which after all is nothing compared to the unmerciful usage of a Bathurst stage-coach. These vans are just large enough for six people; they are set on two very high wheels, and are so arranged that they can be all opened or all closed, just to suit the weather or the convenience of the passengers. The driver sits elevated on a perch in front, whilst the groom who travels with the van stands on the door step behind, so as to balance the weight as much as possible. Every van is painted the same colour - a light yellow or buff, and are all kept in beautiful order and repair, at least such as I saw were. It was my fate to leave in the second lot, which was ordered to be ready five hours after the departure of the first, making it just midnight. ...

...
So that it was a thorough stormy night and no mistake, and to turn out at such a time to encounter the terrors of the desert required almost more than ordinary courage. I did think (don't know what

some of the ladies thought) they might have waited until daylight; but it was of no use, go we must or be left behind - so on with top-coats, extra trowsers (sic), in fact such an increased quantities of extras, that by the time we were all packed in the van, when the fellow popped in a torch to see that all was right, we looked like a coach load of clothes, or a party fitted out for a cruise after Sir J. Franklin.[3] Fortunately, our having been companions before kept us from being strangers now, so that we understood each other, and were fire proof against suspicion of our fellow travellers, who were all military officers, and no ladies fortunately. Soon after 12, our party consisting as before of four vans, and six people in each, started, not without a little backing and filling on the part of the mules, or the horses, or both; for as we were all boxed up (almost air tight) we knew but very little what was going on outside; so we had no alternative but quietly to resign ourselves to the tender care of the Arabs for the remainder of the night; and away they went, soon clear of the town into the sand of the desert with a jump and then a jolt over a few stray sandstones, which shook almost every tooth out of our head, say nothing about a bump which was given in. We, of course, prepared ourselves for a rough passage, especially as it was such a rough and bitter night, and was just about to congratulate ourselves upon our progress when they all stopped, and there arose such a discussion as was really quite alarming, especially at midnight, and that in a wild desert, and after wheeling round and round half a dozen times, they started off a-fresh; could not tell for the moment whether the whole turn out had not been taken in a fit, and what would become of us. At last one of our party was bold enough to pop his head outside to ascertain the exact cause of this rotary motion, and when he popped it back again a paragraph popped in with it, viz., that we had (that is, the Arab had) lost their way, and were quarrelling as to which was right and which was wrong. At last each went his own way, which I thought was much the best, as there would be then a chance of someone being right. They soon, however, got together again, and kept together a few miles, when they stopped again, and just repeated the former dose again; at last it became so frequent that we got accustomed to it, and hardly noticed it. It appears that, in consequence of the darkness of the night, they could not see the road marks, which consist of heaps, of old camel's bones, largo stones, &c.; and the only time they were quite sure they were right was when, they drove right over the top of these heaps, just to remind them they were not far out; and when one or the other had not had a shock from one they got frightened lest they should be out of their track. All this was explained at daylight, when the road was visible, and accounted for what we thought the uneven state of the road. About every eight miles we changed horses, but nobody felt inclined to get out to face such a formidable night, so we still knew nothing of our progress except as we measured distance by the changes, which we supposed to be about every eight miles. At the station we arrived about half-past five in the morning, when we had to get out to feed, and replenish the inner man, which was by this time very necessary, for the intensity of the cold had almost stopped circulation, and it was something doubtful whether all were alive or not. To attempt to picture the frightful dreary scene witnessed on passing from the van to the station, I could not. The wind blew a hurricane, and at times poured with rain; and, at the time we entered the station, we were assailed by a storm of sleet; such a thing I understand, has never been remembered seen in the desert before. The darkness of the night was just breaking for the light of morning, and although you might see a considerable distance, yet there was nothing but the solitary station alone in an ocean of sand. The country is by no means level, and whilst it is not mountainous, yet it is almost as uneven as any other part of the world, being a continuation of sloping hills, plains and valleys. Not the slightest trace of any vegetation or life of any kind beyond ourselves, and what with the frightful wind (which threatened very often to capsize our vans), the squalls of rain,

---

[3] Sir John Franklin, the Artic explorer who had died together with his entire expedition, seeking a Northwest Passage in 1847.

the piercing cold, the heavy clouds, added to which, the wild desolate appearance of everything, left such an impression upon my mind of a winter's night in the desert, that I never shall forget half an hour's active exercise in the hall of the station restored life and animation, so far as to qualify the appetite for its share of the performance; and although the table was adorned with the most tempting dishes, yet there was a good supply of provisions of one sort and another just to check the current of hunger, and stop a gap until we could get something better. Two or three cups of coffee, which was pretty fair, warmed us up, and in the course of an hour and a half we were again ready for another start. By this time it was nearly daylight, but the storm continued, and, notwithstanding the rain, such was the force of the wind, at times, that the sand was driven with such violence against the sides of the van, as to render it impossible to hear each other speak, and frequently the poor horses and mules would turn right round, and would not face it until they were compelled by the lashes of whips, which came rather heavy from the iron hands of the hardy Arabs. With each party is an outrider, who carries a severe whip, and knows how to apply it in case of need, by which means he manages to keep them all pretty well together. As the day advanced, the sun peeped out occasionally, just to throw a light upon the dismal country around; but having had no sleep all night, and nothing over cheering to be seen, we all began to nod very respectfully to each other, and a slight jolt of the van every now and then brought two heads in contact with each other, and knocked all the sleep out for a time. So we went on - a little talk; soon tired, and then another sleep, until we arrived at the second station, which was about eleven o'clock, and about half-way between Cairo and Suez. This was a superior building to the last, and is immediately opposite the Pacha's palace, where he occasionally visits for the purpose of hunting, but what, we cannot imagine. Vultures were the only living animals we saw, and they were very numerous, especially about the carcases of the poor camels, many of which, and the dry bones of many others, lay strewed along the side of the road as we passed along. There is one peculiar fact connected with these birds, which is, that they are never seen unless in the vicinity of their prey; but the moment the camels are abandoned by their drivers, they flock in shoals and commence devouring, and soon leave nothing but the bones, which remain bleaching in the sun. After replenishing the inner man with a feed of curried all-sorts, howlets, and a few other ingredients supplied to stop a gap, we started again with nothing particular fresh except a little occasional amusement arising out of a jibbing mule, or a jumping, kicking horse, which plunge and wheel round or do anything except go ahead. The cold continued very severe, and the wind continued to blow in fearful gusts; and, notwithstanding the frequent showers, we were inundated with sand and fine dust when we left the last station. The palace, which is an immense building, and possesses a most commanding view of the desert (being situated on a high hill), was literally enveloped with sand. From here into Cairo, which is about 45 miles, the road is very good, being kept in repair for the Pacha's accommodation. With the exception of vans going to Suez, to bring up remainder of passengers, and a few camels, we saw nothing worthy of notice - a dry, sandy, barren desert all the way. About three o'clock we arrived at the third station, which is superior to any of the others, and better kept, as the Pasha probably drops in occasionally on his way to the palace. Here we got some beautiful, oranges and other fruits, which had been sent from Suez in addition to the usual display of poultry, curries, &c. Bottled ale, wines, and spirits, or tea and coffee, can also be had at a reasonable rate, and at this station were very good. Half-an-hour was allowed here, so we made best use of the time, and stowed away sufficient stock to carry us into the grand city of Cairo. At these stations, I may mention, the Transit Company provide you with food, but you must pay for what you drink. As the road improved with the horses, we were now drawing fast to the last station where we merely changed horses. Here we had four beautiful grey horses, which took us along in gallant style, and did credit to the Arab blood. As we approached the city, droves of donkeys and their drivers were busy making their way towards Cairo, to be ready to convey us to the different places to be seen in the neighbourhood, so that with these and our four vans, each with four good horses, and going at a pretty smart gallop, the road presented quite a lively and amusing appearance; some of the donkeys kept pace with us for some time, the boys being anxious to be at the hotels as soon as we arrived; however, they could not keep steam up, so that they were very reluctantly left behind. Before we entered Cairo I was much struck with the magnificent heap of buildings which are situated to the right of the road, and form the hospital and barracks for the soldiers - they are really most extensive and substantial structures, and are a quite an (sic) ornament to the entrance. By the road-side in front of them is a most grand and gorgeous tent erecting, said to be for Mehmet Ali. On the opposite side, the spare ground was covered with tents and Egyptian soldiers on parade. If I had not been told they were soldiers, I should have taken them for armed shepherds, as each man was covered with a great light brown shepherd's coat with a cape, and a rough Scotch cap, and had it not been for the military appearance of their muskets should have doubted their being soldiers at all. Here the scene soon changed, with rows of trees on each side the road, which was lined with traffic, people in every colour, shape, and dress; streams of donkeys, carts, and carriages; boys and children in endless quantities gave way to excitement, as we galloped through this extraordinary throng; for I fancied our drivers, like most coachmen, delight to show off when an occasion offers, especially going into or through a town, and with our outrider ahead clearing the way, we arrived at the entrance of Cairo full gallop, and in a few minutes brought up in front of the principal hotel, kept by a Mr. Shepherd (sic), ...

Shepheard's Hotel

Shepheard's Hotel - originalled jointly owned by Samuel Shepheard and a Mr Hill from 1841-45 as the Hotel des Anglais, was better known under the name of its subsequent sole owner (Bird 1957). It was very much the place to stay at least for English-speaking visitors its register being a roll-call of the most important western visitors. Unfortunately the original register was lost when the hotel was destroyed during anti-British riots in 1852. Fortunately, a descendant of Sheapherd, writing the biography of the man and the hotel, had made a hand-written copy which has survived.[4] This register is an impressive list of guests, often detailing their full names, origins and at times information on their dates of arrival and departure and that they might, for example, be leaving to embark on a named Nile boat for a cruise upstream to a named place. For example, in December 1856:

Lord Nemy (sic) Scott   The Saqqarah
Rev. H. Stobort (sic)   9th. December 1856

Shepheard's was the most famous and desirable but it could usually only accommodate those who were staying at least a number of days rather than overnighting before continuing to Alexandria. At the time of TB's stay, because of the Crimean War (October 1853 - March 1856), there were fears the hotel would suffer loss of a key part of its business - leisure tourists, but the war provided alternative business. This is Shepherd himself in a letter dated 'Cairo December 30th, 1854' (Bird 1857: 77):

---

[4] Andrew Oliver, pers. comm. Reading the register is problematic in places as it involves mis-readings by the transcriber and now difficulties reading the transcription.

> "I really never had such a sickening of business as the last month, all quality folks wanting more attention and I in a state that I had ought to have been in bed; amongst the number were Lady FitzClarence & party, Honble C. A. Murray & party on his way to Persia, Marquess of Drogheda, Lord Wharncliffe and Lady, Lord H. Scott, Lord Kerr and a host of others more or less distinguished. We have not had a (sic) empty room scarcely now for more than a month and at the present moment have every bed occupied. There was not nearly so many Nile travellers as formerly, but this is owing to the war which takes away those travellers who travel for curiosity only and who are generally the best customers. However the railroad seems to be changing our trade a little as we have latterly had a good run with the Alexandrian merchants and others who take the opportunity of coming to transact their own business in Cairo themselves. We have had a lot of wounded officers from the Crimea staying with us, ..."

Not just 'quality folk'[5] and wounded officers from the Crimea but as other letters reveal, officers from Indian regiments passing through to Crimea and then in 1855 Shepheard got the British government contract to 'victual' troops passing through to the war (and later benefited again supplying troops going east to suppress the Indian Mutiny (May 1857 - July 1859) (Bird 1957: 78; 97).

When TB arrived in his 'van' in Cairo on 13th February 1854, he recorded:

> *We drove on & on & at last entered by a strong gate into Cairo & drove to the Shephard's (sic) Hotel. Several Englishmen were sauntering at the door. "No room - They couldn't take anyone in". However I suggested that I meant to stay for a fortnight & they soon found me a room. I was taken upstairs at once, & a better room promised me in a day or two. The others went on to search at other Hotels.... Mr Kennedy & Mr Strachan came to try & get beds here & thinking of staying till the next packet. They said they did not care what sort of rooms they had if they c(oul)d be taken & were put into some wretched rooms, with a promise of better. Seaton (sic) Karr & Palmer came in - as soon as dinner was over we 3 had some Tea & then I went to bed & I believe they did the same.*

Having secured accomodation at Shepheard's, TB and these two other passengers from the *S. S. Bombay* – Kennedy and Strachan, spent two weeks on tourism in and around Cairo. By 1854 such tourism by westerners was well-established but increasingly Egypt was attractive for other reasons.

---

[5] In listing these 'quality folk', Shepheard omits a mere vicar like Stobart who was the third (and key) member of the trio with Lords Scott and Kerr.

Chapter 4

Travelling Companions and Encounters from Galle to Cairo

Bowles names sixteen people as companions during fourteen days on the steamer from Galle to Suez and others he met during a subsequent two weeks in Cairo. Several of those named as fellow-passengers cannot be identified and were likely no more than acquaintances but the list is instructive of the diversity of people on this route. Some were military officers with or without families; others were people engaged in trade or commerce; some were clerics like himself.

Some are not identifiable despite frequent mentions: e.g. "Captain Hill, who has his wife, a little person like himself & little girl or girls & an orphan niece of his wife's" is named nine times. Likewise, 'Major Hill', possibly the Captain's father. 'An Engineer' is said to be from Newcastle but given no name, nor is the 'French R. C. Bishop', 'Wood', 'Mr Miller', 'Captain King', 'Captain Baynton'. Mr Kennedy is treated in a later chapter as one of TB's Petra companions.

In roughly the order in which TB names them we may explore those who can be identified.

Rev. William George Tupper
(Westminster, London, 9 December 1824 - at sea, near Malta, 15 May 1854)

TB never travelled with 'Tupper' but mentions him a great deal while in Ceylon and met him again briefly in Egypt. Tupper first appears in the Journal before the current transcribed part when Bowles is at Candy in central Ceylon and mentions meeting Tupper on 17th January 1854:

> ... he told me Tupper was in - I went down to dinner in order to meet him. We shook hands but had to sit on opposite sides of the table, & as he has nearly lost his voice (it is reduced to a whisper) we c(oul)d not talk.

Tupper gets several more mentions while in Ceylon and they met and talked and undertook some tourism together. On one occasion they *'said the Litany tog(ethe)r'*. Tupper was apparently in Ceylon before Bowles arrived, travelling alone and finally left by sea - apparently for Bombay (below), on 28th January a few days before Bowles set off for Egypt. Tupper re-appears in Journal 3 on 23rd February just as Bowles is preparing to leave Cairo and taking the time to leave letters with the British Vice-Consul for Lord Henry Scott and Tupper (above). Two days later, however, he met Tupper in person: while travelling east to Suez on the first stage of the Long Desert Route to Petra, he encountered passengers from a ship newly arrived in Suez and heading in vans to Cairo:

> *At the next station we met 4 more vans & I heard that Tupper was in one of the next. We rode on together again & at the next station about 1/2 past 12 we all dismounted for Tiffin & then finding the vans were expected, I sent our Dragoman on & determined to wait till they came. In about 1/2 an hour they came. ... Here again I heard that Tupper was in the next (the last) batch of vans. I*

> *walked most of the next 10 miles & at last as we got near the 12th station we met large caravans of camels carrying the baggage & goods from the steamer. ... On getting close to the station I saw the vans standing near the door & walked on quickly & found Tupper in the room eating dinner. He was much astonished to see me. Poor fellow he is very hoarse (??). He had not recovered his voice & he says his throat in suffering. I left him eating his food & went out .... Tupper came out & we had some talk & then the vans went on their way. ... Tupper commended my change of plans [to undertake the expedition to Petra and the Levant rather than going home immediately]. I am to meet him at Jerusalem.*

Plainly TB had enjoyed his meeting and conversations with Tupper - enough to take the trouble to have written to him and make a plan to meet again in Jerusalem after he had been to Petra.

Without a name or any further details 'Tupper' might have remained almost anonymous, beyond the possibility he, too, was a clergyman, with a shared religious taste to TB. Happily, Stobart's letter home from Candy of 22nd January provides a crucial detail:

> *... reach Candy about 41/2 pm comfortable quarters at Hotel, find Bowles, also Mr. Tupper, an Oxford clergyman of our day, ...*

Tupper can now be found amongst Oxford alumni:

> *Tupper, William George, 5s. Martin, of St. James's, Westminster, arm. Trinity Coll., matric. 23 May, 1842, aged 17; scholar 1842-7, B.A. 1846, M.A. 1849, warden and chaplain of the House of Charity, Soho, died 15 May, 1854.*

His father, Martin Tupper (1780-1844), was a prominent London physician and Fellow of the Royal Society, while his mother, Ellin Devis Marris (1785-1847), was the daughter of the Lincolnshire landscape painter Robert Marris. These Tuppers were as branch of the prominent Tupper family of Guernsey. William George Tupper's eldest brother was Martin Farquar Tupper (1810-1889), friend of Gladstone, 'poet and writer' (Dingley 2004), prolific author, not least of *Proverbial Philosophy: A Book of Thoughts and Arguments*, which went through 38 editions between 1838 and 1860.

As for Tupper himself and probably the reason Bowles found him engaging, his career and character were summed up in a book on *The Oxford Church Movement* (Wakeling 1895: 47):

> *William George Tupper was the youngest son of the late Martin Tupper, of New Burlington Street; born in 1824; educated at Winchester, and scholar of Trinity College, Oxford; was ordained deacon at St. Paul's Cathedral in June, 1849, and priest in the Chapel Royal, St. James's, in 1850, and very soon became warden of the House of Charity, where he devoted his means and remaining energies to the service of God and the poor. The sea voyage recorded in this book [= Out and Home, see below] was urged on him by his friends and physicians. He started in August, 1853; visited the West Indies, Cape Town, the Mauritius, Ceylon, Bombay, Cairo, etc.; he died in his thirtieth year, at sea, in May, 1854, soon after leaving Malta on his way home, and was buried at sea; it was a privilege even to slightly know him: and who shall measure the influence of such a character?*

Bowles makes no further mention of Tupper after their chance meeting in the desert near Suez and it is unlikely he knew till much later why his new friend never turned up in Jerusalem. Perhaps the poor health implied in his long sea voyage and the ailments Bowles reports and which were still present in Egypt some five weeks after they had first met at Candy, led him to abandon his plans for Jerusalem and sail direct home from Alexandria only to die at sea, aged 29.

Tupper's brother Martin apparently published a 'memoir' about him in 1856 preserving details drawn from the surviving writings which were then - as requested in his Will, burnt (Wakeling 1895: 46; cf. Tupper 1856).

> *... Martin F. Tupper, who in 1856 published a short memoir of him under the title of Out and Home. This memoir contained his journal, three sermons preached at sea, and a few poems. It was a small memorial of a life which, as we well know, and his brother tells us, combined the strictest self-denial with the sweetest cheerfulness; a short life spent in secret good doing; the tongue of gentleness, the heart of love, the mind of sympathy and wisdom - indeed, a character of surpassing beauty, wedded to no common powers of intellect. It was his own request that several packets of his manuscript should be burnt unread, and so all were thankful for the short glimpse this volume gave of a character at once beautiful and of great simplicity.*

Stephen Lushington
(1830 – Puri/Pooree, India, 1860)

Bowles introduces Lushington for the first time on 7th February, almost ten days into the voyage:

> *I awoke early & found we were just entering the Straits of Babel Mandeb [= Bab el Mandeb]. I fell in with Captain Hill later & as we were walking up & down, Lushington joined us & Captain Hill went down to bed. Mr. Lushington is nephew of Lord Cranworth & son of Dr Lushington. He lives near Guildford, or rather near Cobham, at a place, he told me, called Ockham Park. I asked him if he knew Henley Park & the Halseys & he knew them well & told me what I didn't know, that one of the Halseys had come out to India during the last 12 months & was already living on borrowed money to a great extent. He Lushington is going home after 2 years only in India for headache & inability consequently to work. Of course I tried to persuade him to go to Gulley [= James Manby Gully].*

The Lushington family (Taylor 2009; 2015; 2020) had been intimately involved with India since at least the later 18th century. Sir Stephen Lushington (1744–1807) of South Hill Park, Berkshire, was a director and later chairman of the East India Company, and the Company had a ship named for him, the 'Sir Stephen Lushington'. A son, Charles Lushington (1785 - 1866), had been Secretary to the government of Bengal, 1823–27 and his wife, Sarah Lushington (née Gascoyne), wrote an account of their return to the UK via Egypt (Lushington 1829). The records of the East India Company College at Haileybury records several Lushingtons amongst its alumni. Charles' brother, Dr Stephen Lushington, was a key figure in the abolition of slavery and, though he had no direct service in India, two of his sons did. Edward Harbord Lushington (1822-1904) entered service in Bengal in 1841 and rose to be Secretary to the Government of India in the Financial Department (Danvers 1894: 414). His service does not fit with Bowles's description and surely the correct brother is Stephen Lushington (1830-

1860) who had graduated from Haileybury in 1850 and served in India from 1851-1860. At Haileybury he had been a conspicuous prize winner, not least in Sanskrit and Hindi (Danvers 1894: 446). He rose to be Joint Magistrate and Deputy Collector of Dacca. It appears from Bowles's note that Lushington had been in India for just two years unwell and finally unable to work. Whether or not he took Bowles's advice to consult 'Gulley' – presumably the well-known James Manby Gully (1808-1883), famous as an advocate of hydrotherapy ('the water cure treatment'), which attracted such eminent people as Charles Darwin, Lord Lytton and Lord Tennyson, he evidently recovered enough to return to India but died there of 'fever' in 1860 as reported in a letter of 9 November 1860 from his brother Vernon to Elizabeth Barrett Browning (Taylor 2009: 196 and n. 11):

> *I must conclude this with telling you that this has been a very very sad for me & mine. The news came by to-day's mail that my elder brother died of fever in India on the 25th Sept; almost, when I was so happy with you. The first of our ten—we are 10 brothers and sisters—is now taken from us, & we all feel it very much. My brother's life was far away, & in some respects he was himself different from us, yet he was so dear to us. I had to write to you, so I have written, & I cd. not help telling you this sad story of mine.*

Such deaths were not uncommon as his aunt by marriage, wife of Charles Lushington (above) had noted in her book over 30 years before Stephen's death (1828: 5):

> *Among the Europeans in India there are scarcely any old persons, as almost everybody is a temporary resident. Hence, if you search the well-tenanted burying grounds of the large cities, you will discover few besides the graves of the youthful, who have been cut off by some violent disease amid the buoyancy of health, or the tombs of those of middle age arrested by death when just about to reap the fruit of long toil and privation in a return to their native land. It is this which renders our Indian cemeteries so peculiarly melancholy; for though we bow to the decree which summons away the aged and the infirm, yet, humanly speaking, and in our blindness, we are apt to pronounce the death of the young to be premature, and a fit subject of aggravated regret.*

Note: Sir Franklin Lushington (1766-1839), Edward Lear's friend, was from a more distant branch of the family. Lear bequeathed him all his papers and he was responsible almost 40 years later for publishing the section of Lear's diaries which recounted the poet/artist's visit to Petra in 1858, four years after Bowles and with considerably more trouble (Lear (Lushington) 1897).

Dr Patrick Gammie
(Forgue (?), Banffshire, 5 November 1812 - London, 20 May 1887)

Bowles records three conversations with the man he invariably calls 'Dr Gammie'. Helpfully he notes under 8th February 1854 that Gammie had been a military doctor in Australia (*'a good deal in Australia with his Reg(imen)t'*) and they had friends in common there. Gammie also had a major adventure in his past to tell Bowles and Kennedy):

> *After breakfast I sat talking sometime at the Table to Dr Gammie & Mr Kennedy. Dr Gammie was describing his being shipwrecked with troops in 2 transports in the one of the Andaman I(sland)s in the Bay of Bengal.*

Two exasperatingly brief but vital indicators that allow the colourful and adventurous life and career of Patrick Gammie to be reconstructed very fully from a variety of records.

Gammie was born in Banffshire, possibly precisely at or near Forgue, now just across the county boundary into Aberdeenshire (below). The name Gammie is quite common in that area. His family – perhaps farmers, was evidently prosperous enough to send him to King's College, Aberdeen from 1826-1828 (aged 14-16). In 1832 (aged 20) he had become a Member of the Royal College of Surgeons (England) implying he had moved to London to qualify as a surgeon but giving Forgue as his home address. Four years later, 17th June 1836, aged 24, he was appointed Assistant Surgeon of the 80th Foot Regiment and in October 1837 disembarked in New South Wales with detachments of his regiment. He remained in Australia till 1840 when he was sent to New Zealand with the first British imperial troops, as the first military surgeon in the new colony (Lawrenson 2004: 9) and found compiling inquest reports on deaths there (Gluckman 2000: 119-125). He was to remain there – indeed, had a horse *'perform well at 1st Auckland race meeting 1842'* (Taylor 1966: 416-7). By 12th August 1844 he was back in Australia where the entire 80th Foot embarked on four ships for transfer to Calcutta in India. Gammie was on board the *'Briton'* and was soon to involved in the adventure Bowles alludes to (Darvall 1845: 31):

> *The companies two, three, and six were on board the Briton, under the orders of Major, after wards Lieut.-col. Bunbury, and consisted of 311 soldiers, including 12 serjeants and 4 drummers, 34 women, 51 children, and the following officers, namely, Captains Best, Sayers, and Montgomery; Lieutenants Leslie and Freeman; Ensigns Hunter and Coleman; and Assistant-surgeon Gammie, medical officer in charge.*

On 12 November 184 - nine years before he met Bowles, the *Briton* from Sydney and the *Runnymede* arriving from Gravesend in Kent, carrying some civilians and a small contingent of soldiers also bound for Calcutta, sighted one another during a storm in the Bay of Bengal. They were then wrecked close by one another on the same small Andaman Island. Remarkably no one was killed in the wreck though a few did die during the 50 days before rescue arrived. The native inhabitants, famously hostile and believed to be cannibals, did appear from time to time including spearing some of the Europeans. It was partly this wrecking and the reaction of the inhabitants that led to the islands being annexed by Britain and, soon after, being used to house some of the condemned from the Indian Mutiny.

Within a year of the wrecking, the owners of the *Runnymede* had commissioned research on and publication of the event (Darvall 1845). Gammie, then aged 32 and still just an Assistant Surgeon, gets just one further mention, as put in charge of the *'Runnymede'* wreck being used as a hospital ship while the *Runnymede's* own doctor took charge of the camp on land.

Gammie's career subsequently developed in India where he was to spend the next eight years. It was a period of warfare in the Punjab, and Gammie was fully involved as recorded in an obituary (Anonymous 1887):

> *Hart's* Army List *informs us that he served in the Sutlej campaign in 1845-46, and was with the 80th Regiment at the battles of Moodkee [= Mudki, 18th December 1845] and Ferozeshah [21st and 22nd December 1845], and in medical charge of the 31st Regiment at Buddiwal, Aliwal (sic) [= Adiwal, 28th January 1846], and Sobraon (medal and three clasps) [= 10th February 1846].*

Figure 4.1: Indian Mutiny Medal of Dr Patrick Gammie (Courtesy David Galt, Mowbray Collectables www.mowbraycollectables.com).

The service with the 31st Foot was evidently a temporary transfer. On 2nd March 1847 he was promoted to Surgeon but now of the 61st Foot. Then, as of 30th December 1853, just as he must have been preparing to go to Ceylon and join Bowles on the *S. S. Bombay*, he was transferred as Surgeon to the 94th Foot. This regiment was based in the Madras Presidency until embarking for the United Kingdom in March 1854 with service then at Chatham, Windsor and in Ireland. Presumably Gammie was with his regiment but from 30th August 1856 till 6th August 1857 he was on half-pay (Drew 1968: 1, 299, No. 4454). Although not recorded in the standard entries on Gammie's career, he evidently returned to India, presumably with the 94th Foot which had been despatched back to India (Karachi) in November 1857 arriving half-way through the Indian Mutiny. Gammie's military decorations from the Sutlej campaigns were known but in 2017, a coins and medals dealer in New Zealand, auctioned an Indian Mutiny medal inscribed for *'STAFF SURGN. PATK.GAMMIE'* (Figure 4.1) (Galt 2017: 57, Lot 579). He had reached the rank of Surgeon Major Staff on 1st May 1855. On 31st December 1858 he was made Deputy Inspector General Army Hospitals. He evidently returned to Britain after the Mutiny and from 1861 onwards is traceable in Britain through a succession of censuses.

While in India, he had married as recorded in a newspaper:

> *At Calcutta, on 23rd August [1849], by the Rev. Mr. Eteson, P. GAMMIE, Esq., Surgeon, H. M. 61st Regiment, to Mary, widow of the late Captain HOLLINSWORTH, H. M. 10th Regiment, and daughter of the late Captain FRASER, H. M. 80th.*

Mary Maclean Fraser had been born on the Isle of Mull in Argyllshire in 1827. As noted, her father was Captain Fraser of the 80th Foot so she had evidently moved to Sydney when he was posted there in 1839:

Lieutenant Frazer, 80th regiment, Mrs Frazer and 10 children passengers on the 'Marquis of Hastings' from London. Departed London 17 March and arrived Hobart 12 August where the male prisoners were landed.

She married Lieut. Henry Arthur Hollinsworth of the 10th Reg. of Foot on 24th September 1842 in Sydney when she seems to have been c. 15/16 years old. Both the 10th and 80th were posted to India in 1842 where her father was killed on 28th December 1845: he *'gloriously fell while leading his company to the charge at [the Battle of] Ferozeshah'*. Then further tragedy: Hollinsworth died *'on 3 October [1848] at Moultan from the effects of wounds received in battle.'* At age c. 21 Mary was a widow. It seems very likely she would have known Gammie in Sydney and in India where Gammie also served at the Battle of Ferozeshah where her father was killed. Less than a year after being widowed – as was not uncommon in India, she remarried to this older but rising army surgeon, a fellow Scot and probably known to her family: he was 35 [b. 1814] and she would have been 21 [b/1827].

Although Bowles mentions only Patrick, we may wonder whether Mary was also on the *S. S. Bombay* in January and February 1854. Patrick and Mary Gammie seem never to have had children.

Records have Gammie transferring as of 1st May 1855 from 94th Foot to be a 'Staff-Surgeon of the First Class'. In the Census of 1861 Gammie (aged 48) is living in Devonport with four servants and described as 'Deputy Inspector General Army'. In 1871 they are in 14 Stanhope Gardens in Kensington with a Butler and two other servants and Gammie is now 'Insp. Gen. Army Hospital'. In 1881 (aged 68), still at 14 Stanhope Gardens, with five servants now, Gammie is 'Surgeon General H. I. (Not Practicing)'. Gammie died at Stanhope Gardens on 20th May 1887 and Mary at the same address on 7th November 1890. They are buried in the same grave in Brompton Cemetery, a mile from where they had lived for

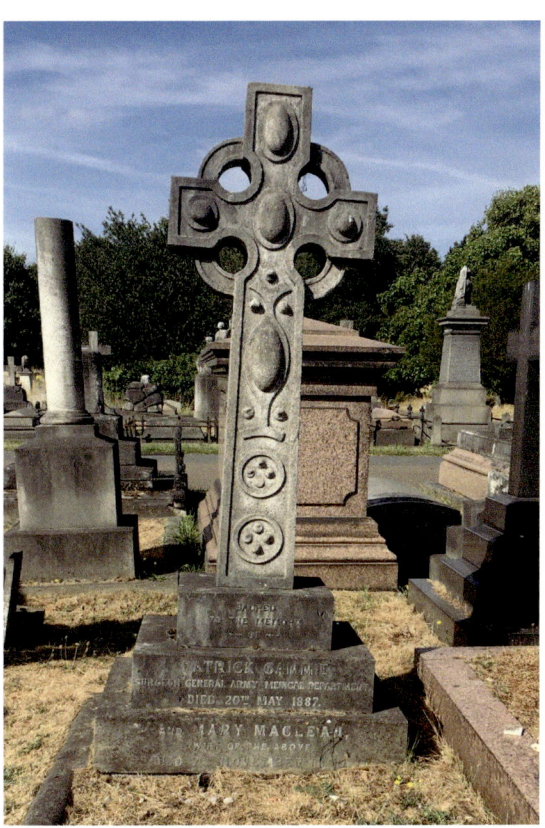

Sacred to the Memory of
Patrick Gammie
Surgeon General Army Medical Department
Died 20th May 1887
And Mary Maclean
Wife of the above
Died 7th Nov. 1890

Figure 4.2: Grave of Patrick and Mary Gammie, Brompton Cemetery, London (Photo: David Kennedy).

some twenty years (Figure 4.2). Appropriately for two Scots, their tombstone is a Celtic Cross – an ancient form but one revived in the 19th century as an assertion of identity.

Both left wills. Gammie's Personal Estate at Probate was a substantial £30,445 –7 -10d. He explicitly identified £9000 for various purposes including £1000 to *'the Minister and Elders for the time being of the Established Church of Scotland in the Parish of Forgue Aberdeenshire'* the interest from investment of this for distribution to the poor of the parish (Gammie Will (COW517740g)). Perhaps hereby identifying this as the parish of his birth (above). He left a further £1000 to his alma mater, King's College, Aberdeen (Gammie Will (COW517740g)), to establish 'The Gammie Bursary', to be awarded to the candidate *'who gains the highest aggregate number of marks in the French and German Papers'* (Aberdeen University Calendar: 39). Finally, he left £1000 to 'Army Medical Officers' Benevolent Fund' (Drew 1968: 1, 299 No. 4454).

Mary left Personal Estate of £17,099-4-6d, diminished by the removal of her husband's £9000 for his bequests elsewhere. Most notably she left bequests to siblings in Queensland, New South Wales and Tasmania, presumably some of the nine other children in her family when they arrived in Sydney in 1839 (above). Finally, she *'bequeathed [Patrick's] portrait, which is now in the R.A.M.C. Mess, to the Mess at Netley'* (Drew 1968: 1, 299 No. 4454). The Royal Victoria Hospital at Netley near Southampton, was, in its day one of the largest building in Britain. It was demolished in the 1960s though the Officers' Mess survives and is a listed building. The portrait is presumably one of the pair she mentions in her will of husband and wife done by the Scottish painter, Henry Farmer.

Walter Scott Campbell Seton-Karr
(Cheltenham, 23 January 1822 – Auchenskeoch Southwick, Dumfries, Kirkcudbrightshire, 22 November 1910)

Seton-Karr appears for the first time on board the *S. S. Bombay* on 30th January 1854, two days after leaving Galle. At the other end, the final mention is in Cairo on 13 February – when TB misspells his name as 'Seaton'. In fact, they have known one another for at least 18 years, as contemporaries at Rugby. Seton-Karr was enrolled there in 1834, aged 12 and Bowles two years later aged 14. This much comes out in Bowles's journal references which suggest an easy familiarity – "Fell in with Seton Karr & stood laughing & talking a great deal of nonsense" (2: 423/713). More specifically (2: 419/709, on Wednesday 1st February):

> *I sat reading Shakespeare afterwards & Seton Karr came & sat beside me with a book & of course we talked for an hour or more, of old Rugby fellows. I have often wondered what had become of so many who were there with me. They seem most of them to have gone to India. He says India seems to swarm with them.*

Seton-Karr was certainly correct about the 'swarming', including by members of his own wider family.

The wonderfully named 'Walter Scott Campbell' – the god-son of the novelist Sir Walter Scott, was the son of Andrew and Alicia Seton-Karr, of Kippilaw, Roxburghshire, but born in Cheltenham. After Rugby, when Bowles went on to Oxford, Seton-Karr enrolled at Haileybury, the college of the East India Company. By 1854 he was already as Bowles notes, fully engaged in administration in India and eventually rose to be Foreign Secretary to the Government of India. Several other members of the wider family were also prominent in the service of the East India Company and its successor.

In July 1856 he married Eleanor Katherine Cust (1823–1903), sister of another great administrator in British India, Robert Needham Cust (Penner 1987; Cust 1899) and aunt of the rakish Harry Cockayne Cust. They had four children, one of home died after a month while in India. He was a prolific writer, not least his biography of Lord Cornwallis. Another exact contemporary from Rugby was this same Robert Cust, his brother-in-law, who dedicated his *Linguistic and Oriental Essays* (1880) to 'WALTER SCOTT SETON-KARR, FOR FORTY YEARS MY FRIEND AND FELLOW-LABOURER IN THE BEST INTERESTS OF THE PEOPLE OF INDIA'.

Seton-Karr wrote a multi-volume 'Autobiography', never published but held now in the British Library. It would be interesting to compare his recollection of the journey with Bowles from Galle to Cairo.

Rev. George Percy Badger
(Chelmsford, Essex, 6 April 1815 – London, 21 February 1888)

On 6th February 1854, the sole day Bowles spent in Aden, he met with one of the most unusual people in his entire trip. In his brief foray ashore while coaling took place, he wanted to visit the Anglican church a design for which had been prepared by his brother-in-law, Henry Woodyer (above, Ch. 1). It was only when he asked directions to the 'English church' that he was told the chaplain's name - 'Mr Badger'. Evidently the name meant nothing to him, and he subsequently struggled to converse with Badger without revealing his ignorance.

> *I am sorry to say I was quite ignorant of Mr Badger's missions to the East & of his work "Nestorians & their Ritual" & could only fish for pegs to hang questions on. It seems that he went out there either by the wish or with the Consent of the late Archbishop. And all the good he was trying & likely to do was overthrown by the annoying interference of the American Independent Missionaries & Mr Layard & then by a member of the C.M.S. a Clergyman who it will hardly be believed associated himself with the Independent Missionaries & communicated with them. So much again for the C.M.S.*

Badger is a fascinating character, far more than TB realised and it is a pity he does not give a fuller picture of a man deserving now of a full biography. TB seems to have been more interested in his host's current religious controversy. Bowles and Badger discussed the latter's problems in his 'missions to the East' and current difficulties in Aden (below) and they had a shared interest in Anglo-Catholicism and evident distaste for the Church Missionary Society (CMS).

At the time of TB's visit, Badger was almost 39 years old, married since 1840, an ordained priest since 1842, a lengthy career in the Middle East and India and already the author of two substantial books (Badger1838; 1852). While Bowles, son of landed gentry, and then educated at Rugby and Oxford, was fairly typical of Anglican clergymen at that time, Badger was from a very different background.

His father Edward Badger was a soldier in the 80th Foot Regiment who had married Ann Percy on 21 December 1807 at Brading on the Isle of Wight. This may be the same Ann baptized at Dulwich College chapel in 1790 in which case she was about 17 when married and Edward about 23. Although Edward is explicitly recorded as '(of the 80th Regt)', that regiment had been overseas since February 1801 in Egypt and (mainly) India and only returned to the UK in later 1818. Of Ann and Edward's known children, Hannah and Sarah were baptized together on 25th April 1813 at Danbury, *c.* 8km/5miles from Chelmsford; and TB's future acquaintance, George Percy Badger in Chelmsford on 6th April 1815. The marriage took place and at least three known children were baptized in the UK while the regiment was overseas, presumably implying that Edward Badger, recorded as 'Sergeant' in the baptismal records, was part of the home depot. A fourth child, Matilda, was baptized at Chatham on 28th December 1817 where the returned regiment was based from 24th August 1817 till 8th January 1818. In 1820 the 80th Foot was sent to Gibraltar and in October 1821 was transferred to Malta. This latter base was the location of a happy event, the birth of a further daughter, Floriana, at Valletta, on 5th August 1822, evidently named from the suburb of Valletta occupied by the British military.

By the time of that last birth, Edward was Quarter-Master Sergeant of the 80th but the new birth and promotion were soon followed by tragedy and upheaval. The records from Malta allow a brief tabulation (https://www.maltaramc.com/regmltgar/80th.html):

- *(Unrecorded date) 1822: Burial of Hannah Badger aged 9 years, daughter of QM Sgt Edward and Ann Badger, in Military and Civil Cemetery Floriana, Malta.*
- *27 July 1823: Burial of QM Sgt Edward Badger aged 40 years, died on 20 July of dysentery on the 13th day of his illness, in Military and Civil Cemetery Floriana, Malta.*
- *19 June 1824: Burial of Florian(a) Badger aged 23 months, daughter of the late QM Sgt Ed. Badger. Buried Malta*
- *11 January 1825: Marriage of Bachelor Acting Cpl William Wade to Anne Badger, widow of the late QM Sgt Edward Badger 80th Regiment.*

The last entry shows the young widow with at least three small children, following a common practice for military widows and swiftly re-marrying to a soldier in her late husband's regiment - a man of a rather more junior rank, so a significant drop in status.

Ann's re-marriage meant the family remained in Malta with important consequences. On 1st January 1835, aged 17, George Badger's sister, Matilda met and married Christian Anthony Rassam (formerly Isa Anton) (1808–1872), from Mosul in what was then Ottoman territory. Rassam was a Christian, born a Chaldaean/Nestorian, educated in a Cairo mosque but converted to Anglicanism. Since 1832 he had been living in Malta working as a translator for the Cairo office of the Church Missionary Society. Soon after his marriage he took part in Col. Francis Chesney's famous Euphrates Expedition (1835-57) and then joined W. F. Ainsworth's expeditions in eastern Turkey and Kurdistan (1837-40). When he returned to Mosul in December 1839 as British Vice-Consul he was accompanied by both his wife, Mrs Matilda Rassam, and his mother-in-law, Ann Wade (formerly Mrs Badger), seemingly now a widow again. Ann was to die in Mosul on 28th April 1844 and was buried there soon after. Matilda Rassam is well-known through her correspondence with the great archaeologist, Henry Layard, who, obliged to remain in the UK, was kept informed about progress in the immense excavations at Nineveh being conducted in his absence by her brother-in-law, Hormuzd Rassam, Matilda's brother-in-law (*ODNB* 2008) (Turner 2001: *passim*).

As for George Percy Badger, his upbringing in Malta made him proficient in Maltese and as a result of a spell in Beirut, he soon became fluent in Arabic. It was as an accomplished Arabist that he was chosen by the Church Missionary Society in 1842 to go to Mosul and investigate the standing of the native Christian population. By then he had married on 8th January 1840 and in 1842 had been ordained in London. The marriage was above his own station, as Maria Wilcox was the daughter of a clergyman. The couple spent two years in Mosul. A contemporary reported that (Ross 1902: 47 n. 2):

> *When Badger returned to Mosul he professed to show people how the English clergy dressed, and went about in a long maroon coloured silk robe with a red shawl wound round his waist, and a red fez on his head.*

Badger became involved in acrimonious disputes while there both with the Catholic and Protestant missionaries. These latter were intent on proselytizing amongst the native Christians while the Anglicans were more content to provide support for these native sects. There was a serious falling out and some regarded Badger as partly responsible for stirring up inter-faith animosities that led to the subsequent massacres of Christians (Waterfield 1961: 213).

By the time Bowles met Badger he had spent a great deal of time in the Arabic-speaking world, become an accomplished Arabist, both literary and colloquial, published a book on Malta (1838) and his two volumes on the Nestorians (1852). In this latter he sets out his personal history after leaving Mosul (Badger 1852: 1, 359-60). In May 1845 he took up an appointment to a chaplaincy in the Bombay Presidency where he worked in the huge Maratha region south of Bombay for 18 months before being outposted to Aden in late 1846. He continues:

> *... a severe illness, which brought me to the borders of the grave and left behind it an inveterate nervous affection, induced the physicians to recommend a change of climate. I accordingly left Aden in March 1849, accompanied by Mrs. Badger, the untiring partner of my wanderings, spent a short time in Egypt, and passed a summer in one of the villages on mount Lebanon. A return of my old complaint, joined with a strong desire to visit the scenes of my former labours, led me to travel farther eastward, and we accordingly proceeded to Mosul, which we reached in safety on the 9th of December [1849].*

Shepheard's Hotel register records Badger arriving from Aden on 6 April and departing 'for Syria' on 26th April 1849. It was apparently during these first years in Aden that Badger met the young Anna Harriette Edwards, later better-known as Anna Leonowens, future tutor to the children of the King of Siam and the subject of many books and movies. He encouraged her in language studies in which she became proficient, and the Badgers included her - just 19, when they made a tour in Egypt in 1849 (Habegger 2014: 67).

As he reported (above), after '*a summer on mount Lebanon*', Badger returned to Mosul in December 1849 where he was again embroiled in the controversies stirred up by the massacres of a few years earlier. Diary entries show him there until November 1850 when he and his wife made a tour through Kurdistan ending at Nineveh in 1851. The Preface of the first volume of his two-volume account of the Nestorians and of his own time living in their territory (1852) is signed and written at the ancient Syrian Convent of Mar Mattai in August 1850. Because of the delay in publication, the Postscript to the Preface is dated *'Aden, Arabia Felix, October 28th, 1851'*. They evidently remained there but there is no mention of Badger's wife in TB's journal.

After Bowles met him in Aden, he had a yet more distinguished career - stellar for the orphaned son of an NCO in an infantry regiment. In the mid-1850s he was playing an important role in Aden negotiating with the tribes. In 1856-57 he was serving in the Arabic-speaking Persian province of Khusestan as part of Sir James Outram's Anglo-Persia War. It is clear from Outram's letters and reports that Badger was appointed not just as a military chaplain but because of his proficiency in Arabic and diplomatic skills. Later still he was sent as an envoy to Muscat then in 1873 as a key assistant and Arabic interpreter for Sir Bartle Frere the British envoy to Zanzibar.

The last twenty years of his life were lived in London. Maria died in 1866 and he married his housekeeper in 1871. He continued to publish extensively, utilizing his linguistic skills and experience in southern Arabia and the Gulf: in 1863 he provided notes for a translation from Italian of Varthema's 16th century account of travels in Arabia; in 1871 it was a study of the Sultans of Oman; later still he produced a massive Arabic-English Dictionary. He died in London in 1888 and is buried in Kensal Green Cemetery. He had no children.

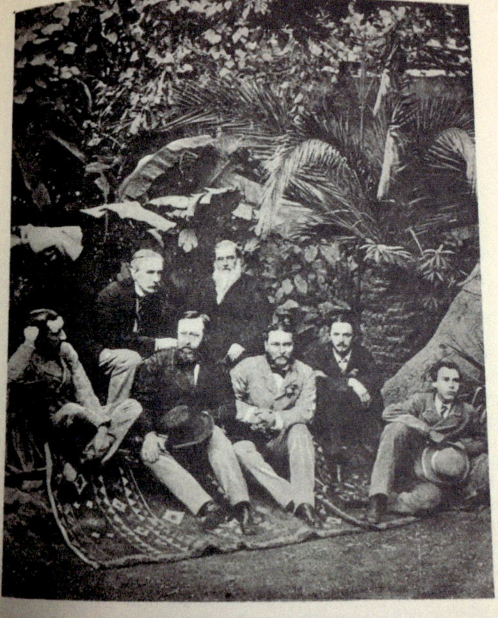

THE FRERE MISSION
Cairo, December 22, 1873

Sir Bartle Frere, Rev. G. P. Badger, Mr. Clement Hill, Capt. Fairfax, Major Euan Smith, Mr. C. Grey, Mr. B. C. A. Frere

Figure 4.3: Badger - back row, right, with long white beard, with the Frere Mission to Zanzibar while in Cairo on 22nd December 1873 (Coupland 1939: facing p. 186).

Only two images of him survive, one of which - an oil painting, is in private hands; the other a grainy image in an old book (Figure 4.3).

NB: Badger appears in several 19th century published books reflecting his role in a series of important activities.

At the time of Bowles' visit, Badger was engaged in controversy locally because of his embracing the contemporary movement to re-introduce catholic doctrine into the Church of England – what Bowles calls – approvingly, 'Puseyite', after the Rev. Prof. Edward Bouverie Pusey, one of the leading figures of what was called the Oxford Movement - the same Movement in which Tupper had been involved.

***

Rev. Charles Tombs
(23 February 1817 – Aden, 22 August 1846)
and
(Later) Major-General Sir Henry Tombs, VC, KCB
(Calcutta, 10 November 1825 – Newport, 2 August 1874).

Bowles mentions each of these brothers in his journey from Galle to Suez. On 29th January, in conversation with Captain Hill he discovered a friend in common – Henry Tombs. A week later, while in Aden, he was taken by George Badger, to visit the grave of Henry's older brother, Charles Tombs.

The Tombs brothers were from Abingdon where the family had been bankers (Hammond 2004). These brothers were two of the seven sons of Major-General John Tombs (1777-1848), of the Bengal Cavalry. They followed their father into John Roysse's Free School (now Abingdon School), founded 1256, eight kilometres from where Bowles was born and 300m from where he is buried. Such rural gentry families would have been well-known to one another as leading members of (what was then) north Berkshire 'society'.

Rev. Charles Tombs (1817–1846), was a graduate of Pembroke College, Oxford (B.A. 1835 and M.A. 1840). He had married in December 1840 to Mary Chatfield, daughter and sister of Anglican clergymen. Soon after, in 1841, he took up a post as Assistant Chaplain at Poona in India. In 1843 he was appointed Assistant Chaplain at Aden., a British possession since just 1839. He died and was buried there three years later, aged 29, leaving a widow and daughter (*The Gentleman's Magazine* 1847: 549). Bowles was unimpressed by the grave – '*a hideous stone box tomb over it*'.

> *At Aden on the Red Sea, aged 29, the Rev. Charles Tombs, Assistant Chaplain to that station. He was the son of Major-Gen. Tombs, of the Bengal cavalry, and son-in-law to the Rev. Dr. Chatfield, formerly Vicar of Chatteris, Cambridgeshire.*

While Charles had been five years older than Bowles, Henry Tombs was

SACRED TO THE MEMORY OF
MAJOR GENERAL
SIR HENRY TOMBS. V.C. K.C.B,
WHO DIED AT NEWPORT
2ND AUGUST 1874, AGED 49 YEARS.
'THY WILL LORD NOT MINE BE DONE.'

Reverse face:
ALSO IN MEMORY OF
HENRY EDWIN. ONLY SON OF
HENRY AND GEORGINA TOMBS.
BORN OCT 27TH 1873.
DIED DECEMBER 2ND 1874.

Figure 4.4: (a) Photo of Henry Tombs in 1857 (Public Domain); (b) Epitaph on his tombstone at Newport.

three years younger. He had omitted Oxford, instead entering the East India Company college at Haileybury and graduating in 1841. He served extensively in India, winning praise for his heroism in a succession of battles. As Captain Hill reported to Bowles on the voyage, *'he is just gone home, with 3 Medals, likely to get his Company at once, & his rank as Brevet Major at the same time'*. The Indian Mutiny provided yet more opportunities for advancement and on 9th July 1857 at the Siege of Delhi, he won a Victoria Cross. Henry Tombs (1825-1874) died young, just 49 (Figure 4.4), but having already had a stellar career including a knighthood and the rank of Major-General.

Henry Tombs married late - in 1869, aged 44. His wife, Georgina Janet Stirling, was a daughter of Admiral James Stirling (1791-1865), explorer of the Swan River in Western Australia in 1827. In part due to Stirling's lobbying, the Swan River Colony (Perth) was founded in 1829 and (then) Captain Stirling became the first Governor of Western Australia. Georgina's mother, Ellen Mangles, after whom Ellen Brook in Perth is named, was the daughter of Captain James Mangles who, together with Captain Leonard Irby, another naval officer, in 1818, had been amongst the very earliest western visitors to Petra and had published a book on their expedition (Irby and Mangles 1823) (Table 6.1 below). Mangles himself visited the Swan River Colony in 1831 and - a keen horticulturalist, cultivated in the UK several Western Australian plants now bearing his name including the 'Kangaroo Paw' plant which is part now of the state floral emblem, bears his name (*anigozanthos manglesii*).

Chapter 5

Thomas Bowles in Egypt

Western visitors in Egypt and the Near East in the 19th century

By the time Thomas Bowles arrived in Egypt in 1854 that country had been a popular destination for western travellers for several decades. The French invasion in 1798 and the British response has brought the country sharply to western attention. The removal of ancient artefacts and then the superb publication of *Description de l'Égypte* (1809–1828) by a team of French scholars, had stirred academic interest and later western visitors often acquired souvenirs and sometimes much more (cf. Stobart, Chapter 2). Indeed, 'exploration' of the ruins of ancient Egypt was no more than treasure-hunting until - precisely at this time, Alexander Henry Rhind brought to his work there the more systematic techniques of excavation and recording he had initiated and developed in Scotland (Gilmour 2014). Although travel to Egypt before the arrival of TB in 1854 was for the relatively wealthy, parts of ancient Egypt itself had already travelled west to the growing collections of the museums of northern Europe - the British Museum, Louvre and the impressive Museo Egizio (as it is now called) in the Piedmontese royal capital of Turin, where artefacts could be seen by a wider public stimulating what came to be called 'Egyptomania'.

Motivations

For many early western travellers in Egypt and the Levant, adventure was a key motivation – novelty in plenty but also danger. After the first flush of travels to Petra, the element of adventure remained but visits became increasingly formulaic, including the frequent upsets and hostilities at Petra. Although antiquarianism was motive enough once the magnificence of the remains became more widely known and travel safer, there were other reasons to visit.[1] Within a short time guide-books were available in several languages – the first edition of 'Murray's' *Handbook for Travellers in Egypt* was published in 1847 with regular updates thereafter (Wilkinson 1847).[2]

Tourism

Tourism was an increasingly popular draw for westerners. Alexandria had attractions invariably starting with the so-called Pompey's Pillar, with visitors able to stay in the westernised Hôtel de l'Europe. Most visitors soon moved on to Cairo - where TB was to spend two weeks and took up residence in one of its hotels for westerners - especially Shepheard's Hotel in the case of most British visitors. There were the outings available to the bazaars, mosques, to the Pasha's Palace at Shobra, the Citadel, to the Nilometer on the island of Rodah or further afield to the Petrified Forest at El-Katameya and - most obviously, the pyramids and Sphinx on the Giza Plateau *c.* 15km/9miles to the southwest.

---
[1] Useful surveys of early western travellers in Egypt and the Levant and their motivations can be found most recently in Oliver 2014 and Sartre-Fauriat 2021.
[2] Written for Murray by Sir Gardner Wilkinson, it was – as the title page explains. 'a new edition' of Wilkinson's previous monograph of 1843).

After exploring Cairo and vicinity, the principal activity of western visitors was to hire a river boat and spend weeks or even months sailing south (upstream) from Cairo. These river boats - of varying sizes and quality, included a crew (with captain - *rais*) and cook with supplies. Families or groups of westerners could band together to rent one. They commonly bore names - sometimes applied by the renter for the duration of the hire. A memorable feature of preparing the older boats was to sink them briefly in the river to drown the vermin and unwelcome bugs.

Journeys could last for weeks or even months depending on how far south they wanted to go. There were numerous attractions to be seen or visited along the banks and passengers often reported going ashore to explore sites and/or on shooting expeditions. The most intrepid would go beyond the Second Cataract and into Nubia (northern Sudan).

Health

Egypt in winter has a wonderful climate and quite soon the well-to-do of northern Europe and the United States identified it as a refuge from the cold, damp climate of home. It became too, a place to recommend for people of a weak constitution - as Algiers was to be later in the century. Lord Henry Scott with whom Bowles had travelled and who was soon to arrive in Egypt himself - and to return at least three times more in successive years, was explicitly travelling for his health, as remarked upon by Stobart in his letters. Stobart and Lord Henry were soon to undertake supervision of excavation at Philae and/or Elephantine on behalf of Alexander Henry Rhind, the latter having himself recently transferred his fieldwork from Scotland to Egypt precisely for health reasons. Indeed, Rhind published a book on the subject (Rhind 1856), adding to two others published at much the same period (Cummings 1839; Patterson 1867). Nor was the concern an exaggerated one: Rhind was to die in 1863, aged not quite 30, while on his way home from Egypt to Scotland. Then there is TB's friend William Tupper: he, too, had left home to travel for his health (above). As noted, he had been unwell when Bowles met him in Ceylon and Egypt, began to fade while in Egypt and died while on the voyage home in the Mediterranean, aged 29.

Jane Eames had the company on the Nile of another young man there for his health (Eames 1855: 53):

> Soon after we came here, J. made the acquaintance of Rev. Mr. Lieder, the English missionary, (well known to strangers visiting Egypt,) and through him and his estimable wife, we had a pleasant companion secured to us for the Nile trip,— Mr. N[icholson], a young Englishman, spending the winter in Egypt for the benefit of his health; and I will here remark, just by the way, that English physicians are now sending their consumptive patients to Egypt, instead of to Italy and the South of France, as they consider the climate here preferable.

Bowles Tourism

TB did not take a Nile cruise. He had spent many weeks at sea, but that is unlikely to have been the explanation - these river cruises were generally relaxing and healthy. He may have been more concerned by cost - as he was travelling at his own expense, and time away from home was now a consideration as he got into the Mediterranean world. Receiving letters at Cairo also induced nostalgia for people and places:

Later in the afternoon, after I had read & re-read my letters, & had sat a long time thinking about them & all in England, wishing I were going on the next morning with the rest of the passengers to Alexandria, ... I went to my own room – read – & went to bed – thinking of England & Home ...

However, as see shall see (below), TB was gradually coming round to the idea of not just delaying his return home but of undertaking a major and rather more challenging expedition.

In the meantime, he was certainly not uninterested in the cultural artefacts of ancient and more recent Egypt: of the 14 full days he had in Cairo before setting off for Petra, no less than eight were spent in tourism (Table 5.1); and when he returned through Alexandria on his way home in May he made good use of his one full day there to visit the most obvious tourist sites.

TB's third journal begins with his arrival at Shepheard's Hotel and he then devotes 64 pages to his two-week interlude in Cairo. The arrival day after the somewhat gruelling journey from Suez was spent settling in and orientation. The second day, *'I slept 2 nights in one & awoke about 7 feeling as if I had done good work during the last 9 hours'* – and in need of a haircut. On the recommendation of an American guest, he hired a dragoman. He had initially thought it unnecessary but was persuaded that unless you could speak the language, a guide was

Table 5.1: Tourism of Thomas Bowles while at Shepheard's Hotel, Cairo and Hotel de l'Europe, Alexandria

Monday 13 February	Caravan from Suez to Cairo through Desert.	458-463 and 1-2
Tuesday 14 February	Cairo. Excursion to Citadel Mosque & Turkish Bazaar.	2-13
Wednesday 15 February	Cairo.	13-16
Thursday 16 February	Cairo. Excursion to Shoobra Gardens.	16-19
Friday 17 February	Cairo. Excursion to Tomb of the Memlooks (Mamluks), Whirling Dervishes, Nilometer on the Island of Roda & Copts Quarter.	19-26
Saturday 18 February	Cairo. Excursion to the Pyramids of Giza, Sphinx & surrounding tombs.	26-42
Sunday 19 February	Cairo.	42-44
Monday 20 February	Cairo. Excursion to Heliopolis.	44-52
Tuesday 21 February	Cairo. Excursion to Mosque of Sultan Tayloon.	52-54
Wednesday 22 February	Cairo. Excursion to Sakkara (Saqqara) & Memphis.	54-56
Thursday 23 February	Cairo.	56-58
Friday 24 February	Cairo. Excursion to Sultan Hassan's Mosque.	58-60
Saturday 25 February	Cairo.	60-61
Sunday 26 February	Cairo.	62
Monday 27 February	Cairo.	62-64
Tuesday 28 February	Depart Cairo for the Desert. Start of journey to Suez.	64-68
Thursday 11 May	Alexandria: English Church, Cleopatra's needle, Pompey's Pillar, Mahmoudie Canal, Pacha's Palace, Library, Roman Catholic Church	282-285

indispensable. The point was underscored soon after when his donkey boy took him to the French rather than British Consul.[3] An initial trip to Bulak, the landing point on the Nile at Cairo, was to collect their luggage sent on by water from Suez, then to the British Consulate to enquire after letters and arrange for permissions to visit places that were not open to the public. Then eight of the next eleven days were allocated to excursions. He began with the nearer sights - the bazaars, the Whirling Dervishes and the Copts' Quarter, but then he undertook a series of tourist visits - by donkey, to the immediate periphery of the old city: Citadel Mosque, Shubra Gardens, Tomb of the Mamluks, the Island of Rodah and the Nilometer, all of it involving passing through the endlessly interesting narrow and crowded streets. In the following days he ventured further afield, to the Pyramids at Giza, the Sphinx and nearby tombs, Heliopolis, Sakkarah and Memphis. In between were further visits to mosques: those of Sultan Tayloon and Sultan Hassan, and to the Tomb of Sultan Qaidbay. Appropriately, he noted that the courtyard of the Citadel Mosque was about the same size as that of his Oxford College, Queen's. The visibility from the top was impressive and he observed that being able to see the Pyramids of Giza which were 10 miles away (actually *c.*14km straight line distance (sld)) would be as if one could see the Radcliffe Library (in central Oxford) from his home at Milton Hill (actually *c.*16.5km sld).

TB was plainly fascinated by Cairo, its bustle and endlessly interesting street life and its people. Likewise, he recorded detailed descriptions of people and places visited, measurements made, and included a few little sketches including a quite detailed one of the Tomb of Mehmet Ali.

It was while he and his two companions (Kennedy and Strachan, below) were visiting the Giza pyramids, the idea had first been mooted - *'in joke yesterday at the Pyramids & then taken up in earnest - viz. the making the journey by Sinai & Petra to Jerusalem & so through the Holy Land to Beyrout & home (!) by Smyrna & Trieste & across the Continent.'* TB was keen to return home after so long an absence but was excited by the possibility and recognized that now - while in Egypt near the start of the route, was an opportunity not to be missed. The idea was firmed up over the next few days, not least through discussion with their dragoman who outlined how he would organize everything for them, and by conversations with another hotel guest, a Mr Ross, who was about to set out on that journey himself. They returned to the subject again and again and Bowles noted the extra time required and that it would cost him £90 more by that Petra and Syria route. By Thursday, he was in a torment of indecision and confided to his journal how this made him ill with worry. However, he soon made up his mind and *'since then I have been at ease & long decided I will not allow myself to discuss the matter again.'* The next day (Friday 24th February):

> *I got up this morning really glad that things were settled. I would have made myself miserable about the delay in reaching home & the expense etc. etc. that I had determination when I made the decision yesterday, that having so decided I w(oul)d not allow my mind to dwell on the question, but enjoy the journey thoroughly, & get what good I can for body & mind out of it & I hope for much for both.*

Having been warned by Mr Shepheard that their local dragoman, Hassan, was too young for this much grander enterprise, he evidently replaced him[4] - though without explicitly

---

[3] In the event, that dragoman fell through and he then joined with two companions in hiring an alternate.
[4] Under 26th February 1854, TB records finalizing accounts with Hassan.

reporting it in his journal with the man he calls alternately Giovanni or John who was to remain with him till he sailed from Beirut ten weeks later.

The last three days in Cairo were devoted to preparations for the Long Desert Route to Petra. The new dragoman, Giovanni, was telling him on 26th to have his luggage ready - some for the journey and some for shipping home: *'my box & canes to Southampton, a much more difficult business than I expected.'* One of TB's final actions before leaving Cairo was to have his hair *'cut - cropped, I might say'*. Oddly considering the imminent departure for several weeks in the desert, the background to his final journal entry was the sound of a waltz being played as a ball got underway in the hotel. By then TB was looking forward to the adventure:

> *This is the last night in the civilised world for some time & the last night I shall spend indoors - I can't say in bed, for I shall have as good a bed in the tent I have little doubt as I have here. But it is anxious being on the eve of starting for the desert. God preserve us & bless this journey to me. It is as if we're treading in his steps. I am very glad. I asked to go. I best now go to bed. To be ready for our first days camel-ing tomorrow.*

Companions in Cairo

Bowles arrived in Cairo with several of the passengers from the *'S.S. Bombay'* but then spent most of the fortnight there in the company of two in particular, Kennedy and Strachan. The former was to accompany him to Petra and on to Beirut and will be treated in the Petra chapters along with all the other travelling companions of that journey.

George Strachan
(Edinburgh, c. 1821- Toxteth, Liverpool, 8 February 1893)

Strachan and Kennedy had been fellow passengers from China (?) to Singapore (?) on the *'Lady Mary Wood'*, arriving in December 1853. Although Bowles does not mention Strachan until he and Kennedy arrive in Cairo seeking a place in Shepheard's Hotel, he must have been on the *'S. S. Bombay'* too from 29th January 1854 when they left Galle in Ceylon until reaching Suez on 12th February. Kennedy, Strachan, and Bowles certainly all stayed together in Shepheard's Hotel for two weeks (13th – 28th February) and often undertook sight-seeing in and around Cairo together. When Bowles and Kennedy set off from Cairo on the Long Desert Route on 28th February 1854, they parted from Strachan who was staying in Cairo and continuing tourism:

> *Away we went for some distance & then Hassan who was with us said Mr Strachan must turn off for the Petrified Forest & we said 'Goodbye'.*

Strachan is not mentioned again. Although they had been together four weeks and in close association, Bowles is as uninformative about Strachan as with almost everyone else, the only hint of his occupation being:

> *After leaving the 2nd mosque we passed a small domed Tomb with a beautifully planned doorway, wh(ich) Mr Strachan drew for me.*

The shipping roster mentioned above calls him 'G. Strachan'. From that, his seeming employment in China and superior drawing skills we may identify him as the George Strachan who had been an architect in China from perhaps as early as 1841 (where he had designed the Clubhouse in Hong Kong in 1845) and then the first professional architect in Shanghai (where he designed the first cathedral and various private and public buildings).

These details in turn allow us to identify him with George Strachan, the 20 year old son of an engineer, John Strachan, and his wife Marion in Edinburgh. He is not on a UK census for 1851 but can be traced again in Edinburgh in 1861 ('proprietor of houses'), at nearby Portobello in 1871 ('architect') and in Toxteth, Liverpool in 1891 ('retired architect'); the family cannot be found on the 1881 Census. He was buried in Toxteth in February 1893 ('architect'). The family had seemingly moved from Edinburgh to Liverpool in the early 1870s.

The most likely marriage record is one for Kensington in 1858. His wife is Mary Dalrymple, stepdaughter of Ellen and daughter of Alexander Dalrymple, 'West India merchant'. George – c. 37 in 1858, was 15 years her senior. Alexander Dalrymple is one of those plantation owners who was compensated after the abolition of slavery in 1832. The Dalrymples were an Aberdeen family, but their daughter Mary was born c. 1835 in Dominica – as was her step-mother before her. By the 1841 census the family was living in London. In the censuses of 1861 and 1871 Mary's widowed stepmother, Ellen, is living in Portobello near Mary and George Strachan.

The censuses and the family epitaph in Toxteth Park Cemetery reveal at least eight children, two of whom died in early infancy. One son - Arthur Claude Strachan (1865-1938), was to become a well-known and prolific landscape artist, seemingly inheriting a talent from his father.

***

William Palmer
(Mixbury, 12 July 1811 – Rome, 5 April 1879)

Bowles mentions a 'Mr Palmer' as fellow passenger from Galle, someone who had recently travelled from India to Australia and who evidently stayed in Shepheard's Hotel as well but is not mentioned again. In contrast is the second 'Mr Palmer' whom Bowles met in Cairo who he - for once, defines very precisely.

Bowles mentions this Palmer twice, first meeting him while visiting the Hotel d'Europe in Cairo on 23rd February 1854 then again when Palmer called on him at his hotel two days later. Although he records him formally as usual as 'Mr Palmer' (Figure 5.1), he goes on to identify the great man more precisely: 'Mr Palmer - William Palmer of Magd(alen) College.'[5]

Bowles was right to be impressed and nervous. The 42 year old Palmer was the eldest son of the Rev. William Jocelyn Palmer and Dorothea Richardson Roundell, herself a vicar's daughter. Of his younger brothers, Roundell became Earl of Selborne, a later Lord High Chancellor; Edwin became Professor of Latin and Archdeacon at Oxford; and George Horsley succeeded their father as vicar of Mixbury in Oxfordshire. William wrote extensively on religious matters and embraced strong views on his high Anglicanism. A few years before Bowles met him, he had tried to join the Greek Orthodox Church and in 1855, the year after their meeting, he joined the Roman Catholic Church and thereafter resided in Rome (Wheeler 2006).

---

[5] By the time Tupper reached Cairo two weeks later, Palmer seems to have moved into Shepheard's Hotel (above, Ch. 4).

By the time Bowles met him in 1854 he had travelled extensively in the Levant from Greece through Beirut, Damascus, Jerusalem and to Egypt in 1849-50 with his brother Edwin, and in 1853-54 was in 'Egypt and the Holy Land' for almost a year. Palmer considered joining TB's developing plans for an expedition to Petra but thought better of it, telling Bowles he might instead go to Suez by 'van' then visit Mt Sinai. In fact he did just that, but went then from Mt Sinai to Jerusalem, by-passing Petra.

William's brother, George Horsley Palmer, had attempted to visit Petra from Jerusalem in 1848 but had given up and travelled instead in the Decapolis and to Damascus. It was not till the next generation in the person of their neice, Sophia, Comtesse De Franqueville, daughter of Roundell, that a member of the family reached Petra (April 1882) (Palmer, S. M. 1883; Ridding 1919).

Figure 5.1: Portrait of William Palmer

Although William Palmer never visited Petra, he did travel extensively in Sinai and will have likely included in his journals his encounteres with other westerners who did visit the ancient city that year, including Bowles.[6]

The Lambeth Library also contains his journals for his visit to Lebanon, Syria and Palestine in 1849-1850 (MS2824-2827); and the journal of his brother Edwin for this same journey (MS2852). Also at Lambeth are extracts from the journal of George Horsely Palmer made by his brother Roundell, and recounting his travels in the Levant in 1848 (MS1906); the whereabouts of the original journal are unknown but in addition to the 'extracts' made by his brother and brief references in the latters published family history, George Horsely's travels in the region can be illuminated through the account published by a travelling companion, Charles Monk (1851).

***

Alfred Henry Pierpont Edwards
(New Haven, Connecticut, 17 August 1810 - Manhattan, New York, 9 January 1857)

One of the first people TB encountered on arrival at Shepheard's Hotel on 13th February 1854 was:

---

[6] Happily his journals survive in Lambeth Palace Library but can only be consulted in person. William's journals are MS 2831-2833 for this visit of 1853-4 running to 850 handwritten pages. The Egyptian part of these extended for four months (19th November 1853 - 20th March 1854) and resulted in a 1200 page, two volume publication in 1861 - *Egyptian Chronicles* (Palmer, W. 1861).

> ... the American, Mr Edwards, who came with us from Singapore. He is going on & on - he had not been able to get to Thebes. It is too late. But he has enjoyed Cairo very much & says he sh(oul)d like to be here longer. He has been somewhere every day & says there is enough to occupy you e(ver)y day between 2 steamers.

It would seem that, while TB has delayed in Ceylon, Edwards had continued and reached Cairo at least a few days earlier. Presumably the *'between 2 steamers'* refers to whichever route he planned to embark on from Alexandria. The following day (14th) Edwards tried to persuade TB to engage the dragoman he had employed for his tourism in and around Cairo. The day after (15th) Edwards has evidently left for the steamer and TB wrote a letter to be delivered to him at Alexandria while waiting to sail.

Not much to go on but fortunately, looking back to an earlier part of TBs journal (not included here), one reads under *'Sunday 1 after Epiphany Jan(uar)y 8th [1854]'*:

> ... I sat talking to Mr Edwards the American - who has been 17 years at Manilla (sic). He describes the island as very beautiful; & Manilla itself a fine town ...

Edwards appears on the final page of the journal as a sole address TB records: *'Alfred H. P. Edwards Esqr, care of Messrs Forbes, Forbes and Co., London'*. There can be little doubt this American is Alfred Henry Pierpont Edwards who would have been 43 when TB met him. He can be traced on a swathe of documents relating to Manila where he was the Consul of the United States on two occasions: 1832-1838? and 1848-1854? (Smith 1986: 159). Consular records for December 1853 record for 'A. H. P. Edwards', *'Resignation - leaving for U. S. William A. Pierce of Salem in charge'*. Exactly in time to encounter TB on the ship from Singapore.

Alfred H. P. Edwards was part of the growing group of American merchants in the Spanish Philippines. On an application he lodged in 1856 he describes himself as *'Importer (Sugar)/Consul (Philippines)'* (Century Association website). He had evidently been in Manila long before his initial appointment as Consul: a passenger list for the ship *Galaxy* records him arriving in New York from Manila on 25th March 1831 when he was 20. There is no reason to doubt that by the time he met TB he has spent 17 years in the Philippines.

The Edwards family was Yankee aristocracy. Alfred H. P. Edwards was the great-grandson of Jonathan Edwards (1703–1758), a notable and controversial clergyman, whose career path culminated as 3rd President of Princeton University in 1758 (dying soon after his appointment, after inoculation for smallpox). At Princeton Jonathan had succeeded his son-in-law, Aaron Burr Snr - who had died in office the previous year.

Alfred's father, Henry Waggaman Edwards (1779-1847), a graduate of Princeton University, went on to a career as a barrister then in politics as a U.S. Congressman, U.S. Senator, and Governor of Connecticut. Alfred's uncle, Moses Ogden Edwards (1781-1862), was a Justice of the Supreme Court of New York. His great-aunt, Esther Burr, was the mother of the notorious Aaron Burr Jnr, 3rd Vice-President of the U. S. A., and the man who killed Alexander Hamilton in a duel. Some of the family was also related to Benjamin Franklin.

At least one other relative of Alfred Edwards was engaged in trade in the Philippines. Ogden Ellery Edwards (1828-1918), another great-grandson of Jonathan Edwards, described on documents as a merchant, also appears on consular documents and went on to be U.S. Consul in Manila from 1854-1860. His eldest child, Catherine Shepherd Edwards, was born in Manila on 24 May 1862.

Finally, an unexpected twist: Ogden P. Edwards of New York, another great-grandson of Jonathan Edwards, is reported as in Cairo in January or February 1854, at broadly the time his kinsman Alfred H. P. Edwards was passing through.

On 16th May 1856, two years after he left Egypt and by then 45, Alfred married 35 year old Mary Griswold (1821-1896), daughter of Nathaniel and Ann Griswold. Presumably related to Charles Griswold of New York, who was to be the U. S. Consul in Manila from 1856-1867. The marriage was short: Alfred Edwards died just a few months later leaving Mary a widow for 40 years. He was buried in New Haven alongside many others of his wider family. They had no children.

***

Mr and Mrs Samuel Briggs
(1776–1868)

Soon after arriving in Cairo, TB visited 'Briggs' then again a few days later went to 'Mr and Mrs Briggs', apparently in search of letters. Briggs and Co could trace its origins back to the arrival in Egypt of Arthur Briggs - the son of a London weaver - in the early years of the Napoleonic Wars. Arthur Briggs enjoyed a long favourable relationship with the ruler of Egypt, Mehmet Ali and carved out an important role as an intermediary between the British government and India. The company offices came to be a useful poste restante address for travellers - in 1846 when TB's travelling companion Henry Stobart, had been in the Levant on his previous tour, as he prepared to leave Constantinople, he had told his mother her *'future letters must be directed to "Messrs Briggs & Cie, Cairo, Egypt"'*. On his first visit TB may only have meant 'Briggs' as short-hand for the business office. However, the Shepheard's Hotel guest book records that Arthur Briggs left that hotel on 17 December 1853 to sail up the Nile to Wadi Halfeh and returned to Cairo on 21st February 1854, three days before TB recorded:

> *After breakfast we went to Mr & Mrs Briggs – where oddly enough I found a letter from Alice dated August.*

That seems to clearly imply he met the Briggses themselves. By 1854, Arthur Briggs was no longer fully active in the business, he was 78 years old and married to his second wife, Camilla (1798–1872). They had no children, but he had a daughter from his first marriage (through whom Briggs was to have two notable grandsons, who were both knighted). By the 1850s the business was largely run by his nephew Samuel Saunders and Henry Ross.[7] However, Samuel Briggs remained deeply committed to promoting the Alexandria to Suez railway and was the principal financial supporter of the Anglican St Mark's Church in Alexandria which reached completion in precisely 1854.

***

---
[7] His published letters are full of interest: Ross 1902.

Rev. Pierce Butler
(Kilkenny, Ireland, 27 February 1826-Ulcombe, Kent 8 February 1868)
and
Capt. Henry Thomas Butler
(1812-Inkerman, Crimea, 5 November 1854)

TB first encountered the Butlers in Cairo at Shepheard's Hotel on 25th February 1854 and they met up again at St Catherine's Monastery at Mt Sinai. 'Captain Butler' is named but later TB refers to '2 Butlers' at St Catherine's. Happily, another traveller's account identifies this pair (Stewart 1857: 1):

> *I embarked at Leghorn on board the French mail steamer 'Bosphore,' on the evening of the 22nd December 1853, and arrived at Malta on the morning of the 26th. Among my fellow-passengers were Mr William Maitland, a partner in the house of Mackillop, Stewart and Company of Calcutta, a most agreeable travelling companion and friend, from whom I parted with sincere regret at Suez; and two brothers of Captain [James Armar] Butler, the hero of Silistria, one a clergyman [Rev. Pierce Butler] of the Church of England, the other Captain Henry [Thomas] Butler, of the 55th Regiment, who was killed, not a year afterwards, at Inkermann (sic), while discharging bravely his duty as a staff-officer.*

Stewart helpfully explains the movements of these Butler brothers prior to TB's first meeting with them at Shepheard's (Stewart 1857: 18):

> *It was our intention to have left Cairo on the afternoon of the 12th of January [1854] but, by the time the Sheikh's contract had been signed, money matters arranged, and the passport returned from the citadel, the day was so far advanced, that it was deemed advisable to defer setting out till next morning [13th January 1854], and Nassar was ordered to have his camels in readiness by seven o'clock. .... We spent our last evening at Shepherd's (sic) in company with the Butlers, who also were to start next morning on a voyage to the Cataracts.*

These brothers were two of the five sons of Lt. General the Hon. Edward Henry Butler (1780-1856), a younger son of the 2nd Earl of Carrick (Irish Peerage). Their father had had a distinguished career in the Napoleonic Wars, serving as a young officer in Egypt under Abercrombie in 1801 and later alongside the Portuguese Army in the Peninsular War (Bromley and Bromley 2012: 1, 124).

The five sons were, therefore, grandsons of the 2nd Earl, nephews of the 3rd Earl and cousins of the 4th and 5th Earls of Carrick. Three of them took up military careers and one went into the church; the youngest, Edward John Butler (1842-1920), the child of a second marriage, was just 12 years old in 1854, did follow a military career but only for a short time, which may be why he was to enjoy the long life denied to his elder brothers.

For the four older sons, tragedy was to strike swiftly and in rapid succession. First to die - as noted by Stewart (above), was James Armar Butler (1827-1854), a Captain in the Ceylon Rifles, educated in Paris and Friburg before joining the army and taking part in campaigns *'against the Kaffirs in 1846 and 1847'* before serving six years in Ceylon including assisting in the suppression of a rebellion (Anon. 1854a: 94). Soon after the outbreak of the Crimean War, while on leave in Rome, he set off at his own expense and joined a few other British officers supporting the Ottoman forces facing Russian siege at the fortress of Silistra on the

Danube (on the present Bulgarian-Romanian frontier). He received a slight bullet wound in the forehead - seemingly not seriously, but died there eight days later on 20th June 1854, aged 27. He was lionised in the British press as 'The Hero of Silistra', the first British soldier killed in the war and figured prominently in a lengthy article in the *Illustrated London News* which also carried a brief biography and testimonial from both Omar Pasha, the Ottoman Commander, and Lord Hardinge, the British Commander-in-Chief (Anon 1854a). He was buried in the Armenian Cemetery at Silistra but subsequently moved to 'Cathcart's Hill' in the Crimea.[8]

Lord Harding had attempted in a letter to the old General Butler to console him (Anon 1854a: 94):

> *I trust that the well-earned fame of one son and the rising merit of the other will ... be a source of consolation to you at this moment of extreme affliction.*

Soon after, however, this other *'rising (son)'* - the 'Captain Butler' TB met in Egypt and at St Catherine's Monastery, Henry Thomas Butler (1812-1854), of the 55th Foot Regiment, was killed at the Battle of Inkerman, on 5th November 1854, aged 42. He had already had a significant career as commemorated in the guide to the Garrison Church at Portsmouth:

> *... served in China (medal) at Amoy, Chusan, Chinhae, (including repulse of night attack) Chapoo, Woosing, Shanghae, and Ching Kiang Foo; also served with the Army of the East, ...*

Then, a third brother, Charles George Butler (1823-1854), a Captain in the 86th Foot Regiment, died of fever at Bombay on 17th December 1854, aged 31.[9] None of these three was married. For the elderly Lieutenant General, the news of the deaths in swift succession, over the course of six months, of three of his four adult sons must have been terrible. He himself died in Paris on 7th December 1856, aged 76. Fortunately, the fourth adult son, the Rev. Pierce Butler (1826-1868), became a Chaplain to the British Forces in the Crimea War, served at Scutari on the Bosphorus, was with 2nd Division at Sebastopol and the capture of Kerch, ... and survived. The entire family is handsomely commemorated on a large brass plaque in the Garrison Church at Portsmouth:[10]

> *TO THE GLORY OF GOD AND IN PIOUS MEMORY OF LIEUT. GENL THE HON: HENRY EDWARD BUTLER WHO SERVED IN EGYPT AND THE PENINSULAR WAR DIED IN PARIS DEC 7 1856 AGED 76 AND HIS FOUR SONS. HENRY THOMAS CAPT 55TH REGT FELL AT INKERMANN NOV 5 1854 AGED 42. CHARLES GEORGE CAPT 86TH REGT DIED OF FEVER AT BOMBAY DEC 17 1854 AGED 31. PIERCE RECTOR OF ULCOMBE KENT SOME TIME CHAPLAIN TO HM FORCES IN THE CRIMEA DIED FEB 8 1868 AGED 42. JAMES ARMAR CAPT. CEYLON RIFLE REGT DIED FROM WOUNDS RECEIVED AT THE GALLANT DEFENCE OF SILISTRIA JUNE 21 1854 AGED 27. THIS WINDOW WAS GIVEN BY NUMEROUS RELATIONS AND FRIENDS.*

The Rev. Pierce Butler, a graduate of Trinity College, Cambridge, was 28 when TB met him travelling with his 42-year-old brother. An obituary reports that after his brother was called

---

[8] James Armar Butler was the author of a journal covering the siege of Silistra (Butler 1854).
[9] Commemorated in the records of Winchester School. Unknown to him, after leaving London for the Danube front, *'he had been promoted a major in the army, and lieutenant and captain in the Coldstream Guards'* (James Armar Butler, *Oxford Dictionary of National Biography*, accessed 17 March 2020).
[10] http://www.memorialsinportsmouth.co.uk/churches/royal_Cgarrison/butler.htm. The brothers are each commemorated separately on other plaques there.

Figure 5.2: Rev. Pierce Butler in Crimea in 1855 (Roger Fenton Collection, Library of Congress. PH - Fenton (R.), no. 252) ('No known restrictions on publication').

away from Sinai to join his regiment, he went on to the Holy Land and Constantinople before returning home (Murchison 1868). Then followed his war service with the Army, during which he was commemorated in at least one photograph (Figure 5.2).

The deaths of all three of his adult brothers then of his father may be why Pierce Butler reportedly then spent five post-war years (1856-1861) travelling extensively 'in America and many parts of Europe' (Murchison 1868: cxxxiv).[11] In 1861 he became Rector of Ulcombe in Kent (a family 'living'), the same year he married Catherine Twisden Smith-Marriott. They had three daughters and two sons and he died in 1868, aged 42.[12]

TB never mentions these two Butler brothers again after their meeting at St Catherine's Monastery nor does he record what they were doing there. Stewart is - again, the starting point for what they were doing in Egypt and Sinai (Stewart 1857: 18-19):

Our esteem for them ripened with the length of our acquaintance, and I regretted not being able to accompany them up the river, as they proposed on their return, following the route I was about to take. Related by marriage to the Rev. Charles Forster, who professes, in his 'Voice of Israel from the Rocks of Sinai,' to have discovered a key to the Himyaritic inscriptions of the Wadi Mokatteb, Captain Butler had adopted, very naturally, the views of his relative, and was resolved to search the Desert for an inscription of twenty lines, in the Sinaitic character, mentioned in that work as having been seen by an Arab merchant, named Cosmas Indicopleustes,[13] in one of his journeys, upwards of 1300 years ago, but never stumbled upon by any one since. He intended visiting the neighbourhood of Tur, as it is known that some of these inscriptions are to be found on the rocks of Ghebel Hummam, not far from that town. I presume he was unsuccessful in his search, otherwise such a discovery must have become known by this time. We parted, in the hope of meeting again in Syria; but, at the moment of embarking at Beyrout, I learned that he had received a summons, after returning to Egypt from the Desert, to rejoin his regiment, then on its way to Gallipoli or Varna, and had been obliged to renounce the Syrian portion of his tour.

---

[11] The European travel included Norway which sparked in interest in a region from which he believed his ancestors had migrated. That in turn led him to prepare a translation and edition of several Danish tragedies and poems (Butler 1874).

[12] Two of his grandsons - brothers, were killed in the First World War: Ralph Twisden Butler was drowned at the sinking of HMS Hampshire on 5 June 1916, aged 20; his brother, Somerset Armar Butler, a Lieutenant in the Royal Flying Corps, was killed on the Greek Front on 16 October 1917, aged 24, and is buried in the Commonwealth War Graves Commission cemetery at Sarigol, Kriston, Greece. A sad echoing of the tragedy in this family two generations earlier (Hillam 2011: 17-18).

[13] A 6th century AD Greek-speaking Alexandrian hermit and traveller.

The *Guide to the Garrison Church at Portsmouth* adds that '*He was the first promoter of the recent and already renowned Survey of the Siniatic Peninsular (sic); ...*'.

Neither of the brothers met by TB published his findings and one was soon dead.[14] However, as Stewart notes, they were relatives (by marriage) to the Rev. Charles Forster (1787-1871), a renowned - and controversial, researcher on places and events in Old Testament history and already famous for the book Stewart mentions (Forster 1851; cf. 1844 and 1853). As a short biographical obituary for Pierce Butler explained (Palmer 1874), he was engaged at the time of his death in 1868 in preparations for a new expedition to Sinai in pursuit of the same material he had been researching with his brother fourteen years earlier in 1854 when TB encountered them. He died on 8th February 1868 the day he was due to set off and the work was taken up by the Rev. Henry Edward Palmer in 1869-70. This latter wrote his own account of his explorations in Sinai (Palmer 1871a; 1871b) but gives scant mention of the work of the Butlers. Not so their relative Charles Forster who frequently credits them in his own study of the important text they had recorded and carried home as transcription and squeezes, photographs of which he now published (Forster 1862). Forster repeatedly thanks and credits the brothers individually and together for their discoveries. In particular (29-30):

But this and all the preceding marks of the common age and authorship of the hieroglyphic and Sinaïtic inscriptions in the scenes of the Exode (sic), are at once corroborated and eclipsed by Mr. Pierce Butler's discovery of a triple inscription on the Djebel Maghara, in the immediate vicinage of the hieroglyphic tablets in the Wadi below; in which a pure Sinaitic inscription, illustrated by the hieroglyphic of an ostrich, stands engraved on the same tablet with two purely hieroglyphic inscriptions; a triple record on the one monument, on the principle apparently of the Rosetta stone.

Not contenting himself with the ordinary information and ordinary phenomena, Mr. Butler, arrived in the Wadi Maghara, cross-questioned and cross-examined his Arab guides as to the existence in the locality of any other inscriptions besides those already known. After much and close inquiry they at length informed him that in a mountain cave, half-way up the adjoining Djebel Maghara, there were writings or inscriptions of the kind he was in search of. He asked them to conduct him to the cave, and they agreed to do so. He climbed with them half-way up the side of the mountain, until he came to where his sheikh pointed out a low-browed cavern. The entrance was between four and five feet in height, but looked so unpromising that he thought himself deceived, and had almost decided not to enter in. However, he wisely judged it better to try; he knelt and entered its mouth, when, to his unfeigned astonishment, he found the entrance on both sides cut into regular planes or tablets; and, upon the right-hand plane, discovered a sculptured triple inscription, two of its columns being in pure hieroglyphics, while the third was in pure Sinaitic characters. The three inscriptions stood side by side on the one tablet; all three cut, not dotted out, obviously with the same graving-tool and by the same engraver.

---

[14] In 1855, the French explorer of Sinai, Lottin de Laval, who had been researching there a few years previously, published his own account in which he makes a brief mention of the Butlers and in a fashion that seems to imply something had been made public - or perhaps it was a letter (Lottin the Laval 1855: 336-7): *On pourrait, à ce sujet, taxer aussi d'exagération M. le capitaine Butler, qui visita ces parages célèbres en 1854, quatre ans après moi; il parle de rochers droits comme des murailles, et les hauteurs des inscriptions varient, selon lui, de 20 à 30 et même 40 pieds. S'il entend par là que les plus élevées se trouvent à 40 pieds de tout endroit accessible et même du sol de la plaine, nous lui répondrons par la dénégation la plus formelle; il s'est trompé.*

The brothers copied the texts and made squeezes (Henry)[15] and casts (Pierce), photographs of which Forster used to illustrate his book. He notes, too, having received communications from both brothers including a letter from *'from my lamented relative'* Captain Butler.[16]

***

Antonio Schranz
(Menorca, 1801–Cairo, 14 July 1865)

While in Cairo on 21st February 1854, TB made the first of two visits to the man he describes as *'a German artist who takes photographs - Schrantz (sic)'* adding, *'He had some beautiful things. I bought 2 drawings.'* When he returned there three days later it was *'about the photographs (sic)'* but he adds that he and Strachan *'looked over some drawings of Petra, Mt Sinai, Baalbec & Upper Egypt'* but then Kennedy joined them and they all left.

Antonio Schranz was one of the leading members of a remarkable artist family - one of *'12 artists, born in four generations in 93 years'* including - very remarkably for the time, five women (Bonello 2017; Schranz 2017; Schranz and Schranz 2017)[17]. Their artist father Anton (the Elder) (1769-1839) had been born in Ochsenhausen in south-western Germany but - after some twenty eight years in Spanish Menorca, had finally settled in Malta in 1817. His three sons all became noted and prolific painters: Giovanni (1795-1882) stayed largely in Malta but Antonio and Giuseppe/Joseph (1803-before 1884) travelled extensively, painting and undertaking commissions. Giuseppe/Joseph had settled in Constantinople c. 1835 after a decade of travelling and in 1838 was appointed to a *'professorship of painting'* at the Ottoman Military Academy of Art (Llewellyn 2017).

Antonio travelled even more: his descendant and biographer, John J. Schranz, calculates *'a conservative estimate of Antonio's travels indicates over 150,000km through vast, strife-riven territories, between 1823 and 1851'* (2016-2018). One of these was in 1837: in his account of his travels in the Levant, Lord Lindsay reports that while at Damascus negotiating - unsuccessfully, to go to Palmyra (Lindsay 1838: 164):

> Mr. Pell, of Devonshire, - Mr. Alewyn, a Dutch Gentleman - and Mr. Schranz, a German artist travelling with them, had just arrived in Damascus from Greece and Asia Minor. ... Schranz was born at Majorca, of German parents, and brought up at Malta; speaks German, Spanish, Italian, and Maltese, (a dialect of Arabic, by which he can make himself easily understood by the natives here), as his mother-tongues, - Greek, and uncommonly good English, - besides being a most accomplished draughtsman.

---

[15] Captain Butler apologised for the quality of the squeezes noting that when they reached the summit of the hill they found their Arab attendants had drunk the water leaving only dregs to work with (Forster 1862: 78).
[16] Other casts and squeezes were made in about the same time by the eccentric Major MacDonald who spent ten years in an ultimately futile attempt to make a living mining turquoise from the sites in the Wadi Mughara once exploited in pharaonic times. His copies were used by Forster - who calls him 'Colonel Macdonnell' and are now in the British Museum.
[17] John Schranz, a descendant, has published a lengthy series of long articles in *The Times of Malta* newspaper about his family (Schranz 2016-2018). Some of that was then incorporated in the chapters he wrote and those by collaborators, in a memorial volume, Bonello 2017.

Schranz was mentioned a little later while the combined parties were in Lebanon but evidently the Pell Party did not continue with the Lindsay one and he is never mentioned again. We know that Pell had been very active in Asia Minor exploring ancient ruins and recording inscriptions (Whitehead 1999) and it may be that Schranz was employed as an artist/draughtsman for that expedition. Certainly he produced sketches and paintings from western Asia Minor dated 1837. As TB notes, Schranz had produced illustrations of Baalbek and there is an original watercolour in the Victoria and Albert Museum in London (Schranz 2017b: 139, Fig. 4)

More important was the commission by Lord Castlereagh to join his party travelling in Egypt, to Petra and in the Levant in 1842. Castlereagh published a two-volume account based on his letters home; in his Preface he notes: *'The plates are selected from a large collection of drawings taken on the spot by one of my companions, Mr. A. Schranz, of Malta.'* He never subsequently names Schranz, but all ten of the illustrations he publishes are signed *'A. Schranz 1843'* (Castlereagh 1847) and include *'Petra, Mt Sinai, ...'* as noted by TB.

Two watercolours not used by Castlereagh may have been for personal pleasure. In 'The Encampment at Hebron', dated June 13, 1842, Antonio has painted himself painting Lord Castlereagh and another western companion - presumably Castlereagh's physician, Thomas Tardrew, with their tented encampment in the background. Many others were still in the family for many years afterwards as his step-son records (Powerscourt 1903: 24-25):

> *In the East Bedroom are a collection of water-colour drawings of subjects in Syria; these were painted for my step-father Frederick, fourth Marquis of Londonderry, when he was Viscount Castlereagh. He went on a tour in the East, taking with him a German painter of the name of Schranz, also a surgeon named Mr. Tardrew. Portraits of the artist, the surgeon, and himself, in Eastern dress, are in the room, painted by Lewis, R.A.; also a portrait of Prince Hassan of Egypt, and of two dragomen, also by Lewis, R.A. The rest of the pictures are by Mr. Schranz*

Perhaps, tiring of so much travelling, Antonio had opened his photographic studio in Cairo in 1848 - a decade before the next such studio. It is interesting that TB was able to look at drawings not just of Petra and Mt Sinai - both of which had been included by Castlereagh, but also Baalbek which was not used - unexpectedly given how gushing Castlereagh was about the ruins.[18] And now Schranz was also taking and offering photographs, a number of which of scenes in and around Cairo, were marketed by him (e.g. Schranz 2017b: 82, Fig. 26).

---

[18] Perhaps because Schranz had made the Baalbek drawing in 1837 when he was with Pell and Lindsay.

Chapter 6

Petra and Visiting Petra in the 19th century: Logistics and hazards of the 'Long Desert Route'

On 28th February 1854, Thomas Bowles set off with his new friend and travelling companion Henry Kennedy on the greatest and most spectacular portion of his long tour. They arrived in Jerusalem - journey's end, on 3rd April 1854, on their 35th day of travel, most of it on foot or riding a camel.

By 1854 visiting Petra was increasingly common but still immensely challenging and not without is dangers. However, almost without exception, those who completed the journey regarded it as having been the most memorable part of their travels. In part that was due to the crossing of the Red Sea and the journey through Sinai with their evocation of the Old Testament and the travails of the Israelites. More important still was the visit to the extraordinary ancient city ruins of Petra in their stunning location. In 2007 'Petra was named amongst the New 7 Wonders of the World and was also chosen by the *Smithsonian Magazine* as one of the "28 Places to See Before You Die"'.[1] How much more so in 1854, long before mass exposure to imagery, the fleets of buses, scores of hotels and tourists arriving through Amman's airport, off cruise ships in the Gulf of Aqaba or as day-trippers from Israel.

***

Petra

The city of Petra had been a fabled place even in Hellenistic times - 3rd century BC, when it first appears in Greek literature as the 'capital' of the Nabataean Arab kingdom. It flourished during the century and half of expanding Roman rule in the wider region (*c.* 63 BC-AD 106) when most of its visible public monuments were constructed during the reigns of successive native Nabataean kings. After the annexation of this Nabataean kingdom by the Roman Emperor Trajan in AD 106, it continued as a prosperous city, part of the newly formed province of Arabia and a regular annual stopping place for itinerant Roman governors on their assize circuit.

After AD 106, Petra was no longer a royal capital, of course, but then or soon after, it ceased even to be the seat of Roman government in the new province. Instead, the already burgeoning city the Nabataeans had developed at Bostra (modern Bosra eski-Sham), in present day southern Syria, became the Roman 'capital' and the place where the major military force of the province - the *Legio III Cyrenaica,* was headquartered in a newly built fortress of 16.8ha. For two centuries, Roman troops - including parts of this legion of citizen soldiers, were posted around the sprawling province from Bostra in the north to as far as Mada'in Saleh - the Nabataean/Roman city of Hegra, 650km/400 miles to the south in the Hedjaz region of north-western

---

[1] https://world.new7wonders.com/wonders/petra-9-b-c-40-a-d-jordan/

Saudi Arabia and 470km/295 miles southeast of Petra. Some troops were stationed in and near Petra. Then, about AD 300, in changing circumstances, an entire small legion was established in a newly created fortress of 4.7ha at Udruh, just 15km/9miles due east of Petra and surely within its city territory. By then Christianity was taking hold and the city soon had its line of bishops and the first of many churches. This last period of Roman Petra is now being revealed through the decipherment of the great cache of Petra papyri recovered in excavation in 1993 (Frösén et al. 2002-2018).

Petra fell to the armies of Islam in the early 7th century AD. Although no longer a city it likely continued as perhaps a scatter of village-like communities exploiting the rich agricultural hinterland at least in the two centuries of the subsequent Umayyad dynasty (AD 640-750) and perhaps beyond as the next Muslim dynasty – the Abbasids (750-969), originated in the small Roman town of Humeima, 50km/30 miles south of Petra.

At the time of the Crusades (1099-1291) and the establishment of western, 'Frankish', states in the Levant, the wider Petra region was annexed as Oultrejourdain - 'Beyond the Jordan'. Major castles were established north of Petra at Shobak, Tafila and Kerak. Petra, too, again became a Christian centre with both a small castle at Wu'eira just east of the city and a fortified post at Al-Habis inside Petra.

Few of the Crusader structures named above were seen by 19th century western visitors to Petra including the two structures at Petra itself. Indeed, for several decades, the castles to the north were outside the zone of safe travel; Karak in particular was notoriously the

Figure 6.1: The Isle of Graia. Coloured lithograph by Louis Haghe after David Roberts who was there in 1839. (Library of Congress) (Public Domain).

Figure 6.2: Lithograph of Johann Ludwig Burckhardt, c. 1820 (© The Trustees of the British Museum)

centre of the predatory Majali family and unsafe for westerners until the beginning of the next century. In the other direction, as they ascended the coastal road to reach Aqaba, western travellers often did remark on the castle on 'Pharaoh's Island', at the head of the Gulf of Aqaba and just off the present Egyptian coastal tourist resort of Taba (Figure 6.1). However, it is doubtful if the substantial castle remains on this so-called 'Ile de Greye/Isle de Graai' - a name seemingly invented in the 19th century, are Crusader; more likely Arab Ayyubid (AD 1174-1263).

Petra disappears from western knowledge for several centuries after the Crusader loss of Oultrejourdain to Saladin in AD 1189. Its mystical - not to say mythical, reputation persisted but even its precise location was lost to western scholarship. A handful of known western travellers in the region passed by in ignorance on the east, not least Ludovico Varthema in 1503 (whose account was to be translated by the Rev. Percy Badger (above): Badger 1863). Finally, on 22nd August 1812, the Swiss orientalist and explorer - an Arabic-speaker, Johann Ludwig Burckhardt (Figure 6.2) (November 24, 1784 – October 15, 1817), 're-discovered' it and wrote extensively of his visit and what he saw. Despite the almost casual nature of his words, it was an astounding report of a remarkable place set within the bowl of mountains with often vertical/near-vertical sides. In a letter to Sir Joseph Banks dated 12th September 1812 and published some years later he wrote (Burckhardt 1819: xlvi):

At the distance of a two long days journey north-east from Akaba, is a rivulet and valley in the Djebel Shera, on the east side of the Araba, called Wady Mousa. This place is very interesting for its antiquities and the remains of an ancient city, which I conjecture to be Petra, the capital of Arabia Petræa, a place which, as far as I know, no European traveller has ever visited.

Although it occasioned immense excitement, it was several years before the next western visitors reached the site (Table 6.1) and just six months after Burckhardt's death. Although he died in 1817 and his accounts were only made public a decade after the visit (1822; 1824), his personal letters and conversations with other westerners he met in Egypt spread the knowledge of what he had seen and how to find it. Most notably, Burckhardt met William Bankes in Cairo in 1816 and the latter was plainly intrigued by what he heard of this remarkable discovery (Lewis: 2004: 136). Two years later, Bankes was one of several men who became the second western visitor to Petra in modern times. Almost all of these new travellers not only

Table 6.1: Early western travellers to reach Petra. The French expedition of 1828 was the first to approach from Egypt which then remained the overwhelmingly commonest route for several decades.

Travellers	At Petra	Route
Burckhardt, John Lewis (Johan Ludwig/ Jean Louis) (1784–1817)	22-23 August 1812	From the North
Bankes, William John (1786–1855) Irby, Hon. Charles Leonard (1789-1845) Mangles, James (1786-1867) Legh, Thomas (1793–1857) Curtin, Henry James (*c.* 1796-1825) Finati, Giovanni (1786-1829?)	24-25 May 1818	From the North
Anson, Hon. Henry (1804-1827) Fox-Strangways, Hon. John George Charles (1803–1859)	1826?	From the North?
Linant de Bellfonds, Louis Maurice Adolphe (1798–1883) Laborde, Simon Joseph Léon Emmanuel, marquis de Laborde (1807-1869)	28 March – 3 April 1828	From Cairo via Mt Sinai and Aqaba

made detailed records of the journey and what they saw but published books which in turn stimulated yet more travellers (Table 6.2).

During the next ten years there was just one set of western visitors but then in 1828 there was a major development. Bankes had met a young French artist in Cairo in 1818 and doubtless told him of his visit. This was Linant de Bellefonds who was long-resident in the region and in 1828 became the joint leader with de Laborde of a major French expedition, which included artists. Due to the fortunate coincidence of plague at Petra when they arrived, they were left alone for eight days recording the site extensively and publishing soon after a large well-illustrated book (Laborde 1830; 1836; cf. 1994[2]). This is what may be called the Pioneer phase; it was soon to give way to what today would be called Adventurer phase and later still to Adventure Tourists.

Many western visitors to Egypt took the opportunity to also visit the Holy Land, so close by. The simplest route was by sea from Alexandria to Jaffa. The more adventurous could go overland by what came to be called the Short Desert Route: from Egypt along the coast through El-Arish to Gaza and on to Jerusalem. From the 1840s onwards, the even more adventurous had the option of the Long Desert Route which would take them to Mt Sinai and Petra. Once begun, it was not long before books began to set out specific guidance.

---

[2] This last item consists of a republication of the original French edition by a French scholar (Augé) together with a descendant of Laborde's travelling companion (Linant de Bellefonds). Helpfully they have prefaced the work with a French translation of Burckhardt's original. The Bibliography of this book also includes the full citations of the other early visitors to Petra and their publications.

Handbooks

For English-speaking travellers there was one of the famous 'Murray's *Handbooks*' which came to be vital preliminary reading and often part of the luggage of travellers.[3] Murray's *A Handbook for Travellers in Egypt* first appeared in 1847 and there were numerous successive updated editions. It is striking that even in the edition available to Bowles and his travelling companions, Murray could offer quite detailed information on what he calls Routes 7 (Cairo to Suez), 8 (Cairo to Mount Sinai) and 9 (Mount Sinai to El Akaba) (Murray 1847: 207-222) but then adds just two short paragraphs tacked onto the end of Route 9 (221-222):

> *In going to Petra (Wadee Moosa) from El Akaba, it is necessary to make an agreement with the Alloween Arabs; but taking advantage of the position of the traveller in these lonely regions, who must pay whatever they choose to ask, or give up his journey, their demands have become so exorbitant, that few will feel disposed to take this route; and it is far better to go from Hebron.*

> *There are two roads from Hebron to Petra (Wadee Moosa); the eastern one by the south end of the Dead Sea, occupies 44 h. 50 m.; the western road, 42 h. 10m. From El Akaba to Hebron, or El Khaleel [= Hebron], is 71 h 45 m.; El Akaba to Jerusalem, 80 h.; but the best road to Syria is from Cairo, or from Suez, on returning from Mount Sinai.*

The second edition of the Egypt handbook as well as the first edition of *A Handbook for Travellers in Syria and Palestine* both appeared in 1858. The former was little-changed as far as crossing Sinai and going to Petra was concerned, but the latter was very detailed - what are now called Routes 1 to 5 (Cairo via Petra to Hebron) are treated in no less than 56 pages, ... but too late for TB (Murray 1858: 1, 9-65).[4]

There had, of course, been many western visitors to Petra between Laborde in 1828 and the first edition of Murray's *Egypt* in 1847 whose published accounts might have been used for the latter but seem not to have been. Between 1847 and TBs visit in 1854 there were many more. Published accounts were often published many years after the journey but as Table 6.2 shows, several would have been available to TB and to his companions if they had planned ahead and knew of the publication. Like TB, however, many trans-Sinai and Petra visitors seem to have seized an unintended opportunity and had to make do with what was available in Egypt.

There is only slight evidence in his journal that TB carried any of these. While still at Cairo he says explicitly he made use of 'Murray' on several occasions in relation to sites visited but while at the Great Pyramid explains:

> *I was all for eating breakfast & sitting quietly down & reading what Murray said, (I had brought his Handbook with me from the Hotel's Library [= Shepheard's 'library']) & then climbing the Pyramid at our leisure.*

---

[3] Digitized copies of almost all Murray's handbooks can be downloaded free as pdf files most easily via archive.org.
[4] The German and English editions of the Baedeker Handbooks for *Egypt* and for *Palestine and Syria* and of Cook's Handbook were 20+ years in the future.

Table 6.2: Western visitors to Petra between Laborde (1828) and Bowles who published extensive accounts of their journeys. Each name usually represents the author only of what was a party of two or more people.

Traveller	Visit	Publication
Stephens	1836	1837
Robinson and Smith	1838	1841
Roberts and Kinnear	1839	1841
Morris, Olin, Formby etc	1840	1843
Millard	1842	1857
Plumley	1842	1845
Measor	1842	1844
Castlereagh	1842	1847
Durbin	1843	1845
Wilson etc	1843	1847
Keith	1844	1859
Bartlett	1845	1849
Martineau	1847	1848
Finn	1851	1867
Lowth	1851	1855
Marsh	1851	1888
Ireland	1852	1859
Stanley	1853	1856
(Stewart - reports his aborted visit attempted just two months before TB	1854	1857)

On the other hand, while on the steamer to Suez he records that *'After dinner I went to my cabin & sat quietly reading Miss Martineau's Egypt ...'* and at Petra itself he twice makes observations about *'Miss Martineau's'* views about specific structures (Martineau 1848). Just before reaching Hebron TB mentions having read a more obscure book, that of John Price Durbin who had been at Petra in 1843 (Durbin 1845). Finally, in Jerusalem he evidently had an opportunity to consult Robinson, Wilson, and Keith (1837, 1847 and 1848 respectively).

It seems clear that TB, like many before and after, could get some advice in Cairo and he makes it clear he was able to compare notes with others planning the journey, most notably with 'Mr Ross' (below, Ch. 7).

The journey that emerged could be divided into sections (Figure 6.3):

Cairo - Suez
Suez - Mt Sinai
Mt Sinai - Aqaba
Aqaba - Petra
Petra - Hebron
Hebron - Jerusalem

Logistics of Travelling to Petra

The broad outline of arrangements consisted of a 'party' - in this case TB and Kennedy, hiring a dragoman and setting out the details in a formal contract arranged under the aegis of their Consul in Cairo and signed by the principals. The dragoman in turn arranged to hire guides/escorts from one of the Sinai tribes and enough people to see to the needs of the party and themselves and to supply the equipment and food and drink necessary for what was normally about 40 days of travel. The precise length depended in part on how long the travellers chose to spend at St Catherine's Monastery at Mt Sinai. More problematic was how long they would have to wait at Aqaba where they had to enlist a new escort and guides from the Edomite tribes to conduct them to Petra and then on to Hebron and Jerusalem, and how long they were permitted to stay in Petra itself. The latter could be stipulated in the new contract at Aqaba, but they were often then at the mercy of the inhabitants at Petra who were unpredictable - indeed, comprised different groups, often competing with one another for what could be obtained from the 'Franks', their catch-all term for all westerners.

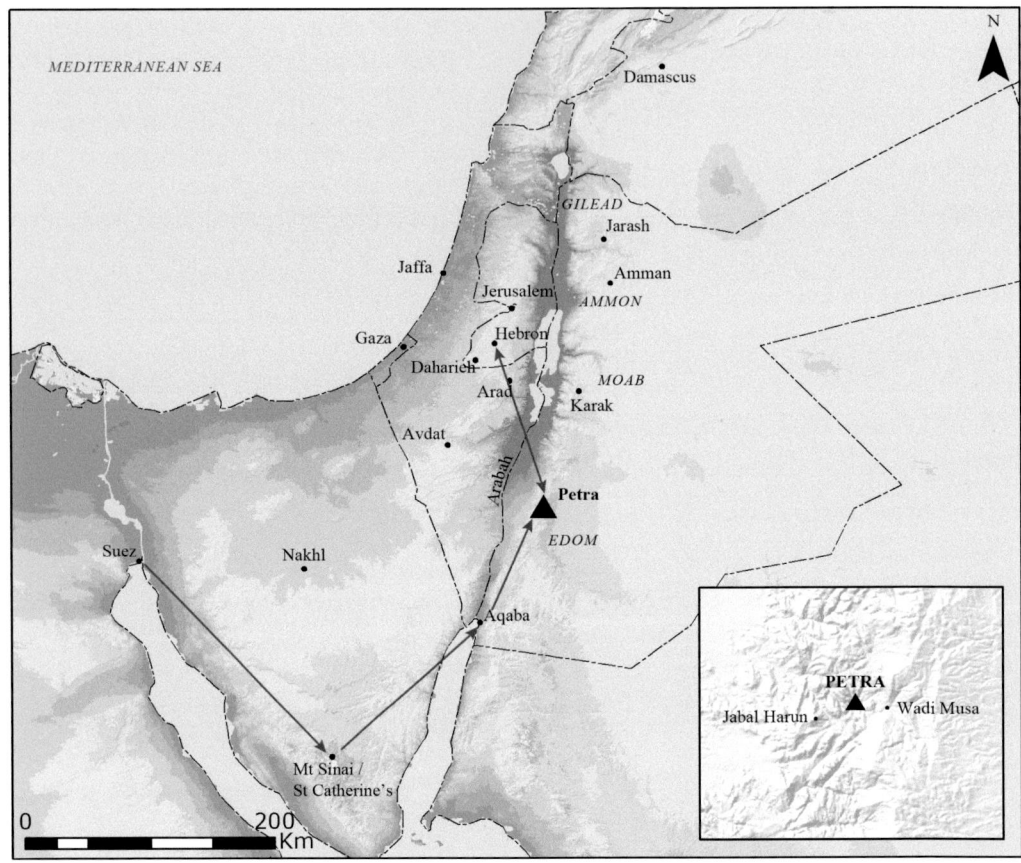

Figure 6.3: The Long Desert Route to Petra (Drawn: Travis Hearn).

Travellers were advised to obtain a letter of introduction from the Monastery of the Sinaites in Cairo addressed to their sister monastery of St Catherine's at Mt Sinai if they hoped for hospitality from the monks, not least accommodation inside its walls. Also, a letter of introduction to the Ottoman 'governor' in the little fort at Aqaba.

'Equipment' consisted of such obvious items as the camels they were to ride and other camels to carry their tents, beds, tables, cooking equipment etc. As there was slight chance of replenishing supplies along most of the route, almost everything they might require had to be obtained in advance. Once guidebooks began to appear, advice was provided on what might be brought from Europe, from Malta or acquired in Egypt. The earliest of these - Murray in 1847 for travellers in Egypt (including those venturing into Sinai) set it out extensively:

pp. 3-6 gives list of items than should be bought in various places from mosquito nets to guidebooks.

pp. 142-145 gives list of goods including food that may be purchased in Cairo and their costs.

pp. xxxii-xxxv gives a corrected list of the above

TB does not say much about the organization and logistics at all and may have relied mainly on Mr Kennedy. Some travellers were sufficiently impressed by the whole enterprise that they published partial lists of what their own great adventure had required. Two who straddle TBs journey are worth quoting - the 20-year-old Hannah Hindley crossed Sinai and visited Petra with her father and two other family members in March 1850 (Hindley 1851: 11-12):

> *I may here mention of what our caravan was composed. Our own party consisted of one gentleman, three ladies, a maid, a dragoman (or interpreter) named Bedair, whom we had engaged in London, and an Arab cook with his assistant. Then we had the camel drivers with their head, the Sheikh, and twenty-two camels carrying the following articles - trunks containing clothes, hair mattresses and tressel beds to hold them, with blankets and sheets, tin wash basins, and all the culinary articles in tin; grids to hold our charcoal fire, a canteen with knives, forks, and necessaries for the table, also earthenware plates, dishes, cups and saucers. We had four barrels and several skins full of water, a hamper of light red French wine, three barrels of bread, a dozen live turkeys, and forty fowls in two hen coops made of palm branches, a thousand eggs, and five hundred oranges. This was provision for five weeks, with a due allowance for accidents. We had three tents, one of which was used as a kitchen and dining room, and we carried mats and carpets to spread in them. About four o'clock every day we stopped, and then it took two hours to pitch our tents, arrange our furniture, and prepare our dinner, during which time we wrote our journals, pressed any pretty flowers we could find, and otherwise spent those two long hours, as best we hungry people could. After dinner we sat reading for a short time, and about half-past went to bed, where we slept most soundly. ...*
>
> *... as we sat at breakfast, the sand was blown on to our bread and butter, and the salt was scattered about the table. We had tea (with no milk), toast, bread and butter, and eggs for breakfast; for lunch, cold meat, bread, biscuits, dates, oranges, and sometimes a gingerbread or a cake, as a dessert. At dinner we had generally two dishes of meat, and one sweet dish. Irish stew came round about three times a week as long as our supply of potatoes lasted, and macaroni frequently made its appearance.*

Rev. Horatio Bonar and three companions prepared for a visit to Petra in February 1856 but were obliged while at Nakhl to accept they could not get there (Bonar 1857: 69-70):

> "*As we moved along I surveyed the whole retinue. There was our dragoman, Haji-Ismael, mounted on one camel, with my two portmanteaus firmly roped to the side of the animal. He was dressed out for the start, with a goodly flowing robe and a head-dress, whose red and yellow stripes glittered brightly to the sun. On another camel were two large square panniers, of palm-branch wicker-work, filled with oranges and lemons, to the amount of at least 1000 of the former and 100 of the latter. On another was our "canteen," that is, two immense wooden chests, containing our dining apparatus, such as plates, knives, forks, spoons, cups, not omitting candles and fenusses, that is lanterns made of linen, apparently saturated with bees-wax or some such substance, which draw out to more than a foot in length when in actual use, but can be contracted or folded together into a very small space for packing. Above this apparatus was placed our kitchen-grate, a long iron box, pierced with a hundred holes in sides and bottom, —its four legs, like signal-posts stretched upwards to the sun. Then came another with an immense wicker-cage, which formed the prison-house of some 100 fowls, all alive, but quite willing apparently, to go to the stake for us, — as four of them are to be called on to do each day at sunset. Balancing these fowls is another cage, with half-a-dozen turkeys, which we are told are to be our "Sunday dinners." On another our bedding is mounted, on another our tents, on another our charcoal, on another our barrels of Nile-water, on another sacks*

containing our camp-stools and table, — our bed-steads, and the pins (watt-watts) of our tents. Most carefully was everything packed up before it mounted the camel. A tribe of wanderers knows how to "pack," as indeed they have no security for any part of their property in moving about save the good packing. They have bags for their money, and sacks for other things ...

On four lighter animals, which were said to be dromedaries, were the four travellers, ...

There were nineteen camels in all, and, including dragoman and servants, about the same number of men, — all of them swarthy Arabs. Notable amongst them is Sheikh Suliman, the Sheikh or chief of one of the Sinaitic tribes, who is our guide and guard. ...

The Journey

Most journies in the 1850s followed an increasingly standard routine against which TBs journal can be measured. The packing of 'luggage' - including livestock, onto camels for the first time was viewed by most western travellers with besumement. The camel owners each tried to keep their share of the luggage low to spare their beast. The result was noise, shouting, quarelling and seeming chaos. To some extent that was repeated each day as consumables declined and loads had to be re-distributed. For that reason, departure was often later in the day and they would then camp for the first night just outside Cairo where the future routine could be trialled for the first time. Tents had to be pitched, beds, tables and stools assembled, animals slaughtered, prepared and cooked. As Hannah Hindley (above) notes, their meals could be quite varied and extensive though they would progressively run out of fresh produce and water became tainted from the containers to the extent that travellers consumed a great deal of 'porter', a dark beer. The two hours or so between stopping and dinner meant that travellers often used that time to write letters or journals. Breakfast was early and - as TB notes, some travellers set off promptly while the camp was struck and animals packed and to move while it was cooler. Within a short time, many travellers discovered a preference for walking much of the route, often then mounting when the camels had caught up. Travel was a time for solitary reflection or conversation - harder when on camelback; or even for reading.

There were adventures starting after leaving Suez when travellers often left the caravan to go ahead by land around the gulf while they took a boat and sailed across to meet it at Ain Musa. Elsewhere they encountered ancient graffiti, some of it going back to quarrying in Pharaonic times. Clergymen in particular but also most travellers to a considerable degree who - as was then common, were steeped in the Bible, constantly viewed places, landscapes and such features as graffiti, through the lens of the Old Testament and the flight from Egypt of the Israelites. The location for the Crossing of the Red Sea, the graffiti in the Wadi Muqqaten and Wadi Feiran supposed by some to have been the records of the wandering Israelites and the holy events associated with Mount Sinai. Not all were convinced of the identification of Mt Sinai but even sceptics could find interest in the early Christian structures at St Catherine's Monastery (Figure 6.4) and with its ancient manuscripts (see now the superb study by Manginis 2016). Most parties paused at St Catherine's for at least a day or two to ascend one or more holy peak. If they had a letter from the headquarters of the monks in Cairo (above) they might take up an offer to be hoisted up and have accomodation inside the monastery. Tourism inside the monastery involved not just the library but the ossuaries for the generations of monks who had died there. It was not uncommon that travellers accepted hospitality but were then outraged and offended when demands were made for payment which they felt far too high.

Figure 6.4: Convent of St. Catherine, with Mount Horeb. Coloured lithograph by Louis Haghe after David Roberts who was there in 1839 (Cleveland Museum of Art Collection) (Public Domain).

Many departed with a sour taste from acrimoniuous exchanges. The duration of this stage of the journey - *c.* 250km/155miles, was fairly predictable: eight to nine days.

After St Catherine's, some travellers headed north to the staging post and Ottoman fort at Nakhl from which they could continue to Hebron and omit Petra entirely or obtain guides there to take them to Petra. Most travellers, however, continued northeast to reach the Red Sea coast of the Gulf of Aqaba. The coast could then be followed much of the way, passing the Isle de Greye (above, Ch. 5), till they arrived at Aqaba, a total of five to six days.

At Aqaba travellers found a miserable collection of a few houses and a small Ottoman fort[5] with a handful of ragged soldiers and a 'Governor' who might - as was the case in 1854, be accompanied by his wife. The place overlooking the gulf lay on a key route. The annual Hajj from Cairo arrived from the west, coming across the Sinai. The governor watched over these passages twice a year - one in each direction. The governors probably looked forward to the passage of a handful of western travellers each year to break the monotony with the intrusion of colourful and unusual visitors including a few women and the opportunity for 'gifts'.

The pleasure of reaching this famous Biblical spot - Edomite Elath and later Roman and Byzantine Aila, soon wore off as there was little to do beyond explore the traces of the ancient town, collect shells and sometimes get fish. It was the more exasperating as the duration was no longer under their control. At Aqaba, the travellers had gone as far as they could with the Towara beduin through whose territory they had been travelling and with guides and helpers from that tribe. Now they passed into the 'care' of the Alawin, a tribe whose territory extended from the eastern side of the Gulf of Aqaba to Petra. Almost every travellers remarked on the striking difference between the 'gentle' Towara and the fierce and aggressive Alawin. At Aqaba, the travellers were far from 'civilization' either to the west (the Nile) or north (Palestine) and at the mercy of the Alawin. The first edition of Murray's *Syria* (1858: 5-6) spells it out:

> *There is just one other tribe of Arabs with whom the traveller may have to deal in his pilgrimage, the 'Alawin, whose sheikh, Hussein, has long claimed the right of furnishing an escort from 'Akabah to Petra. These are a wild and lawless set, far different from the gentle, obliging Tawarah. They are avaricious, disobliging, impertinent, and should thus be avoided if possible: still to attempt to penetrate to Wady Musa by this route without their escort would be madness. In fact, it should be adopted, and strictly followed out, as a general rule, that no traveller should ever attempt to pass through the territory of a tribe until he has secured an escort from it, or has obtained the express permission of its chief.*

It was common when parties arrived at Aqaba to discover the Alawin sheikh was at some distance - or claimed to be. The sheikh was canny enough to delay parties so that those travelling behind could catch up and he could then arrange to conduct a group of parties to Petra. There are indications that he knew what parties had left Cairo or Mt Sinai and could then wait till the last of them arrived.[6] Indeed, several parties often leap-frogged one another

---
[5] The fort is described by many visitors including Jane Eames.
[6] In 1874, the Ridgaway Group were very suspicious and angry (Ridgaway 1876: 120): *It began to be hinted that the sheik knew of the coming of two Englishmen, who were behind us, and that he was simply delaying till they should arrive. It was even surmised that 'Ahmet, our dragoman, was in the game, as the dragoman of the English party is a friend of his, and had requested*

along the way and many might arrive at Aqaba in rapid succession. Then the negotiations began, often noisy and irritable, involving some of the travellers but mainly conducted on their behalf by their dragomen. Some of the latter were seasoned guides and knew what to expect and where the negotiations would eventually end up in terms of not just money but 'gifts' - of cloaks, sugar, tobacco, powder and sometimes even a gun. A contract would have to be drawn up under the auspices of the governor specifying the numbers of camels and attendants, escorts, fees to be paid covering not just the charges made to the Alawin themselves but to cover what the Alawin would then pay the people of Petra for permission to visit. This latter was often a cause of serious trouble as the Alawin sheikhs sometimes sought to enter Petra surreptitiously from the west, hope to remain unseen and hurry their charges away before they were spotted or as soon as they were. Time and again travellers record the ferocious quarrels between the Alawin and the various peoples residing at and around Petra and the obstructions and even threatening behaviour of the latter.

Eventually the group of parties would set off and, as before at previous major starting points, they would encamp just a few hours north of Aqaba, often to the exasperation of the travellers who simply wanted to move ahead after an enforced stay at Aqaba. The groups could be quite large - in 1857, a group which reached Petra in early April numbered 21 including three women; two weeks later another group arrived, totalling 18 including one woman. At least 57 people that 1857 'season' (Kennedy 2018). After Aqaba, TB was to become part of a group of 16 made up of seven parties each with its own dragoman (Table 7.1).

Travellers who had researched the routes into Petra sometimes demanded to be taken by the eastern route which deviated from the Wadi Arabah 8km/5 miles north of Aqaba into the Wadi Ithim. That route took the traveller along the line of the great Roman Highway, the *Via Nova Traiana* which ran all the way to southern Syria, in places over or alongside the Biblical King's Highway. It's attraction was that it brought travellers to Wadi Musa where the modern town and tourist hotels lie and let them then descend via the wadi which led past growing numbers of Nabataean tombs till it reached the Siq, the entry to the narrow cleft that ran through the mountain. The entrance to this passage was crowned by an arch high above their heads. More importantly, its exit after almost two kilometres was directly confronted by the astonishing spectacle of the most magnificent tomb at Petra - the Khazneh. The Alawin, however, routinely refused to take that route or agreed but then took travellers by the western route anyway. The explanation - despite that route being largely through their own tribal terrirory, is that Wadi Musa was occupied by the inhabitants referred to as *fellahin*, because they cultivated crops, lived in stone houses in winter and nearby in tents in summer. There was no way travellers could be kept secret from them and hence the Alawin would have to pay *bakshish* to them.

The western route ran up the broad trough of the Wadi Arabah and would enter through sometimes difficult valleys passing Mt Hor (Jabal Harun) on the way. The duration from Aqaba to a camping place inside Petra was three to four days. Travellers usually wished to stay in Petra for a few days but were often hastened away by the Alawin or by the arrival of the *fellahin* and/or the people who lived in the tombs and valleys. As we shall see (Ch. 8), TB was relatively fortunate, with two full days and a part of a third.

---

him to wait for them. Thus we were wasting and losing, as a company, sixty dollars a day, to say nothing of our time, more precious than gold.

Travellers then retraced their route back out to the Wadi Arabah and turned north. They could expect to reach Hebron after about five days and at that point felt themselves out of the desert. They might have to quarantine at Hebron. They parted now from the Alawin and continued with horses hired by their dragomen. All remarked on the green and fertile landcape, farms and villages. It was then a relatively easy ride via Bethlehem to Jerusalem and the sanctuary of one of the hotels catering for westerners (Gibson et al. 2013) and the support of a European Consul - James Finn in the case of British travellers (Ch. 10).

The overall joirney was costed to be about 40 days; in TBs case he departed Cairo on 28 February and arrived in Jerusalem on 3 April: 35 days.

Conditions of Travel

Most travellers made the journey in winter-early Spring: broadly January to May but most commonly in March and April. The weather could be hot but, more commonly, harshly cold with rain, wind, and sand storms all reported. Travellers often noted scorpions - especially found in the damp and shade under their tent floors. Occasional snakes. Accidents occurred and a few travellers were seriously injured or worse: the American Rev. William Parsons Lunt, died and was buried at Aqaba in 1857, aged almost 52, rather old for so demanding a journey; another traveller that year suffered serious gashes to his face, was sewn up by an accompanying doctor and then had to live off liquids and thin solids sucked through a dry macaroni tube; a third was shot in the face at Petra (Kennedy 2018: 194-197).

This is an American, the Rev. Stephen Olin in 1840 (Olin 1843: 2, 49):

> *A gentleman of the party became seriously ill soon after our arrival in Petra, and was confined to his tent during our stay by a violent fever. We were filled with alarm for his safety; but he recovered so far as to be able to mount his camel when we were ready to recommence our journey, and was soon quite restored. Nothing can be more distressing, or present a prospect more truly appalling, than severe illness under circumstances such as surround a traveller in the Desert, quite out of the reach of medical assistance, in the midst of savages and robbers, where delay and an attempt to advance are about equally dangerous and impossible. I felt truly thankful for the speedy restoration of my fellow-traveller.*

And the gruesome experience of Mrs Dorothy Wordsworth Bolland, aged 31 when she travelled to Petra in 1857 with her husband, the 32 year old Rev. John Bolland. The latter became ill at Aqaba, was too ill at Petra to leave his tent and then died (of pleurisy and/or typhus) a few days later while travelling onwards to Hebron. The travelling must have been excruciating – and included a terrific sand storm which blew down their tents at night. Then his wife had to complete the journey to Jerusalem having seen the body of her husband slung over a camel sent on ahead. He was buried in the Mt Zion Cemetery in Jerusalem (Kennedy 2018: 197-198).

With typical Victorian coyness, travellers seldom mention gastric much less bowel problems or their predictable and uncomfortable consequences. Indeed, they are silent on everyday bodily functions. Even everyday hygiene is almost never mentioned except in the abstract advice offered by handbooks on the usefulness of a collapsable bath. Given the shortage of water, that would be of limited use between Suez and Hebron. Travellers might bathe in the sea and women could do so inside a tent carried for the purpose and erected in the shallows or over streams. Jane Eames reports bathing in the Dead Sea well away from the men but omitted the opportunity at Aqaba and was deterred at Tyre by a crowd of curious natives and her lack of 'bathing wardrobe' (1855: 351-352; 410; 271).

Bodily functions are never mentioned and one feels for the situation of the few women travelling in these parties, and generally overwhelmingly out-numbered by male companions and the all-male guides and escorts. Women travellers to Petra - several of whom wrote accounts, would make an interesting research project. TB was accompanied by two women in other parties (Ch. 8).

TB seems to have maintained a good standard of health though he found the desert air difficult:

I was very unwell in the night. I have found the dry desert air affect me like the air of the sea. I was very unwell all day .... The Desert air makes me quite unwell, unlike?? the Sea air. Most of the day was spent in a very uninteresting way. The air quite thick with sand - & no talking - one's throat got full of sand.

Some of his companions were less fortunate: Bryce and Ward both had health problems and Bryce and both Jane and James Eames were afflicted by boils which must have been a severe source of discomfort. In 1847 Harriet Martineau only found some relief when rain came (1848: 2, 360):

There was then intermitting rain, which settled into a determined downpour at noon. To me, one of the most observable things about this rain was its effect upon my own health. For many weeks I had been very unwell; and, since leaving Cairo, had suffered from a tormenting face-ache. Now, before it had rained an hour, I felt wonderfully relieved; and the benefit of this rain lasted nearly to Damascus, where we had more.

Chapter 7

Companions on the Long Desert Route: 1

Several of the other people who arrived at Petra along with Bowles on 26th March 1854 (see Table 7.1 for details of his journey) set off from Cairo at different times from his party and from each other and leap-frogged along the way, sometimes camping and travelling close together; sometimes separated. It was only when they reached Aqaba, the launching place for their onward journey to Petra, that the various parties came together and were obliged by their Alawin guides and escorts to travel as a single large group (above, Ch. 6). According to Bowles the size of this Anglo-American Group – as we may call it, was sixteen. That is confirmed by Mrs Eames who adds that the group was made up of eight Americans and eight 'English' (1855: 267). One - Mr Rodewald, is said to be German but long resident in New York and was evidently counted amongst the Americans. There were two women, both Americans - Mrs Eames and Mrs Rogers, and three were clergyman - Bowles, Eames and the younger Fenton.

Thomas Bowles records the names of almost all the people with whom he travelled across Sinai to Petra then on the Jerusalem. He seldom provides a personal name - just the surname and sometimes the title. Jane Eames often refers to people but usually only by an initial for their surname - e.g. Dr B or Mrs R. Combining these two allows the identification of the full surname. Bryce's unpublished journal is only occasionally helpful; that of Yeatman is helpful but stops after his journey on the Nile and before he set off for Petra. The Shepheard's Hotel register sometimes provides the initial(s) or full first names. Between these sources we can identify nationalities for almost everyone and provide biographies of TB's companions.

Likewise, we must burrow deep to discover the individual parties within the overall Anglo-American Group which came together at Aqaba. The sixteen people who made up the group from Aqaba to Petra had travelled as a number of smaller parties until that time. Bowles and Kennedy set off together from Cairo on 28th February 1854 but were soon effectively travelling along with others (Table 7.2), a total of four parties: first two Americans - Mr and Mrs Rogers, then two more Americans - Mr Ward and Mr Yeatman, and then two more British - the Fenton brothers. The Eames Party left Cairo on 23rd February 1854 in the company of a party consisting of an American - Mr Fish, an American of German origin - Mr Rodewald, and a Briton - Mr Ross. These arrived together at St Catherine's Monastery at Mt Sinai where they then also encountered three more British travellers - Dr Bryce, Mr Wakefield and Mr Freeman. Although they again moved onwards as parties, they were then all brought to a halt together at Aqaba awaiting the Alawin to conduct them onwards. Mrs Eames noted the group at Aqaba included seven dragomen, confirming that there were seven parties which were to make up the Petra group of 16. (On the other hand, she reports just six cooks and two manservants).

We may now provide brief biographies of the travellers, the seven with whom he travelled most closely as far as Aqaba in this chapter and the other eight in the next chapter, beginning with his own travelling companion, Mr Kennedy.

Table 7.1: Timeline of the Long Desert Route Cairo to Jerusalem (cf. Appendix 2).

Tuesday 28 February	Depart Cairo for the Desert. Start of journey to Suez	64-68
Wednesday 1 March	Desert	68-70
Thursday 2 March	Desert	70-75
Friday 3 March	Suez	75-76
Saturday 4 March	Suez to Ayun Musa	76-80
Sunday 5 March	Ayun Musa to near Wadi Wardan	80-83
Monday 6 March	Wadi Gharandel via Wadi Wardan, Wadi Amarah & Marah	83-87
Tuesday 7 March	Wadi Gharandel to the Sea via Wadi Thall & Wadi Teiyibah	87-90
Wednesday 8 March	Seaside to Wadi Mucatteb (Valley of Inscriptions) via Wadi Shellal, Wadi Sedr, Wadi Maghara (excursion to see Egyptian Inscriptions)	90-96
Thursday 9 March	Wadi Mucatteb (Valley of Inscriptions) to Wadi Feiran	97-100
Friday 10 March	Wadi Feiran to Wadi Shekh	100-102
Saturday 11 March	Wadi Shekh to Mt Sinai via Wadi Ed Deir. Climbed Mt Sinai (Djebel Musa)	102-112
Sunday 12 March	Mt Sinai	112-114
Monday 13 March	Mt Sinai. Excursion to Wadi Er Rahah and 'Smitten Rock'	114-117
Tuesday 14 March	Mt Sinai to Wadi Sahl? Excursion to Tomb of Sheikh El Saleh	117-121
Wednesday 15 March	Wadi Sahl?	121-122
Thursday 16 March	Wadi El Ain (Valley of the Springs)	122-125
Friday 17 March	Along the Gulf of Aqaba	125-126
Saturday 18 March	Along the Gulf of Aqaba to Aqaba	126-128
Sunday 19 March	Aqaba	129-134
Thursday 23 March	Start of journey to Petra	134-137
Friday 24 March	Wadi Araba	137-138
Saturday 25 March	Wadi Gharandel and Seir	138-141
Sunday 26 March	Arrive Mt Hor and Petra	141-145
Monday 27 March	Explore Petra: El Deir, Qasr al-Bint	145-149
Tuesday 28 March	Explore Petra: Al-Khazneh, Siq	149-156
Wednesday 29 March	Depart Petra to Wadi Araba via Mt Hor	156-158
Thursday 30 March	Wadi Araba	158-160
Friday 31 March	Wadi Araba	160-162
Saturday 1 April	Wadi Kurnub and El Tellal?	162-163
Sunday 2 April	Hebron	163-166
Monday 3 April	Hebron to Jerusalem. Excursion to Pools of Solomon and Bethlehem: Church to see Holy Place	166-170

Table 7.2: *The parties making up the Anglo-American Group leaving Aqaba for Petra on 23rd March 1854.*

Party	Members	Number	Nationality
Bowles	Rev. Thomas Bowles and Mr Henry Hyndman Kennedy	2	British
Rogers	Mr John Leverett and Mrs Virginia Beverley Rogers (A)	2	American
Ward	Mr Henry Veazey Ward and Mr Henry Clay Yeatman	2	American
Fenton	Mr Samuel Greame and Rev. George Metcalfe Fenton	2	British
Eames	Rev. James and Mrs Jane Eames	2	American
Ross	Mr Carl [Charles] Reinhard Conrad Rodewald, Mr Robert Ross and Mr Fish	3	A, B, A
Bryce	Dr William Bryce, Mr John Edward Wakefield and Mr Freeman	3	British
Total		16	

Travelling Companions

Henry Hyndman Kennedy
(Newland, Colford, Gloucs., bap. 18 March, 1814 – Fort Augustus [buried in Nairn], Invernesshire, 20 June 1880)

Bowles first encountered 'Mr Kennedy' on board the *S. S. Bombay* from Ceylon to Suez. Although it sailed on 29th January 1854 he only mentions him for the first time on 3rd February: *'After breakfast I sat talking with a man who sits opposite whose name is Kennedy.'* He mentions him several times more during their two weeks on board.

From Suez the passengers were transported in 'vans' to Cairo (cf. Ch. 3) where Bowles reports that Kennedy – now travelling with Strachan who had plainly also been a passenger but gets his first mention in the journal - arrived in a later van and had initial trouble securing a room in Shepheard's Hotel. As we have seen (Ch. 5), these three undertook numerous outings in the city and vicinity during their fortnight's stay. Although Strachan opted out of a grander expedition, Bowles and Kennedy then hired a dragoman to look after them as they formed a party of two for crossing Sinai to St Catherine's Monastery, Aqaba, Petra and on to Palestine. In Jerusalem they were again in the same hotel, frequently went on outings together in the city and to the monastery of Mar Saba and Dead Sea, then onwards in an extended tour of holy sites in Palestine, to Damascus and eventually to Beirut. Kennedy gets his last mention on 8th May 1854 as Bowles sailed from Beirut. By then they had been in close companionship for over 3 months, on a ship, in hotels and - in particular, for several weeks sharing a tent and often going about together.

All told, Kennedy appears by name 91 times. Not once is he given a first name. On one occasion he is part of a group of four described as 'Englishmen' – as he would have sounded to Bowles given his place of birth and upbringing (below). During such a long association and the numerous occasions Bowles records them as conversing and/or walking together, it is inconceivable he did not learn about Kennedy's origins, background, occupation and how he came to be on the *S. S. Bombay*. He records virtually nothing of any of these, nor does he even say anything about Kennedy's character, age, appearance. There are a few hints: Kennedy seems more cautious than Bowles and spends less time walking with him and others while on their desert journey and is described at times as tired – possibly older (Bowles was then 32) and/or unfit. Fortunately, Bowles does provide a clue – just one amongst those 91 explicit namings of Kennedy, which provides the key to a quite detailed biography which can now be written.

On the first stage of the Long Desert Route, between Cairo and Suez, they met a party of westerners going the other way (from a recently docked ship, presumably): *'Mr Kennedy fell in with some man connected with his house in China who was not a little surprised to meet him in the Desert'*.

The implication is that Kennedy belonged to one of the western merchant trading houses, most likely in Hong Kong or Shanghai. *Allen's Indian Mail* for 1854 records five arrivals on one ship from China (?) to Singapore (?) on 29th December 1853, who include *'G. Strachan, H. H. Kennedy'*.[1] As noted, a 'Mr Strachan' had evidently also been a passenger on the *S. S. Bombay* from Sri Lanka and he and Kennedy arrived together in Shepheard's Hotel in Cairo. Kennedy and Strachan subsequently joined Bowles in two weeks of tourism at and around Cairo (Ch. 5).

Census returns in the UK reveal a Henry Hyndman Kennedy, explicitly recorded in 1861 as 'East India and China Merchant'. He was baptised in March 1814 in Gloucestershire (hence Bowles's 'English') but to Scottish parents. He celebrated his 40th birthday while at St Catherine's Monastery, making him eight years older than Bowles.

Plainly Kennedy had the leisure and wealth to undertake a lengthy side trip in Egypt and the Near East, and when Bowles left him at Beirut on 8th May 1854, he was awaiting embarkation for Constantinople.

It is unlikely Kennedy finally reached the UK before mid-summer or autumn. Just a few months later, on 3rd December 1854, he was married in Edinburgh. It was a 'good' marriage. Eliza Mary Ann Gillespie, eight years his junior, was from a notable Edinburgh family. Eliza's father, Alexander Gillespie, the son of a physician in Ayrshire, was a 'D. of Medicine' and surgeon, serving (twice) as President of the Royal College of Surgeons of Edinburgh (as was one of his sons after him). Eliza was baptised in Edinburgh on 31st October 1821 but her own mother, Eliza Mary Shirriff, had been born in Madras where her father, William Shirriff (1759/60–1802), like several of his siblings and nephews, had served in India since 1775.[2]

The Kennedys can be traced in a succession of Census returns – not 1851 when he was still overseas. In 1861, 1871 and 1881 the family of – ultimately, 7 children, live in prosperous houses in London. His last two children were born in Edinburgh and in 1881 those of the family still at home are living in Nairn in Invernesshire. Kennedy himself had died the previous year in Fort William but was buried in Nairn.

---

[1] I am grateful to Andrew Oliver for this vital clue.
[2] In June 1800 William Shirriff was promoted Lieut.-Colonel of the East India Company's 7th Madras Native Cavalry (Annand 1965). Colonel Shirriff is the subject of one of the most important works by the Scottish portrait painter Sir Henry Raeburn, now in the Museum of Fine Arts, Houston, Texas.

We can go further, his will describes him as *'of Pitmuies in the County of Forfar in North Britain of Nairn in the County of Nairn in North Britain and of No. 5 Clarendon Place Hyde Park in the County of Middlesex Esquire.'* Two codicils of the will which were written in 1876 while *'residing temporarily in Dresden in North Germany'*, records him leaving 'under £5000' but this seems not to include £3000 explicitly provided to pay for the *'repair, alteration and improvement'* of *'certain property at Shanghai'* (Kennedy Will COW505220g). He also – shedding light on his interests, leaves to his two youngest sons, *'all my golf clubs fishing rods and guns'*.

Finally, a mention in the will which takes us back to an incident on the steamer from Sri Lanka. Bowles – a Church of England clergyman, was irritated to find the Purser has seated a French Roman Catholic Bishop beside him at dinner. The latter persisted in speaking to Bowles in French which he claimed not to understand. TB fobbed him off at one point:

> *But the French Bishop joined himself to me & asked, a question which I made out with extreme difficulty, if I would play Chess. When I understood him I told him that I couldn't but offered to get him an Antagonist, which he declined & we walked & stood over the gangway a long time, he persisting in talking French & I trying to understand & answer. At last I went below to read & then got Mr Kennedy to play a game of chess with him – which he readily agreed to do. I went up & found him sitting against the Taffrail & came (sic) him below & sat him down to the Chess board – where I settled quietly down to my book.*

In his will, Kennedy refers to his brother, *'Hugh Alexander Kennedy of Reading'*. He is one of Kennedy's older brothers, born in Madras in 1809, subsequently serving as a Lieutenant in the Madras Native Infantry and Captain in the Forfar and Kincardine Militia. More importantly, he was an eminent chess player, founded the Brighton and Bristol Chess Clubs and wrote numerous articles on the subject (Kennedy 1860; 1876). Presumably Henry Kennedy was a player as well and gave the Bishop a run for his money.

The Censuses, birth and death records for Kennedy's parental family are instructive. His father, Alexander Kennedy (1764-1827), MD, FRSE, FSA was born in Paisley, the son of a clergyman. He trained as a doctor at Aberdeen University in 1787 and then served as a surgeon with the East India Company from 1788-1812. He had at least five children: Hugh was born in Madras in August 1809; Thomas in London in October 1810; then Francis, Henry and Patrick at Newland, Gloucestershire in January 1812, March 1814 and April 1815 respectively. The family evidently returned to India where Thomas died in 1818 aged *c.* 7/8. Alexander then worked as a physician in Edinburgh from 1819. So, Henry Kennedy, born to a 50-year old father who had spent a generation in India, seems to have lived in India for at least a few years as a child during which a brother died in 1818. As an adult he returned to the East as a merchant in India and China where he evidently enriched himself.

As a postscript, a bequest in the will of his widow who died 19 May 1917, 37 years after Kennedy, illuminates her husband's tastes beyond hunting, shooting and fishing and explains his *'temporary residence'* in Dresden:

> *"To my friend CLARA DAVIDSON of CANTRAY (sister of the late Major Davidson of Cantray) FIFTY POUNDS STERLING also the following articles presently in my Drawing Room videlicet:- One handsome chinese bowl and stand, two chinese table covers, two very old cabinets full of Dresden*

china and other old china, two Dresden flower bowls on top of said two cabinets. Also the likeness or photograph of myself presently hanging over the Dining Room Mantel piece and bearing date 1884; To my friend Mrs Davidson of Tulloch, Tulloch Castle, Dingwall the following articles presently in my Drawing Room videlicet:- Water-colour picture of myself taken in December 1854, Good frame glass box given me by the Queen of Saxony Dresden 1876 containing a pair of slippers made by herself and a few things from China also a blue velvet picture of the King and Queen of Saxony and his brother."

As Bowles and Kennedy were such constant and close travelling companions for so long, we may wonder if they remained in contact during the years after their return to Britain.

***

Samuel Greame Fenton
(Leeds, 24 December 1821 – Belfast, 6 June 1892)
and
Rev. George Metcalfe Fenton
(Leeds, 24 September 1826 – Christchurch, Hants, 18 Dec 1879)

Apart from Kennedy, no other companions are mentioned as often as the Fentons. The brothers - as we can infer, were introduced without explanation as several parties began the Long Desert Route with the stretch between Cairo and Suez. Thereafter, Bowles names one or both ('young Fenton or 'the Fentons') no less than 43 times. He frequently walks and talks with them but especially the younger brother ('young Fenton') as they cross Sinai and to Petra where they split up. Neither is given a first name and only his reference to young Fenton reading 'the Lessons' in a service he (Bowles) was conducting at Aqaba, hints at his profession. Fortunately, Jane Eames, who frequently refers to 'Mr F.' meaning her fellow American Mr Fish, on one occasion clearly means F(enton) and is explicit (1855: 267):

Sunday, 19th. This morning we had "Morning Prayer," according to the service of the Church of England, in our large tent, the Rev. Messrs. B[owles] and F[enton 'young'] officiating. Including the two clergymen, the congregation consisted of sixteen persons, eight of whom were English, and eight Americans. I think it is a rare thing to see so large a party in the Desert, and for three[3] among the party to be clergymen, and I can assure you it was a scene that will be long remembered by us all, when sixteen persons from the North, the South, the East and the West, knelt down to pray to "Our Father."

It is not even made clear the Fentons are British, but we can infer they are from the tangle of references to names and nationalities amongst the sixteen westerners.

They are fit and adventurous. Young Rev. Fenton joined Bowles at Petra at the eastern entrance to the Siq in climbing up to view the arch from above. After the entire group had completed their tourism at Petra, the brothers left with the others but only as far as the Wadi Araba. Bowles last saw them as he descended Mt Hor after a second ascent and met the brothers ascending:

---

[3] The third was, of course, her own husband.

> *About ½ way down we passed the Fentons to the left down the way we had ascended on their way back to Akaba & we northwards. I was very sorry to say goodbye. They have been very pleasant companions. The usual story in travelling, you make friends to see them again perhaps no more.*

In fact, it was not his last sight of them. Six weeks later, while passing through Alexandria on his way to Malta, he met them again. Although he does provide a little new information, he is characteristically silent about their conversations:

> *The first person I came across in the hotel [in Alexandria] was Paulo who said the Fentons had arrived from Cairo last night having had 2 months in the desert. …. Home to breakfast & much talk with the Fentons.*

Then the following morning:

> *I got up & found the Fenton's having breakfast previous to starting & joined them.*

No indication of where they had been during their weeks in the desert or why or what. No indication of where they were now 'starting' for.

Finally, Shepheard's Hotel register records:

> *Sam. Greame Fenton left Dec, 17th 1853 to 2nd Cataract*
>
> *George Fenton*

The names - especially 'Greame', allow us to identify the brothers precisely. Indeed, both can be traced in the Passport applications index for the period:

> *Rev. George Fenton: 1852, 1853*
>
> *Samuel Greame Fenton: 1851, 1853*

It seems Samuel went abroad in 1851, George in 1852 and then both together in 1853. Perhaps, as with the detail available for Bryce's passport application (below), the original document at The National Archive in London will be more informative.

The names also allow the tracing of the university careers of these two young men in the pages of the *Alumni Dublinenses*:

> *Fenton, George Metcalf, S.C. (Belfast Coll.), Oct. 11, 1844, aged 18; s. of Samuel, Mercator; b. England. B.A. Vern. 1849; M.A. Vern. 1851*
>
> *Fenton, Samuel Graham (sic), Pen. (Belfast Inst.), Nov. 6, 1837, aged 15; s. of Samuel G., Lini Mercator; b. Yorkshire. B.A. Aest. 1849; M.A. Aest. 1852*

And *Foster's Hand-List of Men at the Bar* includes:

Fenton, Samuel Graeme (sic), M. A., Trin. Coll., Dublin, 1852, a student of Lincoln's Inn 17 Nov., 1847 (then aged 25), called to the bar 17 Nov., 1852 (eldest son of Samuel Graeme (sic) Fenton, Esq., of Belfast); born 1822.

The middle name of both Samuel and his father, is correctly 'Greame'. Both were born in Leeds, West Yorkshire where their father, Samuel Greame Fenton Snr was a linen merchant. The Fenton brothers' mother died young in 1826, the year George was born. Their father remarried in 1828 and there is a suggestion this wife, too, soon died and he married for a third time in 1832. Samuel Snr subsequently settled in Belfast to which he moved his linen merchant business and where he was to become Mayor for a year (1852-1853). He died there in 1863.

Samuel Jnr also moved to Belfast - probably to take over the family business when his father died. He married Sarah Pilcher in London in 1861. They had no children.

The Rev. George Fenton married Mary Francis Gregory, daughter of a clergyman, 12 years his junior, in 1861 and married for a second time in 1874, to Mary Anne Amy Armstrong, another clergyman's daughter, and 13 years his junior. He had two sons and a daughter from his first marriage.

When these brothers travelled in Sinai and to Petra in 1854, they were respectively 32 and 27 years old, Samuel was a qualified barrister and George a clergyman. It seems George never had care of a parish - in the Census of 1871 when he was already resident in Christchurch near Bournemouth, he is described as a widower, and a 'clergyman without care of souls'.

It is surprising to find the Fentons leave the Petra Group and return towards Aqaba and then, apparently, spend several weeks travelling in Sinai. As they showed a particular interest in the inscriptions of Wadi Feiran, it is possible they were interested in further such exploration. This was the period in which there was a keen interest being taken in these early Aramaic text and the Fentons would have seen the Butler brothers at Mt Sinai as they set off exploring for the inscriptions some of which subsequently were published in Forster's corpus (Forster 1862) (above Ch. 5). Evidently it was a planned excursion from Petra as TB notes in his journal as they all set off from Aqaba that *'Sheikh Hussein's 2nd son goes as Camel Sheikh to the Fentons, to bring them back from Petra'.*

***

Mr John Leverett Rogers
(Ipswich, MA, bap. 23 October1808 - St Nicholas Hotel, New York, NY 2 December 1869) and
Mrs Virginia Beverley Rogers (née Wood)
(New York, NY, 27 March 1827 - Manhattan, New York, 30 April 1900)

Although he did not yet know their names, Bowles and his travelling companion Henry Kennedy first encountered Mr and Mrs Rogers on 28th March 1854 on the evening of his first day out of Cairo on his way to Suez: *'We got back to the tent where we found a fourth party arrived - their tent sporting the stars & stripes.'*

They continued travelling broadly together, sharing a camp site each night. On the third day, Bowles was accosted by Mrs Rogers - whom he here names for the first time, which he apparently found unwelcome (Bowles 3, 71):

I started on my camel intending to read, but was saluted by Mr Rogers & then found Mrs Rogers was close to me - & I could only ride along talking to her

It may be that the 34 year old life-long bachelor and clergyman - on this first meeting at least, was uncomfortable talking to the 27 year old American woman who seems to have been an assertive character (below).

Bowles and Kennedy then travelled in the company of - if not always as a single party, with these two Rogerses, and with Ward and Yeatman (below) and the Fentons - four Americans and four British. When they arrived at St Catherine's Monastery on 11th March 1854 these eight met up with three more British travellers, Freeman, Bryce and Wakefield. Mrs Rogers was evidently troubled by the cold and at the monastery there was some misunderstanding about the allocation of rooms with Bowles, Kennedy and one of the Fentons chivalrously offering to give up their space for her:

Some trouble about Mrs Rogers. She seemed much upset by the cold & have found there was some difficulty about rooms for all the party. Fenton, Kennedy & I at once offered our room to Mrs Rogers & ordered our things to be removed - assuring them that we had no idea we were keeping them from a room. This they would not hear of. By and by it was all arranged & we each kept our room.

The following day (Sunday 12th March) while most of the assembled travellers set off to ascend the holy mountains, the Rogerses were guided alone up Jebel Musa by Father Petros from the monastery and gone most of the day.

On 16th March, as the now enlarged group of eleven travellers were on the next stage of their travels from St Catherine's to Aqaba, Bowles - who does not normally seem to have had much of a sense of humour, records:

The American lady Mrs Rogers, it appears, gave their Dragoman to understand last night, that it was not right they sh(oul)d be always last [to leave camp in the morning] - & gave strict orders that they sh(oul)d be off first. They weren't to be behind the 2 English parties. So this morning before we were out of bed I believe she was seen all ready to start outside their tent. John [Bowles's dragoman] was all impatient to get us out of our tent & we could not imagine what it all meant - till as we went to breakfast we saw their tents down - & in a very short time their camels were loading. As usual however we with the Fentons quietly walked off the ground, followed however immediately by all 4 Americans, their camels beginning to move at the same time. We then saw how it was & were amused at the way they & their camels pushed on. ... We stopped to luncheon, finding that John had made the usual preparation near the Americans. So that we quickly eat our luncheon & past (sic) on before them. ... when we came out upon a plain beyond which was the Gulf of Akaba - with the Arabian Mountains beyond. We camped about ¼ past 4 after passing along at the foot of the hills on our left for about a mile. The Fentons' tents & ours were all pitched & half furnished before the Americans' camels came out upon the plain. So much for "going ahead"!

Bowles never mentions the Rogerses by name again until a month later (17th April) as the travellers were about to leave Jerusalem for a tour in northern Palestine on the way to Damascus (below). At that point, the new group of parties consisted of Bowles and Kennedy, Mr and Mrs Rogers, Rodewald, the Rev. and Mrs Eames, Ward and Yeatman. They soon split up that same day – *'The Rogerses started with us, but we left them behind.'* Evidently the Rogerses were no better now at making a speedy start and keeping up than in the desert.

Bowles never mentions them again but Jane Eames is helpful both about where the Rogerses had been before the trip on the Long Desert Route and after Jerusalem. She begins with the 11th March, the date and place Bowles (above) had reported the meeting of the growing band of westerners at St Catherine's - in this case, the Eameses were leaving the Convent as the others arrived (Eames 1855: 249-50):

> *I hasten to tell you, that among the party of eight travellers we met this morning, were four of our American friends, Mr. and Mrs. R[ogers], Mr. W[ard] and Mr. Y[eatman], whom we left behind at Cairo, two of whom we hoped to meet in the Desert, but the other two, we had no expectation of seeing whatever. What a place to meet friends and acquaintances, beneath the frowning peaks of the sacred mount! Our greetings were hurried, as they were anxious to reach the Convent, and we were ready to mount our camels, so hoping to meet at Akaba, we said "adieu," and were off.*

She did meet up with them at Aqaba where all sixteen westerners had to await their transfer to the care of the Alawin sheikh. On 20th March she recounts at some length the visit the two American ladies - herself and Mrs Rogers, made to the wife of the Ottoman governor in the little fort (Figure 7.1).

It is a rare account of life at Aqaba for an Ottoman official and his family, almost a place of exile only enlivened twice a year with the passage of the Hajj from Cairo. As it is the observation of a rare western female traveller and would never have been visible to Bowles, it will be quoted extensively in the entry on the Eameses below (Eames 1855: 268-70). One passage worth citing here came as the two western ladies were offered a pipe to share with the Governor's wife:

Figure 7.1: The fort on the shore at Aqaba, residence of the Ottoman governor. The dusty and remote place of 1854 is today a port city and bustling tourist centre of hotels and private beaches (APAAME_20141020_DLK-0040C).

First came a pipe. I begged Mrs. R[ogers], who can smoke better than I, to take it first, and after she smoked a while, I took a few whiffs, and then we gave it to the "Sitteh," [= wife of the governor] who seated herself on the floor, and finished the pipe. Then lemon sherbet was handed to us. After drinking half what was in the glasses, we gave them back, whereupon both the "Effendi" and the "Sitteh" insisted so clamorously, that we should finish the glasses, that I complied with their request, but Mrs. R[ogers], whose glass was much larger than mine, found it impossible to swallow all hers, so she gave back the glass, which after many compliments, the Sitteh drained.

Although it was common for women to smoke a pipe in the Arab world, it was still uncommon amongst western women amongst whom it had traditionally been the preserve of prostitutes and courtesans. However, it was beginning to be adopted by sophisticated society women in Europe - especially after rolled cigarettes became common following the Crimean War when soldiers brought the habit home. It is interesting that both American ladies took the pipe even if it was Mrs Rogers - the younger of the two (below), who is said to have more experience of it (Anon. n.d.; Mitchell 1991).

Arriving at the foot of Mt Hor, Mrs Rogers and two of the men stayed below while all the others climbed the mountain. Again, a few days later (31st March 1854) in the northern Wadi Arabah when the group encountered a difficult terrain and most decided to dismount, '*Mrs Rogers and one or two of the* gentlemen' were the only ones to stay on their camels rather than walk (Eames 1855: 307).

About three weeks later, on the first day out of Jerusalem when the group had initially broken up, Mrs Eames reports meeting up near Bethel with the Rogerses, Kennedy and Bowles and again later near Samaria (Eames 1855: 376; 382-83). They evidently did not meet again on the journey but did all find themselves together once more in Damascus where they joined forces to visit some Damascene houses (Eames 1855: 440-41).

We had heard so much of the unrivalled beauty of some of these palaces in Damascus, we were of course very anxious to see them, to judge for ourselves, but as we had no letters of introduction here, we thought it impossible. However, on going yesterday morning to the hotel to see Mr. and Mrs. R[ogers], they told us there were two or three houses here, which their owners were always happy to have Europeans visit, and the hotel keeper was to send a guide with them, and they invited us to accompany them, an invitation we joyfully accepted.

Once again, Jane Eames in her letter is a marvellous source describing a 'palace', and its inhabitants - in this case a wealthy Jewish merchant. Bowles's description of visiting what may be the same house is a poor one by comparison (under 27 April 1854). Unfortunately, Mrs Rogers evidently did not accompany Jane Eames to the Bath on Ladies' Day (Eames 1855: 445-48; cf. above); it would have been interesting to hear of her reaction.

The Rogerses evidently continued with the Eameses after Damascus along the coast of Anatolia, stopping in several ports including Rhodes. Naturally they viewed it all through the lens of the New Testament and the Apostle Paul. It was only at Smyrna that the Rogerses are again mentioned. Like the Eameses and Rodewald, they were invited to make a visit to an American warship in the harbour (Eames 1855: 474-45; 477):

Some of the officers, however, soon came on board [the travellers' steamer from Beirut], and proclaimed her to be the Cumberland, and finding there were Americans on board, introduced themselves to Mr. and Mrs. R[ogers], Mr. R[odewald] and myself, and inviting us to go on board the frigate, but as the other gentlemen of the party had gone on shore in pursuit of rooms, we politely declined, preferring to wait till they could accompany us. ...

... Yesterday Mr. F[ish] and Mr. and Mrs. R[ogers] went to Constantinople, so we are now quite alone.

As it happened, the Eameses evidently met the Rogerses once more. In a letter dated 'Büyükdere [on the European shore of the Bosphorus 20km to the north of Constantinople], June 4th 1854 [= Sunday]' Jane Eames records (Eames 1860: 55-56):

We spent all the morning of Tuesday [30 May] in the bazaars at Constantinople, in search of a shawl, but I found none to suit, not in quality, but in price. That day [Tuesday 30thMay] we bade farewell to Mr. and Mrs. R[ogers] and Mr. F[ish], who have gone to Greece, and now, for the first time since entering Egypt, we are without companions.

At that point, several months after their arrival in Egypt, the Rogerses left the Ottoman Empire for Greece and presumably, onwards to return to the United States.

Who are Mr and Mrs Rogers? Neither Bowles nor Eames gives either of them a personal name nor an origin beyond that they were both American. There is no hint of age nor of an occupation for Mr Rogers. Once again, the Shepheard's Hotel register provides the crucial key to identifying them: on 3rd January 1854 the register records that *'Mr and Mrs John L. Rogers, New York'* left for Upper Egypt. There can be little doubt that this is John Leverett Rogers, baptised in Ipswich, Massachusetts on 23rd October 1808. Although he appears on successive Census records, there is surprisingly little available about him. The middle name is curiously only ever 'L' in almost every available document including his will. However, other family members recorded in that document have that same name. John Leverett of Ipswich (1662–1724), a grandson of John Leverett, Governor of Massachusetts Bay Colony, was a President of Harvard University (1708-1724). He married another Ipswich resident, Margaret Rogers Berry - apparently a widow and daughter of John Rogers (1630–1684), a previous President of Harvard (1682-84). Presumably the current Petra traveller is a descendant of this Leverett/Berry union from which sprang nine children, six of whom died in infancy.

John Leverett Rogers - is listed as part of a business called 'Downer and Rogers' - and given his full name in the record, in New York which filed for bankruptcy in October 1842. Their business appears to have been as importers of merchandise. Their office was just 300m from that of the Rodewalds in lower Manhattan (below). He evidently recovered from this failure but also - judging by references in his will, made a fortunate marriage. On 25th March 1852 he married Virginia Beverley Wood of Providence, RI, daughter of Silas Wood, originally of Long Island and, like Rogers by that time, owner of multiple properties in Manhattan. She had attended Madame Chegaray's School for Young Ladies in Manhattan, said to be the finest school in the country for young girls. John was then 44 and Virginia was 25. On 1st September 1852 he applied for a passport, and they evidently left for Europe. We may guess at a tour in Great Britain and - more firmly, on the continent as Bowles is recorded as conversing with Virginia about Switzerland. Touring in Italy is also likely. They were in Egypt by 3rd January

1854. Their subsequent travels up to the point at which they left Constantinople for Greece is discussed above.

There is no record of their lives after their European and Eastern tour. John died in 1869 aged 61. They had had no children and Virginia never re-married. She lived on till 1900, dying in New York City, aged 73.

In March 1854 while travelling across Sinai to Petra John was 46 and Virginia turned 27 while actually in Petra. Of all the sixteen people in the Petra group, only Henry Yeatman (aged 22) was younger than Virginia. It is a pity that so little of her character comes out in Jane Eames's letters. Bowles's account of her determination to not be leaving camp after the British each day may hint at a certain immaturity.

***

Henry Veazey Ward
(Sassafras Neck, Fredericktown, Cecil Co, Maryland, 26 September 1806 – Chateau de Coppet, near Ouchy, Lake of Geneva 15 March, 1873)

Ward's broad itinerary can be traced from Cairo on 14th December 1853 to Malta on 17th May 1854, six months later. For at least two months of that he was with or close to Bowles. The latter first mentions Ward as the Petra Group split up in Jerusalem on 17th April. He then encountered him again during the subsequent weeks - in Damascus and Beirut, and at greater length on the steamer from Beirut to Alexandria, then on the next steamer from Alexandria to Malta. He is not mentioned again after they reach Malta on 17th May. Strikingly but as noted with other companions, Bowles never mentions Ward on the Long Desert Route. The only clue Bowles offers to his identity is to say he is an American.

For Ward's identity and the immediate background to Bowles' first mention, we can turn to the Shepheard's register then to Jane Eames.

> *Charles Rodewald, of New York*)
> *Henry C. Yeatman*)   *To Upper Egypt*
> *John H. Pell, of New York*)
> *Henry V. Ward, of Maryland*)
> *Leave Cairo Dec. 14th [1854]*

These are then surely the four Americans the Eameses encountered on the Nile on 14th January (Eames 1855: 148):

> *Our voyage to Wadee Halfeh occupied but two weeks, and I am glad now we went so far, though for a long time we were undecided about going beyond Assouan.*

> *We arrived here this morning about ten o'clock, and saw the American flag waving from a boat behind us, and immediately after breakfast, J. went on board, and found it occupied by four gentlemen, all of whom had left home long before our departure, and they could give us no particular news. Oh! this thirsting for news from home, in a far distant land! Afterwards the gentlemen called on us, and invited us to dine with them; the invitation was given too cordially, not to be as cordially received, so*

about five o'clock we went to their boat, and delightfully the hours passed. Dining out in Egypt, what an event!

She mentions him again on 11th March as her party was leaving the convent at Mt Sinai and new parties were arriving (Eames 1855: 249-50):

... among the party of eight travellers we met this morning, were four of our American friends, Mr. and Mrs. R[ogers], Mr. W[ard] and Mr. Y[eatman], whom we left behind at Cairo, two of whom we hoped to meet in the Desert, but the other two, we had no expectation of seeing whatever. What a place to meet friends and acquaintances, beneath the frowning peaks of the sacred mount! Our greetings were hurried, as they were anxious to reach the Convent, and we were ready to mount our camels, so hoping to meet at Akaba, we said "adieu," and were off.

They did, of course, meet again at Aqaba, travelled together to Petra and on to Jerusalem. As they left the desert and approached Hebron, warned by their sheikh against marauders, they had a brief scare (Eames 1855: 309):

We dismounted, and sat down on the grass amid the bright flowers, and when the parties were all together, the sheikh told them there was danger ahead, for that the sheikh of the tribe through whose country we were passing, had just sent him word, he disputed his right to take us to Hebron, and that he should come out with his Arabs, to fight with our sheikh for the privilege of taking us on. We were warned to keep together, and the Arabs were told to look well to their arms. Mr. W[ard] buckled on his broad red belt, and I surnamed him "the knight of the crimson belt," while Mr. Y[eatman], who is never seen without his gun, I dubbed "the man at arms."

She never mentions Ward again. Bowles reports that Ward was part of his own group leaving Jerusalem on 17th April:

All the party is broken up. The Heneages were gone this morning - for Jaffa & Pagde? too. We go - the Rogerses – Eames - Rudeswaldt (sic) Ward & Yateman etc.

They did not travel together but Bowles met Ward again in the hotel in Damascus on 26th April where he had evidently arrived before him. On 5th May TB met him again, this time in Beirut where he was still in the company of Yeatman. The following day, Ward gives him news about international affairs (*'Japan has consented to open her ports to all nations in 1 year's time'*)[4] from letters he has just received. At noted they were then together on the steamer from Beirut to Alexandria, presumably at the same hotel there, then on again together to Malta - a total of nine days but he adds little beyond that they conversed and Ward lent him an old copy of 'Home News'.

Henry Veazey Ward was 44 when he was at Petra, the second oldest member of the group (Mr Rogers was 45). He originated in Maryland but left the United States in 1826, aged 20, and was to spend most of the rest of his life overseas. Together with a brother, he first moved to the Pacific coast of South America and entered the service of the recently established local branch

---

[4] This is a reference to the threat made by the U.S. fleet under Admiral Perry in 1853-54 to attack Japan, which led to the latter ending two centuries of seclusion and, by the Treaty of Kanagawa (March 31, 1854), allow foreign ships access to their ports.

of the increasingly prosperous and respected London merchant banker Frederick Huth,[5] at first at Lima in Peru then Valaparaiso in Chile. At that time the branches were managed by an American whom Huth and Gruning (his London partner) had enlisted as a local partner, calling the business Huth, Coit & Co. After the initial six year contract (1822-1828), Coit determined to return to the United States and first Augustus Kindermann managed the business - now renamed Huth, Gruning & Co, then after 1839, it passed into the hands of Henry Veazey Ward who had evidently become a trusted and experienced man in the burgeoning trade. Apart from a brief return to Baltimore in 1832 when his mother died, and to London in 1839 Ward was based at Valparaiso, until 1852. The position is summarised by Hawkins in his history of the Huth merchant bank (Hawkins 2015: 34):

> *The initial South American partnership agreement was for five years and, to the disappointment of Huths, Coit did not wish to extend his partnership beyond its expiry in 1828 for personal reasons. ... By 1828 the firm had numerous local and expatriate employees, one of whom, Augustus Hermann Kindermann (c.1797-1852), was considered sufficiently able to take over from Coit and was made managing partner in his place. Also brought into the organisation around this time were Henry V. Ward, who became an expert in textiles, ... A decade later, in 1839, Kindermann was sent by Huths to Liverpool to open a branch there, Ward taking over as managing partner and remaining as such for twenty years.*

In 1852, Ward set off on a tour of Europe, Egypt and the Levant, after which he settled in Boston where he married Caroline Reynolds on 9th April 1856; she was aged 22 and he was then 50. They had a daughter in early 1857 but Caroline died on 13th February 1857, apparently in or soon after childbirth. She is commemorated on a tombstone with touching relief scene of a woman on a bed (Figure 7.2).

On 23rd April 1862, aged 56, Ward married a second time, to Anna Saltonstall Merrill (1828-1901), aged 34, daughter of Hon James C. Merrill, of Boston. They had several children – some at least born in Dresden: Anna Saltonstall, Henry de Courcey, died young (13 July 1865), Marian de Courcy, Robert de Courcey, and Elsa, died young.

---

[5] One of the few non-family - albeit also German, partners in the London bank was Daniel Meinertzhagen, father of the (in)famous, fantasist and fraudster, Col. Richard Meinertzhagen (Garfield 2007).

Figure 7.2: Tombstone of Caroline Reynolds, first wife of Henry Veazey Ward, who seems to have died in childbirth.

While resident in Boston, Ward was appointed as the Consul for Chile. In 1868, having taken up residence in Dresden, he was again appointed Consul for Chile. His move to Dresden was the same year in which another American, with whom he will have served for many years in Chile, was made a London partner.

He died relatively young, aged 66, on 15 March 1873. At the time he was residing in Switzerland at the Chateau de Coppet near Ouchy on Lake of Geneva. He was buried in Mount Auburn Cemetery in Boston.

Part of his estate was in the UK and his will was lodged there with two of the five partners of Huth & Co as executors.[6] The estate 'in England' was valued at not more than £12,000, mainly in shares and investments. He also had estate in Dresden valued at 10,000 Prussian thalers and refers to silver tea sets received at time of his wedding and his first wife's jewellery. The bulk of his considerable wealth as set out in this same English will was in the USA and included properties in Boston and Baltimore; for one of the latter he set aside a huge $50,000 for redevelopment. His numerous bequests were to be funded from income from those investments: eight nephews and nieces were to receive a one-off US$1000 each 'in token of remembrance'. Other bequests - annuities, to wife, children and other nephews, nieces and a cousin amounted to $22,500 per annum. Records from Boston, show his widow and children still living there at 415 Beacon Street long after his death (https://backbayhouses.org/415-beacon/):

> *By the 1894-1895 winter season, 415 Beacon [Street]was the home of Anna Saltonstall (Merrill) Ward, the widow of Henry Veazey Ward. ... She previously had lived at 21 Chestnut. Living with Anna Ward were her stepdaughter, Caroline Elizabeth Ward, her son, Robert DeCourcy Ward, and her daughters, Anna (Anita) Saltonstall Ward and Marian DeCourcy Ward. ... Anna (Merrill) Ward died in April of 1901. ... The Misses Ward continued to live at 415 Beacon [Street] and also maintained a home in Cape Neddick, Maine. Caroline Ward died in April of 1926. Anita and Marian Ward moved soon thereafter, Anita Ward to an apartment at 90 Commonwealth [Ave] and Marian Ward to an apartment at 50 Commonwealth [Ave].*

Her passport applications show his daughter Marian was to become a missionary in Japan and China; Robert (1867-1931) was to make his name as 'climatologist, author, educator and leading eugenics and immigration reform advocate'. According to his entry in Wikipedia:

> *His advocacy for immigration reform and eugenics led him to co-found the Immigration Restriction League which was instrumental in the passage of the Immigration Act of 1924 which reduced Jewish and Italian immigration to the U.S. by over 95% and completely barred Asian immigration until 1952. ... In 1925, he was appointed to the Harvard University Committee on Admission and served on that board until 1931. This committee reduced the size of the freshman class to 1,000 students and arbitrarily reduced the proportion of Jewish members to 15%*[7]

As an odd update, a TV series ('City on a Hill') put together by Bostonian actors Ben Affleck and Matt Damon released in 2019 is fronted by an Assistant District Attorney in Boston

---

[6] In 1862, eight years after Ward's visit to Petra, two of the grandsons of Frederick Huth, aged *c.* 12 and 14 visited the ancient city in the charge of the (then) renowned historian, Thomas Buckle (who died shortly after in Syria).
[7] This cap on Jewish students at major universities in the United States was one of the major factors behind the establishment of the prestigious Institute for Advanced Study at Princeton in 1930.

in the early 1990s called Decourcy (sic) Ward! Hardly a coincidence surely? And Ward is played by a black actor. The events portrayed in this show are based on the actual Operation Ceasefire *'an enormous initiative developed through a collaboration between criminology professor David Kennedy (sic) and ....'*.

There are said to be descendants still living in the Boston area.

***

Henry Clay Yeatman
(Nashville, TN, 22 September 1831 - Hamilton Place, Columbia, TN, 1 August 1910)

Bowles refers to a Mr Yateman (sic) just four times. Yeatman had been travelling with Ward and others in Egypt and now across Sinai; they then travelled together with Bowles from Aqaba to Petra and on to Jerusalem. In practice, however, Bowles only mentions him for the first time on 30th March 1854, a full month after they left Cairo, as the combined party left Mt Hor returning to the W. Araba.

Yeatman stayed in Jerusalem and was then one of the party to the Dead Sea where he joined Kennedy and Bowles in swimming on 11th April. A week later – 17th April, and back in Jerusalem, as the Desert group fragmented still further, Bowles left Jerusalem with *'the Rogerses – Eames - Rudeswaldt (sic) Ward and Yateman (sic) etc'* but he gets no further mentions in the journal until three weeks later, 5th of May 1854 when he is in Beirut:

> *When I got down to dinner I found Briggs & Rycroft & Mr. Ward & Yateman. Much talk - after dinner I adjourned to the roof - where Messrs Ward, Yateman & Rycroft smoked their cigars. I came in & wrote journal.*

As we have seen, Ward receives frequent mentions in Bowles' journal and he is commmonly paired with Yeatman with whom he had travelled in Egypt before the Petra journey. We may reasonably infer that Yeatman travelled with Ward from Jerusalem to Damascus, then Beirut. Ward subsequently sailed on the same steamer as Bowles from Beirut to Alexandria then changed with him to a second steamer to Malta. Whether Yeatman was still with him is not stated.

Bowles may have had more to do with Ward than Yeatman for personal reasons (below). Curiously, Jane Eames, who is usually better at naming her travelling companions and especially the Americans, names 'Mr. Y--' just twice: first, on 11th (?) March, as they rode from Suez to Mt Sinai then on 1st April on the last leg of their journey as they approached Hebron. Once again, as Yeatman was evidently travelling with and probably sharing a tent and dragoman with Ward, this is a curious ommission.

As noted above, the group of four Amercans recorded in the Shepheard's Hotel register as apparently travelling together names him fully but without - in his case alone, reporting his home city/state. He was already travelling with two of those who were to be his companions to Petra - Rodewald and Ward:[8]

Charles Rodewald, of New York)
Henry C. Yeatman   )   *To Upper Egypt*
John H. Pell, of New York.   )
Henry V. Ward, of Maryland.   )
Leave Cairo Dec. 14th [1853]

Fortunately 'Henry C. Yeatman' is well-known. The Yeatmans were a prominent and wealthy family in Tennessee engaged in iron working. Thomas Yeatman (1787-1833) had died in 1833 leaving c.$500,000 to his widow and five children. When the fifth child, Henry Clay Yeatman (HCY), aged four, was allocated his share in 1835, it came to $58,000. In 1852, aged 21, the year he set off for Europe and Egypt, he purchased a one-sixteenth share in the Cumberland Iron Works for $80,000. In 1834, HCYs mother, Jane Erwin Yeatman, still only 41, had remarried. HCYs step-father, John Bell, was a prominent figure in state and federal politics, at one time Secretary of War and also a candidate for the Presidency.

In between, in 1852, HCY had set off for Europe and was to be absent until the end of 1854. After Europe, as recorded above, he had gone on to Egypt, sailed on the Nile with at least two of his companions above as noted by Ditson (1858: 368-69):

A mile or so above the cataracts, and a little below Philae, ... A party thus circumstanced, - Mr. Y. of Kentucky and Mess. P. and R. of New York, - passed us here in an uncomfortable-looking craft, on their way to Aboosymbal. They were the only persons, besides ourselves, who had as yet, this year, arrived with the resolution to traverse Nubia. It was exceedingly pleasant, as one can imagine, to meet our countrymen on the borders of Ethiopia, (they had come up from Asouan by land), and we fired a 'salute' to them, as is the custom in these waters.

This seems likely to be Yeatman, Pell and Rodewald; no mention of Ward. The attribution to Kentucky is surely simply a slip.

After the Nile, Yeatman, Ward and Rodewald set off across Sinai and on to Petra, Palestine, probably on to Damascus then to Beirut as discussed above. He is later recorded as sailing to New York on 15th November 1854. In July 1856 he applied for a passport but his intended destination is not recorded.

On 2 September 1858 this son of a wealthy Tennessee industrial family married into Southern plantation aristocracy. Mary Brown Polk (1835-1890) was the grand-daughter of Colonel William Polk (1758-1834), a leading figure of the War of Independence and in the administration of North Carolina (the part subsequently carved out as Tennessee and in which he became a major landowner). His revolutionry credentials and classical scholarship came together in his naming of his sons (but not daughters): George Washington Polk, Leonidas Polk, Alexander Hamilon Polk, Andrew Jackson Polk, Rufus King Polk, Charles Adams Polk and Lucius Junius Polk (Mary's father). The couple evidently went to Europe on honeymoon as they are recorded arriving back into New York on 2md November 1859 on a ship from Southampton.

---

[8] I am grateful to Drew Oliver for this information.

Henry and Mary Yeatman were resident in the antebellum mansion 'Hamilton Place', built in 1832 by Lucius Polk, Mary's father, and named for his brother who had died young. Mary had been born there in 1835 as were all of her children and many of those of her wider family. They shared this 'Polk' house with others of their extended family including Yeatman's mother and step-father, John Bell. In the Census of 1860 HCY is recorded as having Real Estate valued at $60,000 and Personal Property of $75,000 [or possibly $25,000].

The Civil War transformed their fortunes. Henry Yeatman was soon appointed a Lt. Colonel and one of the aides on the staff of Mary's uncle, Leonidas Polk. The latter was famous - a graduate of West Point in 1827 who had chosen instead to be ordained in 1831, made Bishop of the Southwest in 1838 and first Bishop of Louisiana in 1841. With the secession of the Southern States in 1861 he resigned his priesthood and was appointed a Major General in the Confederate Army. He made little impact as a general and was killed in 1864. Henry Yeatman, too, is hardly mentioned from the years he served. One of his great-grandsons who had a particular interest in the Civil War provided this summary of his military service (Yeatman 1984: 25):

> *An ardent Unionist at the beginning of the secession crisis, he became a reluctant Rebel after Ft. Sumter and Lincoln's call for troops. It appears, however, that he took no active part in the War until 1862, after Federal troops had occupied Nashville. Although he may have been at Shiloh, the first definite mention of his service was in the summer of 1862, at Chattanooga, where he was attached to the staff of Major General Leonidas Polk as a volunteer aide-de-camp, with the rank of lieutenant colonel. In this capacity he served throughout the Kentucky Campaign in the fall of 1862; at Stones River; in the Tullahoma and Chickamauga Campaigns of 1863; and the Meridian and Atlanta Campaigns of 1864. He was present when Polk was killed at Pine Mountain on the Kennesaw Line on June 14, 1864. Shortly afterwards, while attempting to gather Gen. Polk's personal effects, which had been left behind in Alabama, he was stricken with malaria and incapacitated for the rest of the conflict. He surrendered and took the Oath of Allegiance on May 10, l865.*

A few years ago, Henry's English-made Tranter revolver, complete with attachments and in the original presentation case from 'Gardens of Piccadilly', was sold at auction by some of his descendants.[9]

The Civil War proved divisive for Henry and his two (older) brothers. James Erwin Yeatman (1818-1901), industrialist, banker, educationalist and philanthropist, an ardent Unionist, was appointed by the Union government as President of the Western Sanitary Commission at St Louis, whose role was to look after sick and wounded soldiers west of the Mississippi, providing hospitals, staff, medicines. His letters reveal tensions with Henry - 13 years his junior, but a subsequent reconciliation.

Thomas Yeatman (1828-1890), a graduate of Yale, veteran of the Mexican War and attorney in St Louis and then in New Haven, CT, served as an officer in the Confederate Army. However, he seems to have attempted some reconciliation between the two warring sides, acting as an envoy to Lincoln. The proposal was rebuffed. Differences over the war brought breakdown in his marriage - his brother-in-law, John Pope, was a Major General in the Union Army and,

---

[9] https://caseantiques.com/item/lot-297-tranter-revolver-side-arm-of-lt-colonel-henry-clay-yeatman-cased-with-accoutrements/

after a divorce, a great rarity then, he moved to France. There he married Leone Monoury in 1871 with whom he had a further child - his fourth. Together they established a school which educated, amongst others, several relatives including HCY's daughter Mary Polk Yeatman (later Webb). Their grandson, Hippolyte, became a prominent ornithologist in France and married a granddaughter of the engineer, Gustave Eiffel. A great-great grandson, Savin Yeatman-Eiffel is a French film writer-director.

The Civil War severally damaged the family fortunes - even their mansion at Hamilton Place had suffered during a raid by Union troops (one of whom subsequently contacted the family to enquire about the pedigree of one of the horses seized!). On 19th August 1865, just three months after the end of the war, Henry was given an official 'Pardon and Amnesty' for 'taking part in the late rebellion against the Government of the United States' (Figure 7.3). In 1878 the Cumberland Iron Works of Woods, Yeatman and Company was advertised for sale, comprising 60,000 acres of which 14,000 acres had been divided into 94 farms.

It is in this post-war period that Henry and Mary Yeatman had five of their six children - all but two of whom pre-deceased him. Mary died in 1890, aged 55; Henry outlived her by 20 years, being killed by a train near his home in 1910, aged 78. Apparently he had stopped to pick a dog up off the tracks. Both Henry and Mary are buried along with many others of the family in the churchyard of St John's Episcopal Church nearby. Amongst his descendants are grandsons Henry Clay Yeatman II (1916-2013) an expert on crustaceans, and Trevezant 'Ted' Player Yeatman Jr (1915-1996), a Civil War historian and author of a highly-regarded book on the outlaws Frank and Jesse James (Yeatman 2000).

In 1854 when Henry Yeatman set off for Petra he was just 22 years old and probably the youngest of the sixteen who finally reached the ancient city (Appendix 1). That may partly explain why he receives so little attention from the older Bowles (aged 32) and Jane Eames (aged 38). In TB's case, his own high Anglicanism may have prejudiced him towards the child of a presbyterian family. The only hint of his character in this great adventure came on 1st April 1854 towards the end of their journey from Petra to Hebron in one of the only two references by Jane Eames (1855: 309) (above) in which she had dubbed him *'the man at arms'* because he *'is never seen without his gun'*.

Evidently the young Yeatman had a fondness for guns.[10] He would have stood out, too: his passport application of 1856 gives his height as 6' 2".

Just as Henry Yeatman was to find himself on a different side during the Civil War from his brother James (above), so, too, from his travelling companion in Egypt, John Pell (above) who served as a Union Captain in the Minnesota Volunteers.

Henry Clay Yeatman is quite well-known from what can be gleaned on the internet. There is certainly much more both in general terms from the hundreds of letters written by him and Mary held in the Tennessee State Library and - as is revealed from the latter's catalogue: *'Four diaries of Henry Yeatman, written while traveling in Europe and the Middle East during the early 1850s,*

---

[10] Though not the one he carried during the Civil War whose serial numbers seems to indicate manufacture in the 1860s - possibly precisely 1861 (Yeatman 1984: 26). Cf. previous note.

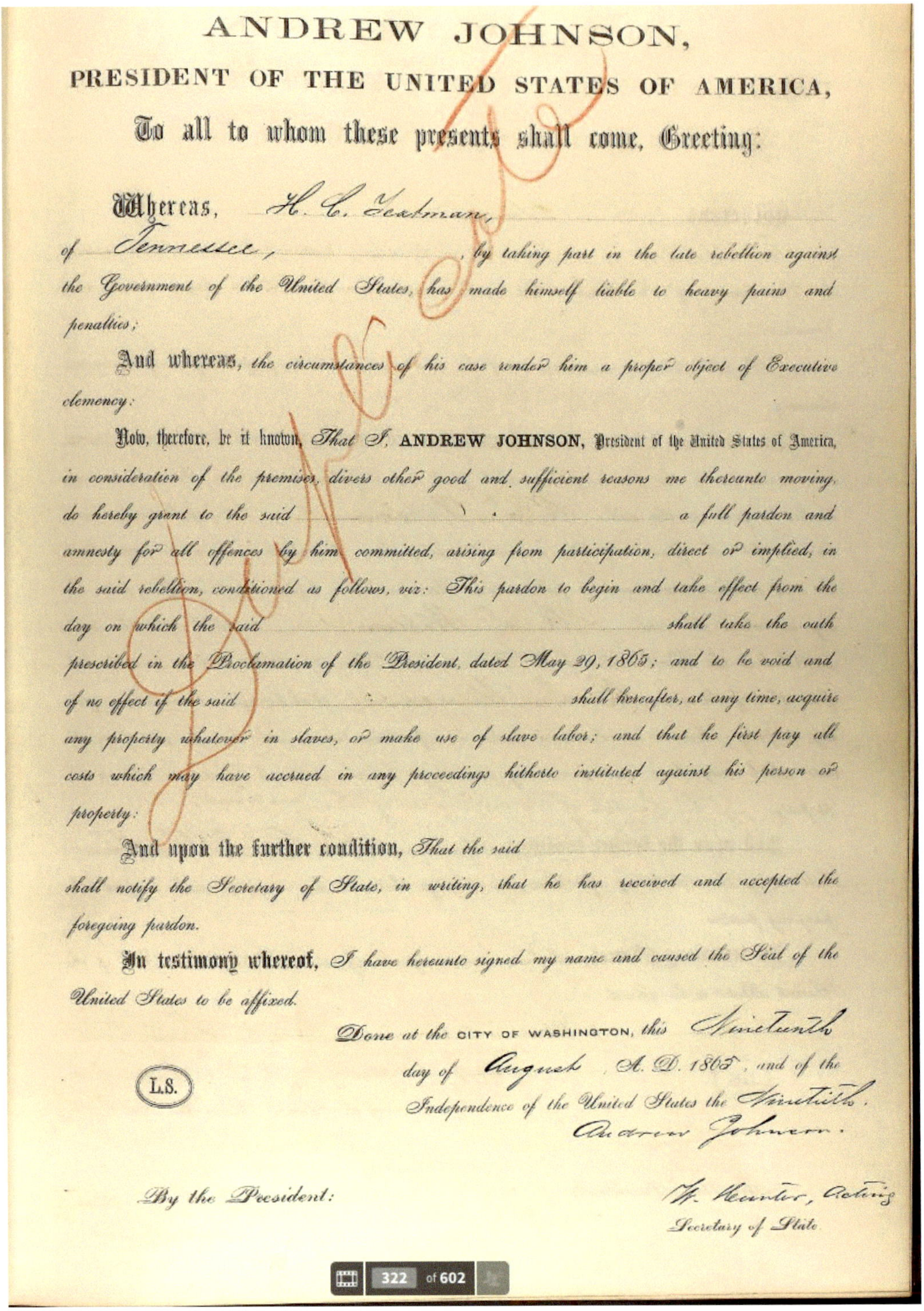

ANDREW JOHNSON,

PRESIDENT OF THE UNITED STATES OF AMERICA,

To all to whom these presents shall come, Greeting:

Whereas, H. C. Yeatman, of Tennessee, by taking part in the late rebellion against the Government of the United States, has made himself liable to heavy pains and penalties;

And whereas, the circumstances of his case render him a proper object of Executive clemency:

Now, therefore, be it known, That I, ANDREW JOHNSON, President of the United States of America, in consideration of the premises, divers other good and sufficient reasons me thereunto moving, do hereby grant to the said ______ a full pardon and amnesty for all offences by him committed, arising from participation, direct or implied, in the said rebellion, conditioned as follows, viz: This pardon to begin and take effect from the day on which the said ______ shall take the oath prescribed in the Proclamation of the President, dated May 29, 1865; and to be void and of no effect if the said ______ shall hereafter, at any time, acquire any property whatever in slaves, or make use of slave labor; and that he first pay all costs which may have accrued in any proceedings hitherto instituted against his person or property:

And upon the further condition, That the said ______ shall notify the Secretary of State, in writing, that he has received and accepted the foregoing pardon.

In testimony whereof, I have hereunto signed my name and caused the Seal of the United States to be affixed.

Done at the CITY OF WASHINGTON, this Nineteenth day of August A. D. 1865, and of the Independence of the United States the Ninetieth.

(L.S.)

Andrew Johnson.

By the President:

W. Hunter, Acting Secretary of State

Figure 7.3: Copy of the official Pardon and Amnesty issued to Henry Clay Yeatman shortly after the end of the 'rebellion'.

contain interesting and detailed descriptions of the places he visited and the things he saw' (see under 'Unpublished Sources' in Bibliography below).

The diaries are indeed interesting (Table 7.3). Unfortunately, they stop abruptly on 26th February 1854 just as Yeatman arrived back in Cairo from his journey up the Nile. They do continue but taking up on 22nd September 1855 when Yeatman is in Washington then continuing with gaps covering a trip to France in 1856. Although the trip to Petra is not included - and may never have been written, the content of Diary 1 which covers the background to that journey is of interest for what it reveals of Yeatman. After leaving Paris in October 1853 he travelled via Malta and Smyrna to Constantinople. Then on to Alexandria to begin his Egyptian tour, by then he had met up with Pell, Rodewald and Ward and shared a Nile boat with them. At least three times he mentions practicing shooting his pistol, he and Pell using their hats as targets. He makes a number of interesting observations including the health of people encountered: eye disease, swarming flies, skin disorders, some with one eye, some blind, discharge from eyes never removed. Almost all the men have maimed themselves: knocking out front teeth, gouging out an eye, cutting off a forefinger of right hand to make themselves ineligible for recruitment into the army.

Table 7.3: Summary of the four Yeatman travel diaries in the Tennessee State Library.

Diary	Date and Place Started	Date and Place Finished
2	22 December 1852 - Paris, France [Cover says 'Volume II: March - June 1853 (sic)]	12 January 1853 - Rome, Italy June 1853
3	1 March 1853 - Rome, Italy [Cover says 'Volume II: March - June 1853 (sic)]	4-5 June 1853 - London, UK
4	6 June 1853 - London, UK 22 September 1855 - Washington City, USA+ 8 August 1856 - Paris# [Cover says 'Volume III: June 1853 - September 1856]	25 October 1853 - Paris, France* 7, 8, 9 October, 1855 - Washington 4 September 1856 - Paris, France
1	25 October 1853 - Paris, France	26 February 1854 - Cairo, Egypt

Chapter 8

Companions on the Long Desert Route: 2

Rev. James Henry Eames
(Dedham, MA, 29 November 1814 – On a ship in the harbour of Hamilton, Bermuda, 17 December 1877)
and
Jane Anthony Eames
(Wellington/Dighton, MA., 21 January 1816 – Boston, MA 8 July, 1894)[1]

Of the sixteen westerners who visited Petra as a group in March 1854, only Jane Eames left a published account. Indeed, over the course of a long life she published several books - appropriate for someone who was one of the first female journalists in the United States (the *Providence Journal* in the 1840s). Most importantly, she was a prolific, perceptive, and amusing letter-writer. Many of her letters sent home during her travels overseas with her husband she collected and then published in a succession of books, each volume of which she called a 'budget'. When TB met her, she was already a published author though he may well never have known that at the time. Their first overseas trip - perhaps a delayed honeymoon, aged 25, was to western Europe leaving the United States in March in 1841 at the conclusion of which she presents some statistics which are worth quoting for an insight into the stamina and style of the writer (Eames 1847: 469-70):

> *We travelled by sea in two different ships seven thousand miles; in steamers, thirteen hundred and twelve;[2] by rail-road, six hundred and seventy-nine; in diligences, thirteen hundred and fifty; in a carriage we hired, eleven hundred and fifty; in coaches, eight hundred and forty-four; and in other kinds of carriages, three hundred; making in all, twelve thousand six hundred and thirty-five miles, of which nineteen hundred and sixty-four miles were in Great Britain. Eight hundred miles we rode on the outside of coaches. Of course I can make no mention of the countless number of miles we rode in hackney coaches and cabs in our excursions around different cities. ... we were gone from home two hundred and forty days, ...*

She was to do so again after her travels in the Europe and the East often in very different circumstances (Eames 1860: 368):

> *We were absent from the United States fourteen months, during which time we travelled nineteen thousand two hundred and fifty-eight miles, viz: thirteen thousand one hundred and eighty-three in steamers, one thousand eight hundred in our boat on the Nile, six hundred miles on camel, five hundred on horseback, in diligence and other carriages, three hundred and forty-five, and two thousand eight hundred and thirty miles by railroad. Of course, I could form no conception of the number of miles I rode on donkeys in Egypt, and in hackney coaches in Europe. During this period, we slept sixty-two nights in steamers, seventy-two in our boat on the Nile, fifty-nine in a tent, two in a diligence, and two in railroad cars, avoiding, as much as possible, night travel on land, which is very exhausting to the strength of travellers.*

---

[1] I have been unable to locate any portrait or photograph of either of these interesting people.
[2] These distances by sea are positively heroic for the Rev. Eames whose obituary observed *"It has been his misfortune, at all times, when at sea, to suffer from sea-sickness."* (Anon. 1878). He was to die just as he neared the end of another sea voyage.

For present purposes it is the letters written during the 14 months James and Jane Eames spent in Britain, 'the East', Constantinople then Germany that are of relevance here (Eames 1855; 1860).[3]

Although she frequently 'names' travelling companions, nationalities and sometimes records characteristics, she almost invariably does so in reducing them to initials - 'Mr. R---', 'Dr. B---', etc. Happily, combined with the unpublished travel journals of Bowles, Bryce and Yeatman, almost all can be identified, and fuller biographies teased out.

As Jane Eames' letters are readily available to read in entirety, they will be used here either to assist in identifying others, their characteristics and - through quotations, events in which she and her companions to Petra shared. As women were a very small minority of travellers to Petra and their accounts rarer still, her letters are especially informative for the experience of women and their encounters both with non-western men and some of the latter's womenfolk. In this instance, her letters are worth quoting for episodes especially involving just her and her sole female companion, Virginia Rogers.

James Henry Eames was born at Dedham in Massachusetts in 1814, one of nine children, but his family moved during his childhood to Providence, Rhode Island. He graduated from Brown University in Providence in 1839 and received a Doctorate of Divinity from Norwich University, Vermont, in 1862.[4]

> *He was ordained deacon in December, 1841, and presbyter in 1842; was rector of Ascension Church, in Wakefield, for about four years, when he took charge of St. Stephen's Church in Providence, remaining there until 1850, and then engaged in missionary labor in Rhode Island; became rector of St. Paul's Church, Concord, N.H., in 1858, and held that position until his death, which occurred in the harbor of Hamilton, Bermuda, December 10, 1877. For many years Dr. Eames was chaplain to the asylum for the insane, and performed a large amount of missionary work in New Hampshire. Three times he travelled in Europe, and spent part of several winters in Bermuda.*

Although James Eames seems a rather quiet undemonstrative character *'courtly and dignified in manner, sweet and gentle in disposition'* - his posthumous reputation benefitted from a literary wife who is the likely author of the 56-page 'In Memoriam', including numerous testimonials (Anon (Jane Eames probably) 1878).

On 1 October 1839 at Providence, RI, the 25-year-old James Eames married 23 year old Jane Anthony, one of eleven known children - seven of whom died in infancy, of Sally and merchant Hezekiah Anthony. The latter was a wealthy merchant in Rhode Island who evidently underwrote significant parts of the costs of the travels by Jane and James. She was to dedicate her *Budget of Letters from the East* (1855):

TO MY FATHER,
WHOSE CONSTANT CARE AND AFFECTION,
HAVE BRIGHTENED MY EVERY PATH IN LIFE,

---

[3] Twenty years later she published a further selection of Letters from a series of lengthy winters spent in Bermuda for health reasons (Eames 1875).
[4] https://www.biblicalcyclopedia.com/E/eames-james-henry-dd.html. Cf. Catir 1964: 74-80.

AND BY WHOSE LIBERALITY,
I HAVE BEEN PERMITTED TO VISIT FOREIGN LANDS,
This Work is Dedicated,
WITH A DAUGHTER'S LOVE AND A DAUGHTER'S REVERENCE.

She evidently inherited a great deal of money: (*New York Times* obituary 10th July 1894):

Mrs. Jane Anthony Eames, widow of the Rev. Dr. James H. Eames, for many years rector of St. Paul's Episcopal Church, Concord, N.H., died Sunday night of paralysis, in Boston. She was one of the most widely known women in the State. She was born in Providence, and belonged to the Anthony family of Rhode Island. Mrs. Eames had traveled extensively and had written much for the press. She began newspaper work on the Providence Journal more than fifty years ago, having been one of the first woman journalists in the country. During her life she gave to religious and charitable objects about $50,000. She left a large estate, nearly all of which will go to Episcopalians and benevolent objects in New Hampshire. She was seventy-eight years old.

In addition to her own lively and interesting writings, we have the testimony of one of the other Americans with whom she travelled to Petra, Henry Clay Yeatman who had previously encountered them on the Nile (above, Ch. 7) (Yeatman Diary 1: 44B):

Mr & Mrs Eames - from Providence, Mass & Mr Nicholson [=G. N. Nicholson, of England] an English gentleman far gone in consumption who travels with them, dined with us today. Mrs E. is a very smart clever little yankee woman & her husband quite an agreeable person. I think she is the head of the house.

That fits rather well with her own statements that *'you well know that I am generally not backward in speaking my mind'* (Eames 1847: 302) and - when writing of their prospective dragoman for the Long Desert Journey, *'True, he has his faults, and when he has done wrong, and has not suited me in everything, I have not hesitated to give him "a piece of my mind" on the subject'* (Eames 1855: 339-40).

They were to have no children and Jane outlived her husband by 17 years. Understandably it was Jane, the forceful and well-known journalist and author, who merited several obituaries and entries in literary records. A flavour of these can be illustrated by quoting her account of one event in which neither TB nor any of the other men, were able to participate.

On 20th March she recounts at some length the visit the two American ladies - herself and Mrs Rogers, made to the wife of the Ottoman governor in the little fort on the shore at Aqaba (above Ch. 7). It is Aqaba as seen by the rare western female traveller (Eames 1855: 268-70):

The Governor here, has made visits to us all. I told Hassan [their dragoman] this morning to ask him if a visit from Mrs. R[ogers] and myself to his wife would be agreeable, and on being assured that it would be, we went this afternoon, escorted by his excellency, or the "Effendi", as he is called, to pay our respects to the highest lady in station at Akaba. The Governor has rooms in the Citadel or fort, a large, castellated looking edifice, immediately in the rear of our tents. Passing through two immense gateways, where the soldiers, in anything but "uniform" costume, presented arms, we entered a quadrangle, where were one or two guns, not calculated, judging from appearances, to do much execution. In this court is a large well of good water, the only place where it can be obtained for some distance, I believe. Ascending a flight of stone steps, hens and chickens fluttering before us, we were shown into an ordinary looking room, where "Sitteh[5] Mohammed," (the wife of Mohammed, received us according to the Eastern mode, kissing our hands, and wishing us "peace". There was another woman in the room, with a baby in her arms, but neither Mrs. R[ogers] nor I, could muster Arabic enough, to find out who she was, except that she was not one of the Governor's wives, "Sitteh Mohammed" being the only one he had. The "Sitteh Mohammed" was dressed in rather a scant robe of Syrian silk, open from the neck to the waist, thus exposing the chest a good deal. She seemed, like our common mother, to be on "hospitable thoughts intent,"[6] for she bustled about continually, to do honor; in her way, to our visit. And not only she, but the Effendi himself, in a robe of Syrian silk, and a crimson mantle over it, with the assistance of a Nubian slave, was occupied all the time of our visit, in going back and forth to wait upon us. First came a pipe. I begged Mrs. R[ogers], who can smoke better than I, to take it first, and after she smoked a while, I took a few whiffs, and then we gave it to the "Sitteh," who seated herself on the floor, and finished the pipe. Then lemon sherbet was handed to us. After drinking half what was in the glasses, we gave them back, where upon both the "Effendi" and the "Sitteh" insisted so clamorously, that we should finish the glasses, that I complied with their request, but Mrs. R[ogers], whose glass was much larger than mine, found it impossible to swallow all hers, so she gave back the glass, which after many compliments, the Sitteh drained. Then commenced a great talk, but neither of us could make out what it was about, excepting that it concerned something to eat. After fruitless attempts to make us understand, the Effendi and the Nubian boy, went out of the room, and soon we heard a clatter and a shuffle on the stone steps, and the door opened, and in walked the Effendi with a large knife in his hand, followed by the slave dragging a sheep! And then it turned out, that they had been begging us to allow them to kill a sheep, and cook some part of it for us! In vain we thanked them, and declared our own dinner was waiting, they seemed determined we should eat something, so eggs were brought out, and the Sitteh, pointing to the fire of coals, where the coffee was being made, begged she might cook some for us, but this too we declined. Coffee was then brought, which you know I never drink. I sipped a little, and then handed back the cup, but no! I must drink it all, so making a violent effort, I drained the cup, which fortunately was small. After this, bread was offered us, with many assurances that we should find it very nice, so I gave them to understand, we would not eat it there, but take it to our tents, which proposition was graciously received, and then we rose to take leave, but we were urged so violently to stay longer, we once more seated ourselves. Fortunately, I had become so heated with my efforts to swallow so much sherbet and coffee, I could not stay longer with comfort, and we soon after came away, inviting the Governor's wife to visit us to-morrow in our tents. All the while we were there, the other woman either stood up beside us, or sat on the floor, and every time the Governor came in, she would hastily cover her head and face, though she made no scruple at nursing her baby in his presence. The baby was quite a pretty looking child, with the rims of its eyes stained black, and if it had been clean, I could have kissed it; as it was, I kept it at a reasonable distance.[7]

***

---

[5] 'Lady'
[6] John Milton *Paradise Lost*, V, l. 331.
[7] She had previously noted the social value of smoking (Eames 1855: 108): 'I have thought many times to-day, (notwithstanding my objections to tobacco in all its forms,) "blessings on the man who invented pipes." Nothing can be more awkward, than a visit from strangers, with whom you can have but few thoughts and associations in common, and with whose language you are so unacquainted, as to be obliged to resort to an interpreter. You sit in silence, it would be 'awkward silence' elsewhere, but as the pipes are brought, there is an excuse for not speaking; it is indeed "the pipe of tranquillity." Fortunately, the tobacco we brought from Cairo is so good, our visiters (sic) are glad to smoke some time, and as they seem to be better satisfied with smoking than talking, why should we be discontented with their choice? And here I may add, for your further information, if any of you should ever come to Egypt, when your tobacco is praised, take it as a gentle hint to give your visiters(sic) some. Do not think they would feel offended, even though the visiter(sic) rejoices in the high-sounding title of Governor. Ten chances to one, if you do not offer it, they will not hesitate, after they have left you, to ask your dragoman for some, or to send back one of their servants, with a polite request.'

'Mr F[ish] from Alabama'

In his Journal, TB mentions Fish just twice, once towards the end of their trip to Petra (30 March 1854):

> *I had a walk in the morning with Freeman & then with 2 of the Americans - Yateman (sic) & Fish. They are certainly very unlike Englishmen.*

... and a second when he is travelling in Palestine between Nablus and Jenin (19 April 1854):

> *We passed several villages. At the top of one of the hills or passes - we saw the Mediterranean & then descending again amongst a great many olives, we sat down to luncheon. We found Mrs Eames & Mr Rudeswaldt (sic), who had come a shorter way & while resting after luncheon, Mr Eames, Fish, & Adam came up. They had stayed behind to ascend Gerizim. They too had come the same shorter way & so had missed Samaria.*

Jane Eames in her letters has two people she refers to as 'Mr. F.' but it is easy by reference to TB's parallel account, to differentiate between F[reeman] and F[ish]. The latter, a fellow-American, travelled in the company of the Eameses from her first mention of him in Cairo when they agreed to travel together to Petra (Eames 1855: 210. Seems to be 21st February 1854):

> *... so we have made all arrangements for Hassan to take us, and we hope to get away early next week. Our party consists of Mr. R[odewald], a German by birth, though a resident for several years in New York, with whom we dined at Assouan, another Mr. R[oss], an Irish gentleman, Mr. F[ish], from Alabama, and ourselves. Hassan takes his son Ali with him, and a cook, so that our temporal wants will be well attended to.*

... until they finally parted company in Constantinople three months later when he left with the Rogerses for Greece (Eames 1860: 55-56):

> *We spent all the morning of Tuesday [30th May 1854] in the bazaars at Constantinople, in search of a shawl, but I found none to suit, not in quality, but in price. That day we bade farewell to Mr. and Mrs. R[ogers] and Mr. F[ish], who have gone to Greece, and now, for the first time since entering Egypt, we are without companions.*

As he had previously split with the Eameses at Smyrna and gone on ahead to Constantinople with the Rogerses, it maybe he found their company more to his liking or convenience.

Despite the numerous references, Fish remains opaque. One may suspect he was of a similar age to the Eameses and Rogerses - late 30s/early 40s; carries a Bible (which Jane Eames has him reading in the Dead Sea). He may be bookish and interested in the natural world as implied by Jane Eames who twice refers to him with pen and paper and once in search of geological specimens.

Without a first name or even initial, Googling any combination of Fish, Alabama, Egypt, Petra, Jerusalem, Syria, Smyrna, Constantinople, Greece returns nothing of use. The unpublished journals of Bryce are no help; perhaps those of Ward when available will provide a clue.

***

Charles Rodewald = Carl Reinhard Conrad Rodewald
(Bremen, 7 March 1811 - Kehrsiten, Switzerland, 11 August 1884)

A graffito in the Khazneh at Petra was read as 'C. Rodewald 1854' (Euting in Brünnow and von Domaszewski 1904: I, 193). It is still visible despite having been written only in charcoal - the first letters are unclear but *'----wald 1854'* is certain (Figure 8.1). This is surely the 'Charles Rodewald, of New York' who, together with three companions was recorded in the Shepheard's Hotel register as having left Cairo for Upper Egypt on 14 December 1853 (below).

Bowles mentions him five times as 'Mr Roudewald (sic)' - and with other spellings. In the first of these, on 6th April 1854, without any further introduction, he arrives at Bowles' hotel in Jerusalem to take him to a meeting. It is only on the second mention - on 17th April, that he mentions that Rodewald had been one of the group of 16 at Petra:

> *All the party is broken up. The Heneages were gone this morning - for Jaffa & Pagde (?) too. We go - The Rogerses - Eames - Rudeswaldt (sic) Ward & Yateman etc.*

That in turn allows him to be identified in the pages of Jane Eames (1855: 210), in a letter dated 'Cairo, Feb. 18th. [1854])':

> *... so we have made all arrangements for Hassan to take us, and we hope to get away early next week. Our party consists of Mr. R—, a German by birth, though a resident for several years in New York, with whom we dined at Assouan,[8] another Mr. R—, an Irish gentleman, Mr. F—, from Alabama, and*

Figure 8.1: The graffito in charcoal of '[Rode]wald 1854' in the Khazneh at Petra (Courtesy of Prof. Fawzi Abudanah).

---

[8] Eames 1855: 148. Letter dated: 'On the Nile, Jan. 13th.' but the meeting having taken place on Saturday 7th January 1854.

ourselves. Hassan takes his son Ali with him, and a cook, so that our temporal wants will be well attended to.

The two others are Mr Fish (above) and Mr Robert Ross (below).

The Rodewald family is well-served on the internet with a major genealogical site following members generation by generation. The family's origins lie in northern Germany. Parts of the family rose to prominence in Bremen as merchants and bankers in the 18th century and then in the 19th century some pursued their business overseas: some can be traced sailing to Baltimore but key members seem to have gone to New York where 'Rodewald Brothers' were in business on Beaver Street in Lower Manhattan, just 250m from Wall Street and already a thriving business area. Carl was a popular name in the family including those who migrated to the United States. However, Jane Eames is again the guide as she records that Rodewald was older than Ross who - as we shall see, had been born in 1813 (below). Only one candidate - Carl Reinhard Conrad Rodewald, fits the evidence.

Carl Reinhard Conrad Rodewald (1811-1884) was born in Bremen and - although he was, after a lifetime overseas, at the end of his life resident again in Bremen, died in the little Swiss town of Kehrsiten on Lake Lucerne. He was the third of seven sons and one daughter of Johann Friedrich Arnold Rodewald and Anna (née Quentell) Rodewald, all born in Bremen. Three of the sons included 'Carl' as the first of their three personal names but two were known by one of the other two personal names - Augustus and Ferdinand[9] respectively. Only Carl Reinhard Conrad was actually known as Carl. Most of the brothers can be traced as adults in the Americas - principally New York, Baltimore and New Orleans, all presumably because of the role of those ports in the cotton trade. Carl Reinhard Conrad, however, is found much further south. In 1841, aged 30, he married Leonora Dam(m)asia Zimmerman in Buenos Aires, Argentina. Once again, the link is likely to be cotton and Leonora is presumably a daughter of an ethnic German settled there. The obvious candidate is Johann Christian Zimmermann (sic) (1786-1857), a merchant in the city who served successively as Consul of the Hanseatic cities of Bremen and Hamburg and later also Vice-Consul of the United States in Buenos Aires, then Consul in Bremen and Hamburg of Argentina from 1828-1847. Both his wives had the surname Halbach Schmidt, a German family resident in Buenos Aires. The first died in 1824 the year Leonora Dammasia married her fellow German Carl Reinhard Conrad Rodewald. They had two sons, both born in Buenos Aires - one died there relatively young in 1867; the other died in Bremen in 1910. Leonora Dammasia Rodewald died in Madeira on 30 November 1852 and was buried there.

Carl Reinhard Conrad remarried - to Adelheid Gildermeister of Bremen where her father, a lawyer, was a state Senator and her brother was to become a Senator and Burgomaster. The first of three sons and two daughters was born on 2 April 1856. All were born in Bremen and all but one died there. It would seem that when Carl Reinhard Conrad Rodewald arrived in Egypt in 1853, he was recently widowed from Leonora but remarried some time before June/July 1855 and presumably after his travels in the Levant, to Adelheid. The eldest son of this second marriage, Carl Adolf Ferdinand Rodewald, was to marry his American-born cousin

---

[9] Ferdinand outlived his elder brother by 22 years and his gravestone in Liverpool calls him Charles Ferdinand Rodewald.

once removed, Anna Fredericka Rodewald - a daughter of the New York Rodewalds, and settle in the United Kingdom where he died in London in 1953.[10]

Returning to Charles Rodewald (aka Carl Reinhard Conrad Rodewald) in the Levant, as noted, Bowles first mentions him on 6th April 1854, a week after they had completed the Long Desert Route from Cairo via Petra to Jerusalem. They had plainly travelled in - at the very least, close proximity for the early part of the journey and were certainly part of the same group of 16 from Aqaba to Petra then of 14 from Petra to Jerusalem. Of his 15 companions, Rodewald is one of just two Bowles never mentions during the journey itself - the other, Henry Veazey Ward, is also an American. Rodewald was at least ten years older than Bowles.

After the Petra group broke up in Jerusalem two sub-groups emerged: Rodewald evidently joined the American group of the two Eameses, the two Rogerses, Ward, Yeatman and Fish. Bowles nevertheless encountered Rodewald several times after Jerusalem, on the last – 6th May 1854, they were in Beirut.

For characterisation of Rodewald, we can turn to Jane Eames who explains his nationality and that they *'had dined [together at] Assouan'* some time before. In an earlier passage, without naming them, she noted that the dining party consisted of four Americans (Eames 1855: 148):

We arrived here this morning about ten o'clock, and saw the American flag waving from a boat behind us, and immediately after breakfast, J. went on board, and found it occupied by four gentlemen, all of whom had left home long before our departure, and they could give us no particular news. Oh! this thirsting for news from home, in a far distant land! Afterwards the gentlemen called on us, and invited us to dine with them; the invitation was given too cordially, not to be as cordially received, so about five o'clock we went to their boat, and delightfully the hours passed. Dining out in Egypt, what an event!

It is certain that these are the four Americans recorded in Shepheard's Hotel register[11]:

Charles Rodewald, of New York)
Henry C. Yeatman) *To Upper Egypt*
John H. Pell, of New York)
Henry V. Ward, of Maryland)
Leave Cairo Dec. 14th

All but Pell were then to join the Petra group. Jane Eames then mentions the 'Mr R---' who can be identified as Rodewald rather than Ross, on at least 17 occasions, mostly after they had reached Jerusalem and travelled onwards together. Early on, however, she describes his crossing of a branch of the Red Sea (Eames 1855: 222):

... while I was talking, the American R, quite a stout man, had mounted on the shoulders of an Arab, who was struggling through the water, waist deep, with his burden. Then followed the other, Mr.

---

[10] On a personal note, Cosmo Rodewald (1915-2002), the only son of this marriage of Adolf and Anna, and, therefore, a grandson of Carl Reinhard Conrad Rodewald/ Charles Rodewald, was my tutor in Greek History at the University of Manchester in 1971-3 where he was a long-time academic and author, fondly remembered by his students (See Sekunda 2007).

[11] As so often, I am grateful to Drew Oliver for this information.

R., and, as I found it useless to delay any longer, my turn came next, and borne between two men, to whom I clung with the tenacity of a drow(n)ing man, I was carried through the sea, and safely landed on the shore of Asia. What an introduction to that consecrated soil!

Rodewald was later part of the Eames sub-group which made an expedition from Jerusalem to the Dead Sea. On 17th April they set off again, this time quitting Jerusalem for good (Eames 1855: 372):

To-morrow morning [= 17th] we are off, and I look forward with much interest to the journey. In addition to Mr. R[odewald] and Mr. F[ish], Mr. A[dam according to Bowles], an English gentleman, invited by Mr. R. to join us, goes with us.

The Eameses split up into different parties as they progressed through Palestine with Rodewald sometimes away from Mrs Eames. However, they finally reached Beirut together - where they again met Bowles, so that Rodewald could catch a specific steamer. In the event he changed his mind and, together with the Eameses, sailed for Smyrna on 13th May with stops at Lattakia and Rhodes. At Smyrna, these Americans were invited to visit the American frigate *Cumberland* in the harbour but declined. That same day (17th May 1854) Jane Eames recorded (1855: 477):

On the day of our arrival, we parted with our pleasant companion, Mr. R[odewald], who had been with us more than three months. He has gone to Malta, and we may scarcely expect to see him again, this side the Atlantic ocean, but if our lives are spared, we hope to meet sometime in the "new world." Yesterday Mr. F[ish] and Mr. and Mrs. R[ogers], went to Constantinople, so we are now quite alone.

At that point Rodewald, having been first encountered leaving Shepheard's Hotel for Upper Egypt on 14th December 1853 then meeting the Eameses at Aswan in mid-January 1854 moves out of sight in mid-May 1854 sailing for Malta. As noted above, whether or not he first returned to New York, he evidently went to Bremen then or soon after and remarried before June/July 1855.

***

Robert Ross-of-Bladensburg
(Westminster, 19 February 1813 - Leghorn/Livorno 28 November 1859)

Brünnow and von Domaszewski (1904: I, 193) provided till now the only record of a visitor to Petra by the name of Ross. They cited a short article in a German periodical in 1855 which pointed to the visit being in Spring 1854 (Blau 1855). Amongst the "Inschriftenaus Petra" in the article are three which Blau credits to someone he refers to several times as 'L. Ross' and on the first occasion calls "a brave English traveler". Later he says that when Ross came to Constantinople, he gave the records of the inscriptions to him (Blau). The article itself is dated 'Constantinople den 3. Sept. 1854'.

Although Blau gives his initial as 'L' that seems to be incorrect, a confusion with the contemporary German (but half-Scottish) Ludwig Ross (22 July 1806, Bornhöved – 6 August

1859, Halle), a former Director of Antiquities for Greece (1834-36), then Professor of Archaeology at University of Halle (1845-59), a well-known explorer of classical Greece and publisher of many Greek inscriptions and several books on the antiquities of Greece.

Brünnow and von Domaszewski (1904: I, 193) could cite only 'L. Ross' and 'C. Rodewald' under 1854 for western visitors to Petra. We can place them both in the large group of sixteen in the ruins from 26 to 29 March as revealed in the Bowles Journal and the Letters of Jane Eames but with a correction to one name.

Bowles refers eight times to one companion as 'Mr Ross', starting in Cairo with the latter being partly responsible for his own decision to undertake the Long Desert Route via Petra to Jerusalem. Their paths crossed from time to time as they made their separate ways to Aqaba. Thereafter they all travelled together to Petra then on to Jerusalem where Bowles mentions 'Mr Ross' for the last time. Although he seems clearly to be British, Bowles gives no details – on first name, age, occupation, character, although he implies he is fit and energetic.

Jane Eames is much more helpful. In Cairo as the Eames Party prepared to set off in February 1854, she says:

> *Our party consists of Mr. R—, a German by birth, though a resident for several years in New York, with whom we dined at Assouan, another Mr. R—, an Irish gentleman, Mr. F—, from Alabama, and ourselves.*[12]

Fortunately, Mrs Eames differentiates between the two 'Mr Rs' and we can then follow them individually in her later pages. The first Mr R is Mr Rodewald discussed above. The second 'Mr R.' is Ross who is characterised as the younger of the two and is found doing athletic things such as climbing to the top of the arch at the entrance to the Siq at Petra. Rodewald is not only older but apparently quite overweight (above). Neither is given a first name (as noted above, the third member of the trio is Mr Fish.)

Mrs Eames subsequently met Ross in Constantinople in July 1854 which is where Blau says he was given the Nabataean inscriptions by his 'L. Ross' (above). We can now add that in the Bowles Journal we read:

> *Mr Ross had come so far to show us a Tomb opposite the Theatre in wh(ich) he had found inscriptions.*

At Constantinople he was mixing in eminent diplomatic circles. Crucially for identifying Ross, Mrs Eames observes that:

> *This morning, at the request of Lady G., Mr. E. read prayers in her drawing room, and though the congregation was small, it was a select one, consisting of Lord and Lady G., the wives of two officers in the English army, Mr. M. and his wife, occupying for many years a prominent station in the British possessions in North America, the family of the American Ambassador, and our fellow traveller in the Desert, Mr. R., son of General R. A more pleasant circle than this, it has not often been our lot to*

---

[12] In his unpublished travel journal Dr William Bryce (below), born in Ireland, also refers to Ross as a fellow countryman in the context of talking about St Patrick's Day.

meet while travelling, and it is one of my sources of regret at leaving this place, that I must say good bye to so many agreeable acquaintances.

Finally, in September 1854, Mrs Eames, in a subsequent collection of letters from her travels, records meeting at Dresden,

... our fellow-companion in the Desert, Mr. R. to say which was the most surprised at this unexpected meeting, would be exceedingly difficult, and the way we talked over, during the evening, our mutual adventures, since we parted on the Bosphorus, was not slow.

Who was Mr Ross? He has the leisure and wealth to support travels from at least Egypt in February 1854 to Dresden in September 1854. The Eameses stayed in a hotel in Vienna Ross had recommended to them. He has the standing to mix in elevated circles in Constantinople. Finally, and most important, he is Irish and the son of a General – indeed, apparently a general well enough known (to an American) to be referred to by Jane Eames as 'son of General R.'

While Ireland produced many high-ranking soldiers including generals for the British army, only one 'General Ross' fits the dates.

Major-General Robert Ross, born at Rostrevor in County Down in 1766, served extensively during the Napoleonic Wars as recorded on a monument in his honour at Rostrevor in Ireland:

HILDEN 1799, ALEXANDRIA 1804, MAIDA 1806, CORUNNA 1809,
VITTORIA 1813, ORTHO 1813, PYRENEES 1813, BLADENSBURG 1814, BALTIMORE 1814.

The last two are significant: in 1813 Ross was the commander of British land forces on the Eastern Seaboard of the United States during the War of 1812. He defeated American forces at Bladensburg, *c.* 10km northeast of Washington, on 24th August 1814. The same day the British troops occupied Washington and – in retaliation for the *'wanton destruction of private property along the north shores of Lake Erie'* in Canada including Port Dover and York (predecessor of Toronto) the previous year, burned several public buildings in the capital, including the White House and the Capitol.

General Ross subsequently advanced on Baltimore but was killed by American sharpshooters on 12th September 1814. He was buried in Halifax, Nova Scotia but commemorated by an impressive monument at Rostrevor, a large plaque in St Paul's Cathedral in London and a portrait in the Rotunda of the Capitol in Washington.

At the time of his death, despite his long and distinguished career and high rank, he had received no honours. Posthumously, his family was granted the right to change their surname to 'Ross, of Bladensburg' and a coat of arms was designed to include an arm grasping a broken staff with a banner of the (then) 15 stars and stripes of the United States.

General Ross had married in 1803 and at the time of his death, had four surviving children:

1. Andrew (1802-1822)
2. David (1804-1866)

3. Elizabeth (1811-1827)
4. Robert (1813-1859)

Andrew and Elizabeth were both long dead by 1854; David would have been 50 which would not fit well with Mrs Eameses description of him as the younger of the two Mr Rs. More likely our Mr Ross is Robert, 41 in 1854. Fortunately, a footnote in Porter's book on his years in Damascus (Porter 1855: 1, 14) confirms the identification:

On the 19th of May 1854 I visited the fountain of 'Anjar and the ruins near it with two friends.[5]
5. Robert Ross, of Bladensburg, Esq., and Edwin Freshfield, Esq.[13]

At the moment only a few further details can be added:

Born in London (Westminster) on 19 February 1813 eighteen months before his father's death; graduate of 'Trinity' – apparently Dublin rather than Cambridge (1831-35); admitted to Inns of Court Dublin (1833) and in London (1835); died in 1859 and buried in the (New) Cimitero Inglese at Livorno (Leghorn), Italy. Interestingly, his sister Elizabeth had died at Siena in 1827 and is buried in the same cemetery (Milner-Gibson-Cullum and Macaulay 1906: 63):

Elizabeth Ross of Bladensburg,
daughter of Major General Robert Ross of Rosstrevor, Ireland,
born May the third 1811, died at Sienna April twenty seventh 1827
before she had attained the age of sixteen, in the full assurance of hope and faith in our Lord Jesus Christ.

Ross's father, the future Major-General, had been in Egypt himself 60 years earlier at the time of Napoleon's expedition. Ross's older brother David had undertaken with the famous David Urquhart, a tour of inspection in the border areas of the soon to be independent Greece in May to July 1829 (Urquhart 1838). He was evidently appointed an attaché at the British Embassy in Constantinople soon after and remained for several years (*c.* 1829-1834) (Chesney and O'Donnell 1885: 177; Bolsover 1936: 451). He was an authority on the Ottoman Empire, spoke and read Turkish (Ross 1836) and also recorded numerous Greek and a few Latin inscriptions from visits to coastal cities of southern Anatolia (Whitehead 1998; 1999, 2014). Half a century later, David Ross's son/Robert Ross's nephew, Lieutenant (later General Sir) John Foster George Ross, of Bladensburg (1848-1926) was deeply involved in the continuing problems of the dissolving Ottoman Empire in the Balkans when he served as an assistant commissioner inspecting the Serbian-Turkish border in 1878 (Watson 1909: 98-101) and soon after in 1881 was British financial commissioner in Constantinople.

Robert Ross seems never to have married and died aged 46.

***

---

[13] Edwin's cousin, Douglas William Freshfield, a famous mountaineer, was to travel extensively in Southern Syria and the Hauran in 1868 (Freshfield 1869; Fisher 2001) and was to then engage in an acrimonious argument with this same J. L. Porter, a great man who was not pleased to be challenged by the much younger man.

Dr William Bryce
(Killaig, Co. Coleraine, 28 April 1821 - Edinburgh, 21 February 1914)

Although Bowles and Bryce will have met as the various parties passed and re-passed one another between Cairo, Suez, Mt Sinai and Aqaba, it is only at the last of these on 19th March 1854 that he is mentioned for the first time. Bryce is named when all of the westerners then congregated at Aqaba came together for a Christian service. The service was conducted by Bowles and with a lesson read by Fenton; the third clergyman, Eames, played no direct part although the service was held in his tent. As was to be the case in most of the six subsequent references, Bryce is paired with Mr Wakefield and both are described by TB as 'Dissenters of some sort'.

'Dr Bryce' is a medical doctor and a few days later he was called upon to treat one of their beduin escort for a scorpion bite (below). A month later when TB was in Damascus, he records that Bryce and Wakefield arrived, having crossed the Jordan below the Sea of Galilee and travelled on its eastern side to Damascus. Although they had had no trouble, they reported that another party of travellers, mainly natives of the region, on that same route and encamped close to them, had been robbed and stripped. A week later, Bryce and Wakefield, now accompanied by Freeman, caught up with Bowles once more, this time at Beirut. Bryce is mentioned twice more: on a ship from Beirut to Alexandria, Bowles was joined in exercise around the deck by Bryce and a few days later still, they are both now on the same steamer from Alexandria to Malta. They seem to have gone their own ways at Malta which they reached on 17th May 1854. It is a pity TB says so little about Bryce as he is an interesting man from a talented and soon-to-be distinguished family.

Bryce had been at Malta before. Jane Eames recorded in a letter home dated January 1854 and written at Qasr Ibrim on the Nile in Nubia (Eames 1855: 133-34):

> As we came down the hill, we heard the report of a gun, and on looking down the river, saw a boat with an English (sic) flag, coming up. As our boat did not return the salute, we concluded Mr. N. was not up. After we got into the small boat, to be rowed back to the dahabieh, we met the English boat, and saw two gentlemen walking on the bank. We exchanged salutations, but to my great regret, they had no newspapers, and thus we are as ignorant as ever, of what is going on in the great world. What a singular meeting this was! The two gentlemen were our fellow passengers in the Indus, from Gibraltar to Malta. There they left us, and I am sure I never expected to meet them again, when lo! who should appear, but Dr. B[ryce]. and Mr. W[akefield]!

As we shall see, Bryce and Wakefield were travelling together, and they had been on the same steamer from Southampton as the Eameses. When Jane Eames met them at St Catherine's monastery, they were travelling now with a third Briton, Mr Freeman who seems to have formed a party with Bryce and Wakefield for the Long Desert Route and then beyond at least as far as Beirut (Eames 1855: 244):

> ... we found a party had just arrived, consisting of our friend, Mr. F[reeman], and Dr. B[ryce] and Mr. W[akefield] who came out from England with us, so we went in to see them.

Later still, when the group reached Petra and most set off to ascend Mt Hor, Mrs Eames joined in (Eames 1855: 280):

With my guide on one side, and kind Doctor B., who, being accustomed to mountain climbing, politely offered his services, on the other, I went up with very little difficulty, pausing only once to rest, although the ascent occupied an hour.

Bryce and Wakefield - and perhaps Freeman, too, parted from the rest of the Petra Group at Jerusalem, and Mrs Eames never mentions them again. Fortunately, Bryce can be identified, and a reasonably full biography constructed. Indeed, his travels in the East were recorded by him in detail and the National Library of Scotland in Edinburgh contains the four volumes of his *'Journal of a Visit to the Middle East'*.

William Bryce was born at Killaig, Co. Coleraine on 28 April 1821 and died in Morningside, Edinburgh on 21 February 1914. Although he probably sounded (Ulster) Irish, he was in fact a Scot. His father, a graduate of the University of Glasgow, the Rev. James Bryce (1767-1857) (= James Bryce I)[14], was from Airdrie in Lanarkshire and his mother Catherine Annan from Auchtermuchty in Fife. The family moved to Killaig in Co. Coleraine in 1805 where James Bryce took up a position as minister of a new presbyterian church for adherents of the Anti-Burgher sect.[15] He was to die in Killaig 52 years later having seen all six of his sons born there. The parish was one of the poorest in Coleraine and one suspects much schooling was done at home and included at least one subject not likely available locally (Fisher 1927: I, 4):

From this poor home among the flats of Antrim, where Greek was the only luxury, there issued in course of time a remarkable brood of sons.

Obituary notices for William Bryce seem vague, thin, or confused but this one seems broadly correct in saying (Anon. 1914: 553):

"Dr. Bryce was brought up at Killaig, and began, at a little later age than most men, to study medicine, first at Dublin and then at Edinburgh. He graduated M.D. at the last-named university in 1853, gaining a gold medal for his thesis."

It is certainly true he graduated from Edinburgh University and not till July 1853, by which time he was 32. We may suspect after he finished his schooling he initially enrolled at (Trinity College (?)) Dublin[16] but dropped out and only returned to studying several years later, but this time at Edinburgh. Whether the gap came between the two universities or was immediately after schooling, several years of his early adult life are unaccounted for. In view of Bryce's expertise in 'mountain climbing' mentioned by Jane Eames (1855: 280) and her record of overhearing a conversation between Bowles and Bryce about Switzerland (1855: 306), it is tempting to speculate about a lengthy series of tours in Europe before commencing his interrupted medical training. Certainly, mountaineering - and a keen interest in geology,

---

[14] To reduce confusion, I have numbered as I, II and III the three successive Jameses, father, son and grandson.
[15] The timing may have been due to disputes within the Anti-Burgher church which resulted in a split the following year.
[16] His younger brother Archibald enrolled at Trinity College in 1845, aged 21.

botany and landscapes, ran in the family: his nephew James (= James Bryce III) (below), reminiscing about his Uncle William, wrote (Fisher 1927: I, 28-29):

> *The summer of 1850 included a short visit to the Lake Country where my Uncle William was then living at Kendal.[17] We ascended Helvellyn for the first time and were delighted with Windermere, and Grasmere, and Rydal, and the view from the top of Scouts Scar to the west of Kendal. The beauties of the Lake Country made a profound impression on me, and Wordsworth's house was pointed out where he had died that year. It was the first thing that brought into my consciousness the great poet who has been so much to me ever since.*

And later (Fisher 1927: I, 32-33):

> *My Father [= James Bryce II] enjoyed it [= the island of Aran in the Clyde Estuary] especially because its geological features were so interesting; and I enjoyed it specially because my Uncle William Bryce, who was then studying medicine in Edinburgh, had taught me the elements of botany and set me to collecting plants and forming a herbarium. He had me to stay with him for some days in the May of (I think) 1853, though perhaps 1854, and insisted on my accompanying him on botanical rambles in the country round Edinburgh. I remember being a little reluctant the first day, but by the end of it he had got me interested, and on the remaining days I was as keen as he could wish. Few people realise what an enormous difference to the life of a boy or a girl may be made by the implanting of a taste at the right moment, and by persevering even if the first seed sown does not seem to be taking root. My uncle took me out a second, a third and a fourth day on short excursions, and by the last day the seed had taken root so deeply that the taste has remained with me ever since and constituted one of the chief pleasures of my life. It fell in happily with the passion for mountain climbing and exploring every new region where one happened to be, and from that time on I never saw a mountain or a wood, or a common, or a river bank without searching for uncommon plants; and even on railway journeys I tried to catch (when the speed was not too great) the plants that grew along the line. Without those four days I should never have taken to botany, never have formed the habit of closely observing nature, and should have lost half the pleasure of foreign travel.*

A lengthy obituary for William's son Thomas Hastie Bryce (1862-1946), also a medical doctor, includes the following:

> *"It is worth noting the early history of Thomas's father, Dr William Bryce. After taking his degree at Edinburgh, he set out for the Crimea to join the Army Medical Service, but on his arrival at the front learned that peace had been declared. So he assumed the dress of a Syrian and spent over five years travelling and observing in Syria and in Palestine, earning the friendship of the natives by giving medical advice. He returned to Scotland in 1861 and set up in practice in Dalkeith; later he transferred his practice to Edinburgh where he made a reputation as a homoeopathic physician."*

The dates do not square up - he graduated in July 1853, the Crimean War began 3 months later. And ended in March 1854. We know from his own *Travel Journal* that he set off from Edinburgh on 23rd August 1853, collected Mr Wakefield, the young man for whom he was to be companion and medical support on the planned tour, and can be traced then through the journal with occasional corroboration in the pages of both Jane Eames (above) and Bowles.

---

[17] Perhaps engaged in research: this medal from Edinburgh University was one of the four medals awarded that year for a dissertation on 'The Medical Topography of the Lake District of the North of England.'

The latter part of Bryce's *Travel Journal* is incomplete after he reached Jerusalem but - as noted above, TB's Journal records him in north-western Jordan, Damascus and Beirut and then on the same ship as Bowles from Beirut to Alexandria and then records him as a fellow passenger on the next leg of the journey by sea from Alexandria to Malta which they reached on 17th May 1854 and seemingly parted. TB continued to the UK, but we don't know where Bryce was between May 1854 and the signing of the Treaty of Paris ending the Crimean War on 30th March 1856.

If we accept the notices of his life written many years later (above), he had spent the five years 1853-1858, travelling in the Ottoman Empire. That is accurate as far as the first year is concerned. However, on 29th July 1857 he was in Scotland where he married Bessie Darling Hastie at Dalkeith, Midlothian - he was 36 and she was 25; their first child was born on 23rd December 1860.

A possible solution is that Bryce dropped off Mr Wakefield at Malta and continued his travels through the Balkans and Ottoman Empire before offering his services for the war. Then home in 1856 and marrying the following year. The earlier trip with Wakefield and a new trip after Malta might come to 3-4 years in total.

The Censuses and birth records give the names of six children:

- Mary Darling Bryce = Dalkeith, 23rd December 1860
- Thomas Hastie = Dalkeith, 20th October 1862 - 16 May 1946
- Catherine Annan = Dalkeith, 29th May 1865
- Bessie Darling = Dalkeith, 12th November 1867
- William Hastie = December 1870 (?)
- Reuben John = Edinburgh, 28th July 1874

In the Census of 1911, asked for details of the number of children born live, the Bryces reported eight.

He seems to have initially practised medicine in Dalkeith - where he acted *'also as surgeon to the Duke of Buccleuch's Militia there'*, but by 1868 he was settled in Charlotte Square in Edinburgh where he remained in practise.[18] As it happens, the Royal College of Physicians of Edinburgh has two items of William Bryce's which their entry summarises as:

> *British Passport of William Bryce, MD given in London on the 10 October 1853 stating intent to travel to Egypt, Syria, and Turkey. Also included is a small homeopathy case with small glass vials of various homeopathic remedies, possibly also belonging to Bryce but unknown. Items were in envelope labelled 'Dr Williamson'.*

It is likely the *'homeopathy case'* mentioned above is Bryce's. The obituary in the *British Medical Journal* in 1914 notes that (Anon. 1914: 513):

---

[18] One of his brothers, Dr Archibald Hamilton Bryce, ran the Collegiate School on the same square.

Dr. Bryce had leanings to homoeopathy; he laid great stress upon the importance of thoroughly mixing the constituents of prescriptions, and he was ever ready to take a critical position in respect to drugs (such as calomel) which were enjoying what he regarded as too high a degree of popularity.

Sixty years before, Bowles explicitly says that when Bryce was called on to treat one of their beduin escort on the Long Desert Route for a scorpion (sic)[19] bite, the man:

"... was taken to Dr. Bryce at once - who lanced it, put on a tight bandage up the leg - & got one of the Arabs to suck the wound. He them rubbed in sweet (?) oil - applied caustic - & gave the man a dose of Calomel. Strongish measures for a man who had never tasted anything stronger than coffee."

The source of his attraction to homeopathy may have begun in his student days at Edinburgh. One of those he would have encountered there was William Henderson (1810-1872), Professor of Pathology.[20] The 1840s was a decade in which homeopathy was attracting adherents including Henderson. It led to bitter attempts to displace him from his position with one of his most implacable opponents dismissing homeopathy as *'a system of consummate charlatanry'*. Henderson weathered the storm and just two years before Bryce graduated, it was said that of the 48 graduates of 1851, at least eight openly embraced homeopathy.

William Bryce's talents and intellectual interests were wide-ranging: mountaineering, geology, botany, homeopathy, and travel. The latter is self-evident, and his journals are filled extensively with observations on the rocks and mountains encountered.[21] On a lighter note, his *BMJ* obituary noted that, *'He was a keen golfer, curler, and bowler.' 'The cause of death was acute pneumonia'* (Anon. 1914: 553).

Bryce was one of seven brothers, all of whom made their mark in various professions and all of whom lived to advanced ages (Fisher 1927: I, 8):

Ulster Scots are proverbially a tough race, but surely few even among the stalwart families of Ulster can claim to eclipse the Bryce record for longevity. The subject of this biography [= James III] died with his zest for life still unsatisfied and insatiable at eighty-three. His mother lived to within three weeks of ninety. His father showed every sign of a long career when he was killed at a little over seventy in what appeared to be the bloom of robust manhood. His uncle Reuben John failed to reach ninety by three months. His uncle William, botanist, physician, and golfer, died at ninety-three, his uncle Archibald at eighty. His aunt Catherine lived to be ninety, his grandfather, the Minister of Killaig, was active at ninety and preached two sermons on the Sunday before he died.

James Bryce II, William's brother, a renowned geologist, alluded to above, only lived to *'a little over seventy'*, being killed by a rock-fall while *'geologising'* near the shore of Loch Ness. His son, the subject of the biography by Fisher, James Bryce III, William's nephew, was an outstanding figure of late Victorian/Edwardian times: prolific author of books on subjects as wide ranging as from *The Flora of Aran* through *The Holy Roman Empire* to *Transcaucasia and Ararat* and official reports on the Ottoman treatment of the Armenians and German war crimes in Belgium

---

[19] In his diary, Bryce says the bite was by a 'serpent' and he describes his treatment in detail.
[20] I am grateful to Jenny Dawe, Editor of *The Grange Newsletter*, for drawing my attention to her article (Dawe 2019) on Henderson who is buried in Grange Cemetery in Edinburgh and the underlying publication by David Boyd (Boyd 2005).
[21] Bryce's brother.

during the First World War. He was Regius Professor of Law at Oxford, Chief Secretary for Ireland, Member of Parliament for 27 years and Ambassador to the USA (1907-1913). After the last of these, he was elevated to the House of Lords as James, Viscount Bryce of Dechmont.[22] Almost exactly 60 years after William Bryce had been at Petra, this nephew, and his wife Marion, would be amongst the last visitors to Petra in late April 1914 before WW1 broke out with Turkey on 5th November 1914. Their immediate predecessors at Petra in February that same year were the redoubtable Lady Evelyn Cobbold and (not yet famous) T. E. Lawrence.

***

John Edward Wakefield
(Kendal, 8 August 1830 - Great Malvern, 30 July 1858)

Very little can be said about Wakefield from the pages of TB's journal. As noted in discussion of Bryce, he is mentioned just three times: along with Bryce at Aqaba (both described by TB as 'Dissenter'), and then with Bryce again at Damascus and Beirut; Jane Eames recorded meeting them both on the Nile and recognised them as fellow passengers from Southampton to Malta, then again at St Catherine's monastery but not thereafter (above). Finally, the transcription of the Shepheard's Hotel Register includes under early December 1853:

William Bryce, M.D.)   to Upper Egypt
John W. Refield.)

Andrew Oliver (pers comm.) plausibly suggests this is 'John Wakefield'

As we now know from Bryce's *Travel Journal,* he and Wakefield had travelled together from the latter's family home in Cumbria to Malta where they split from several of Wakefield's family travelling with them and went onwards through Greece and Smyrna and finally - for the benefit of Wakefield's health, to Egypt. They had cruised on the Nile then joined with Freeman to form a party for the Long Desert Route and gone on as a party through Palestine, to Damascus and to Beirut. Bryce's *Journal* does not cover this last section so there is no reference to what became of Bryce who certainly continued onwards to Malta at least.

Armed with the fuller information than available in TB's journal, a brief biography can be set out. John Edwards Wakefield was the third of William and Susanna (née Birkbeck) Wakefield's four sons. The Wakefields were a landed and banking family in Kendal (Westmoreland/now Cumbria) where, together with another local family, the Crewdsons, they ran the local Kendal Bank. A daughter of this latter family - Rachel Crewdson, had married Henry Fox, a woollen manufacturer in Wellington (Somerset). The Crewdson and Fox families were both Quakers. The four brothers took different directions: one into banking, another into agriculture. John Edward went into medicine - or rather, he undertook study and appears in the Scottish census of April 1851 as *'student of medicine'*, presumably at the University of Edinburgh. His address is at 9 Drummond Place, a large Georgian house in the Edinburgh New Town which he is sharing with his mother and two younger brothers.[23] This is probably where Wakefield met William

---

[22] Derived from Dechmont Hill in Lanarkshire, from which the family hailed in the 18th century.
[23] An interesting feature of this house concerns the later owner, 'an advocate called *Charles Scott* [d. 1892, aged 65]. A classical scholar, he had almost every room in his house decorated to look like the walls of a Roman villa'. Only one of

Bryce who had completed his medical studies at Edinburgh where he graduated in July 1853, awarded one of the four medals that year for - perhaps significantly, a dissertation on *'The Medical Topography of the Lake District of the North of England.'* In short, Bryce and Wakefield were students of medicine at the same period although the latter evidently never completed. Bryce's *Journal* shows him beginning the preparations for his grand tour on 23rd August 1853 just a few weeks after graduation and recording in his first sentence that he is fulfilling an undertaking to accompany an invalid to Egypt. As noted above, Bryce's Passport application specifies travel to Egypt, Syria, and Turkey.

Bryce's preparations included a short visit to Kendal where he stayed with the Wakefield family - apparently the oldest brother, William Wakefield, 'of Birklands [Old Mill]' on the outskirts the town. Later he records arriving in Southampton where he met the travel party consisting of Wakefield's father and his younger brother George Henry Wakefield. Also there were Mr and Mrs Henry Fox and their daughter Rachel Crewdson Fox and Charles Henry Fox, presumably a cousin. Jumping ahead to beyond the grand tour, John Edward Wakefield married Rachel Crewdson Fox on 13th September 1854 in Wellington, Somerset. She was 20 and he was 24. Wakefield died young - not quite 28, on 30th July 1858 while at Great Malvern - presumably taking a health cure at the spa there. He and Rachel had a son and daughter both of whom married and had children

When Bryce and Wakefield left Southampton on 22nd October 1853 they were 32 and 23 respectively. More will surely be available from Bryce's *Travel Journal* when it has been examined more closely and issues of copyright have been resolved.

***

'Mr Freeman'

This is the last of TB's Petra Group and the most opaque. Bowles refers to 'Mr Freeman' eight times; Jane Eames frequently mentions 'Mr F' but only twice does she mean Freeman rather than her fellow-American Mr Fish; Bryce has numerous references to Freeman both on the Nile and later as part of the group to Petra. Not once does any of them give him a first name and even his British nationality has to be inferred. Now we can add the evidence of the Austrian diplomat, Baron Andrian-Werburg.

Putting together the various references, 'Mr Freeman' is listed as a passenger sailing on the steamer *Indus* from Southampton to Alexandria on 20 October 1853 (*Allen's Indian Mail* 1853: 659). The same listing includes 'Mr and Mrs Eames' and Jane Eames noted that when she later met fellow passengers at Mt Sinai (Eames 1855: 244):

> ... found a party had just arrived [at St Catherine's monastery], consisting of our friend, Mr. F[reeman], and Dr. B[ryce]. and Mr. W[akefield] who came out from England with us, so we went in to see them.[24]

---

the rooms survives in that form but it is very impressive. https://holeousia.com/2020/11/04/he-had-almost-every-room-in-his-house-decorated-to-look-like-the-walls-of-a-roman-villa/

[24] That is misleading as we know from Bryce's *Journal* that he and Wakefield sailed on the *Tagus* and only transferred to the *Indus* at Gibraltar where they became fellow passengers with the Eamses and Freeman.

In Egypt he stayed at Shepheard's Hotel which records a set of names who all left the hotel on 13th December 1854 for a cruise on the Nile:

William Bryce M.D.
John W. Refield (sic) = John Edward Wakefield
... Andrian
... Freiman (sic)
Fletcher

'Freiman' is interpreted as Freeman and supported by Bryce's note that Freeman subsequently shared a boat on the Nile with a German Baron. 'Andrian' in the list above is Viktor Franz Freiherr von Andrian-Werburg (1813-1858). Happily, Andrian kept an extensive journal and quite recently it was published as a transcription and with a commentary including his months in Egypt.

Unhappily, despite now having the journal of a man who spent two months sharing a Nile boat with Freeman, we still do not get a forename. He first appears on 1st December 1853 at Shepheard's Hotel (Andrian 2011: 2, 693):

I've probably already found a travel companion, a Mr. Freeman, a very well-mannered-looking young Englishman, who lives here in the hotel ...

These two then undertook some local tourism together around Cairo and finally left on 13th December 1854, joined now by a second Englishman (Andrian 2011: 2, 696):[25]

... I've gained a second travel companion, a M. Fletcher, a young, very well-bred-looking grass devil and lieutenant, an acquaintance of Freeman's, ...

This the Fletcher in the Shepheard's register (above). There may be some way of identifying Freeman by first identifying Fletcher, *'his acquaintance'*. Andrian is of assistance as he later records (Andrian 2011: 3, 35):

There were a lot of boats, about 15, English, American and French, including Kennard, his friends Wilmot and Stephens, a cousin and namesake of Fletcher and 2 of his friends in the Coldstream Guards, Bouverie and Tower, young London Elegants in a magnificent boat.

Bouverie and Tower, both officers in the Coldstream Guards, were soon on their way to the Crimean War where the former was killed later that year. Andrian later differentiates between the Fletcher cousins with the one on the other boat being Fletcher Senior - presumably older. A search produces an 'H. C. Fletcher' sailing from Southampton on 20th November 1853 on the steamer *Euxine* bound for Alexandria (*Allen's Indian Mail* 1853). Fletcher Junior is explicitly said by Andrian to be a lieutenant and it is likely both Fletchers were officers along with Bouverie and Tower. Fletcher Senior cannot yet be identified - he is last encountered in the Andrian diary as a fellow-passenger from Beirut to Smyrna along with Bouverie and Tower, but the last two continued to Constantinople to their regiments and the war while Fletcher

---

[25] Andrian records that he briefly considered other potential co-passengers on his large boat, including the French traveller Charles Didier but describes him as 'indecisive, stingy, and at the same time inclined to become pompous' (Andrian 2011: 2, 693).

is explicitly said not to have done. Fletcher Junior, Freeman and Andrian's companion on the Nile is surely Henry Charles Fletcher (Marylebone, London, 28th April 1833 - Spencer House, Putney, 31 August 1879). He was the son of of Major-General Edward Charles Fletcher from a family hailing recently from Kircudbrightshire in Scotland. In December 1853 - mid-February 1854 Fletcher was not quite 21 having been in service as a lieutenant in the Scots Fusilier Guards as early as 1851 when he was 17.[26] Despite Andrian's earlier characterisation of both Freeman and Fletcher as 'well-bred' (above) - even using this English term in relation to Fletcher, by the time he was back in Cairo on 15 February 1854 he was to have formed a less favourable view after two months in their company (Andrian 2011: 3, 45):

> *The peace of life here does me good after the two-month trip to the Nile, but especially that I am finally alone and in control of my movements. I did get along fairly well with my two traveling companions, but they were extremely insignificant people with whom there was absolutely nothing to talk about but empty stuff. Freeman a man for whom the word snob seems to have been invented, disagreeable and vulgar in many respects, Fletcher a young, insignificant, ignorant fellow, yet not possessing even the good nature and good humor of his age. If I were to make this trip again, which I'm sure will not happen, I would probably do it alone, but otherwise I would at least do it with Englishmen. People of other nations, unless they are very agreeable, are utterly obnoxious.*

One may suppose that Andrian himself is something of a snob. The problem is more likely to be that the 40-year old Baron had little in common with two very young Englishmen. He says as much in relation to Freeman elsewhere in his diary implying again that Freeman was young and inexperienced. So, still no name and clear identity but there is the likelihood Freeman was in his very early 20s in 1853-54, making him younger even than Wakefield (23.5) and Yeatman (22.5) and perhaps explaining why none of the rather older diarists found much to say about him.

On his voyage with Andrian on the Nile he joined the latter on some shore excursions but one enigmatic one stands out. While on the Nile at Aswan on 11 January [1854] (Andrian 2011: 3, 18):

> *In the evening Freeman and I went on a voyage of discovery into amorosis (sic), and had a peculiar meeting with two black beauties.*

After returning to Cairo, he joined up with Bryce and Wakefield to cross Sinai to Petra: they seem to have arranged three tents: one for Bryce and Wakefield, one for Freeman and one a shared meal tent; they also seem to have shared a dragoman. After Jerusalem, Freeman - having originally asked to travel with Bowles and Kennedy, evidently continued instead with his Desert companions Bryce and Wakefield: all three seem to have gone together to Damascus, then are met as a group in Beirut on 5th May. His last mention is on 9th May in TB where they are together on the steamer from Beirut to Alexandria.

Without a first name it is not currently possible to say more. Nine Freemans are recording applying for British passports in 1853. After eliminating the woman and the doctor and two who were only applying after we encounter him on the steamer from Southampton, we are left with possibilities: Charles, Charles Edward, Edward B, Alexander Daniel and William Henry.

---

[26] Census of 1851 has him resident in the home of his uncle, James Fletcher, an 'East India Agent.' He had apparently enlisted as an Ensign in November 1850.

Chapter 9

Thomas Bowles at Petra

"Oh master," said Giorgio (who is prone to culinary similes), "we have come into a world where everything is made of chocolate, ham, curry powder, and salmon"; and the comparison was not far from an apt one. (Lear 1897: 422)

Petra lay near the mid-point of the journey from Aqaba to Hebron, the parts divided into usually four and five days respectively. How much time was spent in Petra varied not just with what was contracted for but what the inhabitants of the valley and the villagers outside would permit and the willingness of their Alawin escorts to support them. In TB's case, his group evidently contracted to spend a full day on the site but were able – very unusually, once there to negotiate a second day.

TB's Party reached Aqaba late afternoon on Saturday 18th March 1854, after the Ross Party - which seems to have arrived the day before, but ahead of the other five parties, two of which he is pleased to note had set off from St Catherine's before them. Even after all the parties had assembled there was a delay. The Alawin sheikh who was to take over and provide escort, camels and guidance for this next stretch, was in the desert - 24 hours distant they were told. The sheikh would have been in no hurry - as always, he would have been well-aware how many parties were strung out along the route from Suez and would wait till all had arrived and he could assemble a single group.

In the event, TB's Party had four full days at Aqaba after the day of arrival then finally departed on the morning of the sixth day, Thursday 23rd March. Sheikh Hussein (below) arrived on Monday 20th leaving the assembled group to get to know one another, make the most of the location on a beach for bathing, and to explore what there was in the tiny settlement and its vicinity. For the first time they could identify as a unitary group - eight Americans and eight British, including two women (both American) and three clergymen, all but one (Rodewald) being native English-speakers.

Accounts by travellers in other years of their enforced stay in Aqaba often include detailed information about the negotiations with Sheikh Hussein. TB says little about the negotiations: the travellers were all invited to join the sheikh and his son and there was indeed some wrangling but soon it was agreed to leave the dragomen to conduct their affairs.[1] Oddly given his concern for costs, TB never records how much they were to pay Sheikh Hussein. Likewise, while at Petra itself the group experienced almost none of the harassment or even importuning reported by so many other visitors (below).

After the long trek from Suez, often far from water, TB records bathing in the sea at Aqaba at least once each day. He used the enforced leisure also to write up his journal and to take long walks

---

[1] This is Dean Stanley who had been at Petra the previous March (1856: 85 n1): *'I have purposely omitted all account of the often repeated, though to those concerned always interesting, negotiations with the old chief himself at 'Akaba'.*

and conversations - most commonly as on the way from Suez, with one or both Fentons, but usually the younger, a fellow-clergyman. He also had a long conversation with the self-effacing James Eames and seems to have been surprised to learn he, too, was a clergyman. Although he surely visited the Castle, he never mentions it - not even that their arrival was greeted by a 'salute' from Eames's revolver and another from the governor at the castle as reported by Bryce. We depend on Jane Eames for an informative and amusing account of the visit she made jointly with Mrs Rogers to a meeting with the Governor's wife and the Governor himself (above, Ch. 7).

TB spends a great deal of time exploring further afield. Although he would have been well-aware that this was the supposed location of Old Testament Ezion Geber, unlike Jane Eames he never mentions it nor any of the traces of ancient settlement other western travellers noted. Instead he climbed up and round the rugged mountains on the eastern side of the settlement.

The group finally left Aqaba on Thursday 23rd March 1854 under the guidance and escort of Sheikh Hussein of the Alawin (below). It was an impressive cavalcade - Bryce records that there were 70 camels carrying travellers, servants, escorts and baggage. When they camped, we may calculate for the travellers probably 15-25 tents for sleeping and cooking.

The Alawin[2]

All western travellers arriving from Cairo had to go via Aqaba,[3] the little settlement at the head of the Gulf of Aqaba - the eastern prong of the Red Sea. The Sinaitic beduin - the Towarah, who could guide and escort travellers that far were then obliged to hand over their clients to a new tribe, the Alawin who are heard of as early as Pococke in 1738 (Pococke 1745). It was a recognised feature of tribal societies that one tribe would not pass through the territory of another - or only with permission and by paying a fee of some kind. At Aqaba, the Alawin were not interested in a fee alone for permission, they wanted total control of the next stage, not just permission to pass but provision of their own camels and drivers, escorts and guides. Numerous western travellers who had little negative to say about the 'gentle' Towarah beduin who had conducted them from Cairo, were shocked by the wild and often aggressive conduct of the Alawin, but at times also admiring and impressed. Several described their appearance: hair in long locks, sun-blackened, often ragged and immensely proud of their camels.

They were proud and warlike, wily in negotiations, and frequently accused of brazen duplicity: formal contracts were often brought up for re-negotiation before the caravan set off or along the way; solemn promises to some parties to enter Petra from the east via the Siq were routinely and brazenly broken; and the number of days spent in Petra was often cut short or spoiled by the Alawin trying to urge or frighten the travellers to move on. Personifying the tribe was, for a long generation, their leader, Sheikh Hussein (below) who alternately appalled or impressed (despite themselves) the travellers.

---

[2] Bowles spells the name of the tribe 'Aloines' and I have noted no less than ten alternative spellings in reports by other travellers to Petra. 'Alawin' is the preferred form in most accounts then and now. There is a very useful modern literature on the Alawin and Howeitat and on the people of Petra (e.g. Canaan 1930; Lewis 2004; McKenzie 1991; Miettunen 2008; Van der Steen 2013). Much of this is based on the published reports of western travellers in the 19th century and may be supplemented and refined now by the increase in availability of unpublished journals - a task for another day.

[3] A tiny number diverted from St Catherine's Monastery to Nakhl on the Haj Road from Cairo to Aqaba and attempted to go from there but fewer still were successful.

The Alawin were one of three main components of the large Howeitat tribe (below) whose territory extended from Al-Wejd, on the Red Sea coast of the Hejaz, *c.* 400km south-southeast of Aqaba, to the region just beyond Petra, *c.* 150km to the north-northeast. The tribe was not under a single authority but marked by alliances and fractures, and constant shifts. The Alawin component, for example, had the monopoly of the revenue from conducting western travellers and that was regarded with envy and sometimes more, by those Howeitat who lived southeast of Petra, presided over by a sheikh in this period who was the cousin of the Alawin sheikh, and could block or demand a fee if the westerners tried to enter Petra from the east, down the Siq. Quite simply, tribes and sub-tribes, believed their relatives in other branches owed it to them to share at least some of their good fortune. Hence the insistence of the Alawin to normally approach up the Wadi Arabah and into Petra from the west avoiding their cousins to the east. The animosity and dissension of these two components was no less serious because their sheikhs were kinsmen for much of the mid-19th century (below).

Rationally, the Alawin would have contracted to bring westerners into Petra from the east, down the Siq. They would have shared their fee with their cousins of the Shera'a region southeast of Petra and alerted the *fellahin* and *bdul* to their approach and again share the fee. That might well have involved the westerners paying a larger fee, but they would have had the enormous benefits of security, freedom from harassment and threats, approach by the most stunning route and ability to stay as long as they wanted. The Alawin, however, were often more interested in keeping their fees to themselves and westerners were were complicit because keen to keep fees as low as possible.

The difficulties were made worse by the relationship of the Alawin with the people who resided in and near Petra. Although the Howeitat laid claim to this immense swathe of territory, they, too, had to conduct delicate relations with various groups within that area, in particular for present purposes, the people who lived *inside* the ruins of Petra and the farmers/semi-farmers - the so-called *fellahin*, who lived in seasonal villages at the springs nearby, principally at Wadi Musa/Elji, still so-called and location of the modern town, tourist hotels and Visitor Centre.

The relationship might vary depending on the relative strength of the different parties and their sheikhs at any one time. Contracts drawn up at Aqaba commonly stipulated that the monies paid to them by the westerners included that they would undertake to pay the inhabitants in and near Petra for the right to enter their territory. A common theme, however, was the attempt of the Alawin, to evade the payment and keep it for themselves. To that end they tried to avoid the inland route to avoid their Howeitat cousins and to avoid the eastern entry to Petra where they could not help being seen by the *fellahin*. They almost invariably entered from the west where they might arrive unseen by the inhabitants if they were present and escape longer detection by the *fellahin* in Elji/Wadi Musa a few kilometres beyond the Siq entrance or those in the small villages to the north such as Debdebah.

Dominating the Alawin and the man most encountered by western arrivals at Aqaba for several decades, was Sheikh Hussein Ibn Injad. All were struck by his confidence, cunning and easy duplicity. At least one sketched him (Formby 1842) and another had him photographed (Figure 9.1).

The characterisation of him by Lord Castlereagh who met him in June 1842 is one of the fullest (Castlereagh 1847: II, 34-44 passim):

Akaba is a wretched village situated on the sea-shore, exactly at the opening of Wady Araba. When we reached our tent, and were arranging matters, the Governor, or Nazir, marched in, and sat down with half a dozen of his attendants, and with him came the notorious Sheikh of the Alouins and his son Mahommed. They are both remarkable people. The Sheikh dark and scowling, with pointed features and restless eyes; the boy very intelligent and distinguished.

… The Sheikh having requested a second interview, I received him and his son in my tent, making them sit on each side of me, and desired that no one else should enter. After pipes and coffee, the Sheikh quietly asked when I meant to go, …

Figure 9.1: Engraving from a photograph taken in Cairo in 1855 (Prime W. 1855: 96).

… While this debate was going on, it was curious to watch the play of Hussein's face, and the restless eye which never for a moment was fixed upon an object. He looked at everything in the tent, but never in my face. The young chief, who was on my left, watched everything I did, never took his glance from me, and evidently was as deeply interested in the matter as if he had been thirty years older. Mohammed's countenance, without being handsome, is very remarkable, and one not easily forgotten; possessing singular brightness about the eye, and a sinister smile. I heard him laugh once, but it was a joyless sound; the echo of another's hilarity rather than a demonstration of his own. His figure is perfect; tall, slender, and very muscular. We had seen him in the morning as he was going into the sea to bathe; his red caftan falling off his shoulders, and his long tresses of hair, like a woman's, hanging over it. After another ineffectual attempt on his part to induce me to accede to his proposal, another Sheikh, by name Aboudjazi, made his appearance, and took his seat in the tent in a manner very different from the well-bred coolness of Hussein.

Here we see three of the key players: Sheikh Hussein of the southern Howeitat, the Alawin; his son and ultimate successor, Mohammed; his cousin, Sheikh 'Arar ibn Jazi of the Howeitat group north of the Alawin and east and southeast of Petra. A third division of the Howeitat further north, rather more opaque, are the Abu Rashid branch.[4]

The Inhabitants of the Ruins

Western visitors were seldom interested in the identity of the people they encountered at Petra itself. They usually knew the names of tribes: the Towarah in Sinai, the Alawin bringing

---

[4] There is now an extensive article (in Arabic) on Sheikh Hussein (Abudanah 2016).

them from Aqaba, the Tiahah south of Hebron. But at Petra the people encountered were often described by the colourless terms 'Arabs' or 'beduins'. Indeed, they were often unaware they not only had a name but that there were two distinct groups. Most immediately were those who inhabited the actual ruins, living in some of the caves/tombs. These are the *Bdul*, whose descendants were moved out of the ruins in 1985-86 and now occupy the new, purpose-built village of Umm Sayhun ('Bedul Village'), 2km northeast of the Colonnaded Street.

The *Bdul* are an enigmatic group, poor and living a simple existence. What little is known of their practises can be summed up by Canaan writing some 75 years after TB's visit (1930: 80):

> The Bdul *rarely leave their district. In the winter season they spend between two and three months in the caves of Petra; in the spring they encamp around the wadis, while the summer is spent at the tops of the high mountains of Petra or on one of the surrounding ridges. A few of them live in el-Beda. They are the poorest of all the bedouin tribes of this district of Trans-Jordan. Their children mostly run about quite naked.*

In short, as most western travellers were at Petra in the March-May period, they would not normally have encountered the *Bdul*.

In contrast are those inhabitants who - when they are defined at all by more than 'Arabs', are called by the Arabic term *fellahin* (pl. of *fellah*), peasants, farmers engaged at least in some part as tillers of the soil, in contrast to the beduin who were herders of animals. Of course, like peasant farmers everywhere, they often also had their flocks but it was the fact of their growing of crops and hence fixture to a place that distinguished them.[5]

At Petra *fellahin* were found not *inside* the ruined city itself but nearby in some small villages - though that term is misleading. Closest and the one referred to by several visitors, was Elji (see Figure 9.2),[6] the modern Wadi Musa - or rather the small village that lay on the southern side of the wadi opposite where the modern town lies on the north side. Normally western visitors only encountered these *fellahin* if they arrived from the east - a rare occurrence (above), or if they went up the Siq then followed the tombs along the widened part of the wadi up towards Elji. However, the *fellahin* were also sometimes encountered inside the ruined city in the form of shepherds or children tending flocks, who then alerted the adults in the village. In addition to Elji there was a second village of *fellahin* at Debdebah, *c.* 6km northeast of the central ruins - more distant but easier of access. There were other small villages further north still, but these were the two with whose inhabitants western visitors commonly came in contact.

These *fellahin* seem not to have resided in the stone houses at Elji or Debdebah; the buildings at the former are described as intended for storage with just a handful of people living there. The people were semi-nomads but within a restricted orbit. Several travellers refer to them being in permanent camp on the hills just north of Elji, perhaps shared with the *fellahin* of Debdebah. Indeed, the *fellahin* are said by several visitors to be divided into four sub-tribes

---

[5] Not immediately relevant at Petra but notable are those people who were semi-nomads - annually going in winter into the nearby desert with their flocks (but not, principally, camels) and returning in summer to the place where they had sown seeds.
[6] Even in the 19th century some travellers referred to the village as Wadi Musa rather than Elji.

Figure 9.2: Elji as recorded on a photograph by Charles Hornstein from a visit in 1895 (Courtesy Palestine Exploration Fund).

(whose names were recorded many years later by Canaan (1930: 61)), each with its own sheikh and perhaps spread across two further 'villages' beyond Debdebah.

Travellers frequently describe these *fellahin* as fearsome in appearance and behaviour, threatening and hostile, and it is easy to imagine that when travellers entered the ruins it was most likely some of these people they encountered, armed and aggressive, rather than the distant and impoverished *Bdul* who are more commonly characterised as ragged and near naked, and possessing few weapons. Some of the *fellahin* sheiks were named - indeed, infamous, although several earlier commentators were under the impression that Muqabil Abu Zaitoun and his successor Suleiman Abu Said, were *Bdul* rather than Liyathneh (below).[7]

The numbers of inhabitants were quite considerable. In 1843, Rev. Dr John Wilson was told by the 'sheikh of the *fellahin*' that (1847: 331):

> There are 500 of us able to bear arms under Sheikh Suleiman [= Wadi Musa], and 500 under Sheikh Aubed [= Debdebah].

---

[7] Lewis 2004: 14, n. 14 *'Some recent writers, as well as some of the 19th century, travellers have described Abou Zaitoun as sheikh of the Bdoul. A careful reading of almost all the 19th century accounts has led the present writer to conclude that this was not the case and that Abou Zaitoun was in fact the sheikh of the Liathna, as stated above.'* My own reading - now supported by a wider range of contemporary writings than available to Lewis, had already led me to the same conclusion.

The numbers are suspiciously round and similar, and imply - once families are included, a settled population of 2000 and more in each place. The known ruins at each place would suggest rather smaller numbers but the extent of the stone structures is misleading. The nature of the 'villages' is unexpected: even Elji is said to be just 20-30 stone buildings and used largely for storage rather than accommodation and the ruins at Debdebah as seen today are similar in extent.

An indication of the actual scale of population in these places is provided by the poet and painter Edward Lear who had a very difficult visit to Petra in 1858. His visit was undertaken from Hebron and had evidently not involved the *fellahin*. They soon mobbed and manhandled Lear's servants and even him. As his visit progressed more and more angry *fellahin* arrived (Lear 1897: *passim*):

> *What a scene! Groups of nine or ten Arabs, in all upwards of one hundred in number, are around the tents; ... On regaining the ruined terraces above the stream in the valley, I was sorry to find nearly double the number of Arabs I had left there gathered round the tents, not fewer I suppose than two hundred in all. ... so I was compelled to stand still while they rushed by me singly to the number of one hundred and fifty or thereabouts, on their way to their brethren at the cave's mouth. ...*

These are more realistic numbers - several hundred people in total between these two villages. A further check is provided from 1870. Edward Palmer visited Petra in 1870 and was taken past the stone buildings of the village of Elji to the *fellahin* camp on a hill an hour and a half to the north - that might be *c*. 5-6km, to a place he says is beside ruins which may be those of Debdebeh which is almost 6km north of Elji (Palmer 1871b: 52-53):

> *Passing the village of Eljí, we ascended the hills for an hour and a half, and at last reached the camp of the, Liyátheneh, which consisted of about a hundred tents arranged in a great square, with a double row forming a street on either of two of the sides. As we neared the place we were met by a party of men from the camp, the Sheikh Silmán [= Suleiman Abu Said?] amongst them, ... In the mean time a great and hideous row was going on between old Hamzeh and the sheikhs of the fellahín about the amount of black-mail which we were to pay. Towards the end of the afternoon we took a walk to the eastern end of the camp, where, at the head of a valley, is a well, and the remains of a ruined village called 'Ain and Khirbet D'háah [= Debdebeh?].*

Once again, the numbers of tents suggest a total population for *fellahin* in the two principal villages - perhaps in the four sub-tribes as a whole, at several hundred.[8]

The *fellahin* had a name - *Liathneh*[9] and an identity. A few visitors, who took the trouble to examine and describe them, characterised them as being different to the beduin Alawin. Despite the shock the Alawin presented to travellers (above), those who encountered the *fellahin* found them yet more savage in character and thought them different in appearance and manner to the beduin. One traveller thought even the Alawin sheikhs were fearful of the *fellahin*.

---

[8] Canaan 1930: 65 computes precisely for the late 1920s, four sub-tribes totalling 470 armed men and 61 horsemen.
[9] And variant spellings: Liathna, Liyatheneh, Lyathene...

Figure 9.3: Muqabil Abu Zaitoun (Sheick Yomgebel Abouseeton) (Formby 1843: Title page).

Wilson was told they did not inter-marry with the 'Arabs' (i.e. beduin) and concluded, too, their names were distinctive to themselves (1847: 331). Speculation - fostered to some extent by the *fellahin* themselves, concerned whether they were descended from a Jewish community, from the Nabataean-Roman inhabitants or from the Crusader settlers.

The characterisation of the better-known sheikhs of these Liathneh, is generally unfavourable. Ruthless old men, often marked out by a red robe. The two best-known in this period were Muqabil Abu Zaitoun (Figure 9.3) and his successor Suleiman Abu Said (above).

Thomas Bowles inside Petra

TB was very fortunate in his timing and encountered no impediment to extensive tourism - just three years later a succession of westerners were treated with considerable hostility, indeed, violence and gunfire (Kennedy 2018). It might have been different and TB records that:

The Fellahhines have come down in great numbers & there were violent debates. The Dragomen said there w(oul)d be a fight. Sheikh Suleiman, one of the Fellahhine Sheikhs said that were it not for his respect for Sheikh Hussein he w(oul)d kill every European here this year.

TB's Anglo-American Group reached Jabal Harun on Sunday 26th March 1854. He and others climbed up to the shrine of Aaron then proceeded into the city ruins to find their camp. Unlike so many other visitors who were compelled by the hostility they met to leave early -often after a single day, TB's group then had two full days and then left on the morning of the fourth day (Wednesday 29th March), again, in a few cases, climbing to the top of Jabal Harun. Indeed, the second full day was only requested and agreed after they arrived and realised how extensive the attractions were.

Jabal Harun and the Shrine (weli) *of Aaron* (Figure 9.4), was the first of several named places TB visited. During the 19th century many travellers were either forbidden to go up to the shrine or were forbidden to enter it if they did. TB and several of his companions ascended twice and explored the exterior of the shrine but found the door to the interior locked. The shrine itself is Muslim but closer inspection in more recent time has revealed it is constructed over

Figure 9.4: Jabal Harun: the plateau with the Byzantine Monastery and the Muslim shrine on the peak above (APAAME_20051002_DLK-0086).

a larger building, apparently a Christian chapel. Despite the extensive excavation undertaken on the plateau below, the recent Finnish excavations were unable to explore this Christian structure further. The 19th century visits to the *weli* and what has been drawn of the traces of the underlying Christian building are now set out in detail in the superb Finnish publications (Fiema and Frösen 2004: esp. Ch. 2). TB seems not to have noticed the traces of the earlier structure.

Neither did TB remark on the extensive ruins on the plateau below. A closer look might have suggested its purpose as mosaic tesserae are abundant and there were remnants of the curving walls of the apses in the church within the complex (Fiema and Frösen 2004: esp. Ch. 5).

TB found the camp pitched 'facing the Grand Temples', his name for the Royal Tombs on the eastern cliffs of the city (Figure 9.8). Like many visitors, TB regarded the monumental tombs as temples (or, in some cases, palatial homes). In fact they are all tombs (McKenzie 1995). It was also a common belief that at least the grandest of these *must* be Roman; likewise the theatre. In fact, detailed research has dated the theatre and almost all the principal rock-cut tombs to the Nabataean period. The first tomb all visitors explored in detail - what TB called 'the Great Temple', and sometimes revisited, was the *Khazneh* - or, *Khazneh Faroun* - The Treasury or Pharaoh's Treasury (Figure 9.5). Indeed, the Arabs at the site were convinced the urn in the broken pediment, was filled with gold and the pitting from bullet holes records their attempts to shatter it and disgorge is riches.

Even those visitors inclined to sneer at the debased or poor attempts at Classical architecture, agreed the Khazneh was a wonder, TB visited it three times, remarking:

> ... the Temple is as fresh & sharp as the day that it was cut. We sat & enjoyed it. Even with its' awkward broken or interrupted pediment it is beautiful. I can't describe it so as to help my memory when I forget it, if I ever do. ... It grows on you more & more. ... This morning the sun was partly on the Temple - & the rich rose colouring was much more beautiful than it was the Evening before. I tried to make a rough sketch of it, swiftly in order to <u>learn</u> it. ... It is more & more beautiful from every new place you see it from - perhaps most, when you catch bits of it between the dark rocks of the Es Sik.

Another visitor who was there three years later and rather sceptical of all the wonders he had seen in Europe and Egypt, nevertheless, wrote in his letter-book (Robertson MSS 183 On-the-Desert Thursday April 23rd 1857):[10]

> *However much we were tried by the villany of the Bedowins and had our tempers sorely tested by their rascality, my visit to Petra will be one which I will never forget, I consider it the gem of the Desert. Without this I would have (I am sorry to say) voted this whole trip a grand bore, but this I had looked forward to with anxious anticipation, I had all my life head of this city of rock and considered it one of the most singular places now in existances and not withstanding all my extravagant ideas and my daydreams of the grandure of this ancient place. I can now say that in none of them was in the least disappointed, I was highly pleased with everything and in many I was astonished, the wildness of the surrounding mountains, the romantic beauty of its secluded glens, and the magnificence of its excavated structures, would well compensate a man for the fatigues and hardships of this tiresome desert. Nothing in all my life that will in any wise compare with it. Egypt has its stupendous works, Greece Rome their monuments and works of art but Petra combines them both and in her uniqueness she defies the world.*

The Khazneh was explored in detail. Not just the simple chambers on the walls of which scores of visitors inscribed or wrote in charcoal their names and sometimes their home place and a date (Brünnow and von Domaszewski 1904: 192-24). (Mr Rodewald was to add his name in charcoal, the only one of TB's companions to do so (above, Ch. 7)). Now, the huge build-up of material washed down the Siq has been excavated in front of the tomb to reveal tomb facades at a lower level. The tomb is likely to be the most photographed of all the structures at Petra, not least in the famous view, glimpsed by those arriving down the Siq.

TB went up the *Siq* on the second day, accompanied by Kennedy and the Fentons. This is the route by which almost all tourists arrive today but it was rare in the 19th century when the Alawin were anxious to avoid both their Al-Jazi cousins and the *fellahin*/Liathneh (above). Many travellers nevertheless took the opportunity while at the Khazneh to walk all the way up the *c.* 1.5km natural passage; a few went beyond to where the Wadi Musa broadens out as it gets near Elji. Unlike the open track one sees today, the Siq in the 19th century was overgrown with oleander bushes, full of boulders, rocks and gravel washed down from its eastern (higher) entrance. Some travellers found themselves not only struggling through bushes and rubble but against a stream - this was, after all, the Wadi Musa channelled through the gorge and ultimately out into the open area in front of the Khazneh then on down alongside the main street.

---

[10] I have retained his spellings and errors.

Figure 9.5: Khazneh from above (APAAME_19980520_DLK-0200C).

Despite the obstacles, travellers - including TB, often noted the survival of paving in places (Figure 9.6) and the water channels (which they usually call 'aqueducts' cut into the side walls of this narrow canyon). TB also noted the niches cut into the walls and spotted occasional inscriptions. At the top, TB and 'Young Fenton' climbed up to examine the *Arch* that spanned the entrance high above and noted also the grave slots cut into this high surface invisible from below. The arch was intact in 1854 and sketched and painted by several visitors. The central voussoirs seem to have fallen in 1896 leaving it as one sees it today. The Siq is a stupendous natural wonder. Even without the archaeological traces and despite the crowds of tourists (and horses and carriages) today, it is one of the most memorable places to visit.

Unseen by many travellers, even those who went beyond the eastern entrance, was the remnant of an ancient dam. Following a flood in 1963 when 22 tourists and a guide were drowned, the dam was rebuilt. Its purpose now, as in antiquity was to bar and deflect flood water which would otherwise have been a constant hazard every year and potentially very damaging in the city centre. The deflected water was turned northwards into a minor narrow wadi and then into a man-made tunnel, still in use. Many travellers never noticed it, but a few entered and walked through to where it continued in a narrow wadi course, ultimately emptying out into the city centre through the Wadi Mattaha and then continues westward to the Wadi Siyagh .

Figure 9.6: The paved road in the Siq. When Thomas Bowles was there, the Siq could be a veritable river in winter, deep in boulders and gravel and thick with oleander bushes, hiding the roadway (APAAMEG_20070409_DLK-0024).

Figure 9.7: The Obelisk Tomb in the Upper Wadi Musa (APAAMEG_20070409_DLK-0150).

Beyond the Arch over the entrance to the Siq, the Upper Wadi Musa offered further attractions to the more adventurous travellers who went on. TB was one of those and he duly noted the most striking of the tombs - the *Obelisk Tomb* (Figure 9.7):

> ... we found a large Tomb - or perhaps 2 Tombs one above the other. The lower had pillars & a pediment & from the Terrace above this rose again other doors & windows - ending above in 4 pyramidal stones.

Back inside the ancient city, TB focussed on the most noticeable and accessible structures. First was the *Tomb of Unayshu* shown to them by Ross who was exploring separately. It lies opposite the theatre and is distinguished by the Nabataean inscription it bears and from which it gets is name: "Uneishu, brother of Shuqailat, son of ..." (Zayadine 1974; cf. Wenning 1990). Beyond lay a succession of very grand tombs on the east face of the cliffs, the so-called Royal Tombs (Figure 9.8). These must be amongst the most photographed tombs at Petra, particularly striking as the light of late afternoon reaches them.

The *Theatre* - not amphitheatre as many visitors erroneously name it, remains impressive, cut deep into the rocky face and exposing existing tomb openings like theatre boxes (Figure 9.9).

Figure 9.8: The Royal Tombs: from the left, the Palace, Corinthian, Silk and Urn Tombs (APAAME_20160918_DLK-0367).

For many years it was labelled as 'Roman Theatre' but excavation has revealed it as Nabataean, probably late 1st century BC/early 1st century AD; a remarkable addition to the architectural furniture of this Nabataean city and indicator of its cultural influences. It is thought to have been able to seat about 5000 people. Its scale and context enchanted numerous werstern visitors - but TB says nothing. Jane Eames on the other hand (1855: 290):

> *Within a semicircle of hills, stood the ancient amphitheatre (sic), the seats rising up in thirty-three rows, all hewn out of the solid rock. In some places the rock is crumbled away, but we had no difficulty in climbing up to the highest row of seats, and there we sat down, to gaze around us at our leisure, …. This amphitheatre was immense, and capable of seating more than three thousand persons. A double row of columns ran along the front, but they have fallen to the ground, leaving their bases to show where they once stood.*

The *Triumphal Arch* standing on the main *Colonnaded Street* near the *Qasr el-Bint* (Figure 9.10) was already collapsed but clearly identifiable. Originally it had a tall central arch and two lower ones flanking on either side. The rumps of the columns that once flanked the street are often still preserved and behind those in the shade would have been rows of shops. The ornamental fountain was not seen by TB. The 'Palace of Pharaoh's Daughter' was certainly a temple and virtually the only major building still standing. Probably built in the 1st century AD on an earlier structure then remodelled in the Roman period.

Although almost all the tombs at Petra belong to the Nabataean period, one of the monumental ones is different. This is what TB calls "the Temple with a Latin Inscription", which is in fact

Figure 9.9: The Theatre (APAAME_20151005_DLK-0429).

Figure 9.10: Aerial view of the Colonnaded Street, Great Temple, Temenos Arch and Qasr el-Bint al-Pharoun (APAAME_20051002_DLK-0102).

the monumental *Tomb of Sextius Florentinus* (Figure 9.11), of one of the early Roman governors of the province of Arabia. T(itus) Aninius Sextius Florentinus was in office governing Arabia in December AD 127 and evidently died soon after - a successor is attested in November AD 130. His son arranged for him to be buried in style in a place flanked by the grand tombs of Nabataean kings. It is one of the few tombs with an inscription and - even rarer, it is in Latin (rather than Nabataean or Greek). Few visitors ancient or modern go beyond the Royal Tombs to visit this one but TB did ...and then he went on a little further.

The inscription is high above the door and always been difficult to access and read and TB made no attempt. In fact, it is a fascinating record of the career of a man who would be almost entirely unknown ... but for the accident of his death while in office and the pretensions of his family who commemorated him in this remarkable way. As a scholar of Greek and Latin, TB would doubtless have been delighted by its content. It reads (*IGLS*XXI; IJ IV: #51):

Figure 9.11: Tomb of of the Roman governor of Arabia Titus Sextius Florentinus.

T(ito) Aninio L(ucii) f(ilio) Pap(iria tribu) Sextio Florentino, IIIviro aur(o) arg(ento) flando, trib(uno) mil(itum) leg(ionis) I Minerviae, quaest(ori) prov(inciae) A[c]haiae, trib(uno) pleb(is), leg(ato) leg(ionis)VIIII(sic) Hisp(anae), procos(uli) pr[ov(inciae)] Narb(onnensis), leg(ato) Aug(usti) pr(o) pr(aetore)prov(inciae) Arabiae, patri piis[sim]o, ex testam[e]nto ipsius.

Titus Aninius Sextius Florentinus, son of Lucius, of the Tribe Papiria. Member of the Board of Three for Minting in Gold, Silver and Bronze, Military Tribune of the Legion I Minervia, Quaestor of the Province of Achaia, Tribune of the Plebs, Legate of the Legion IX Hispana, Proconsul of the Province of [Gallia] Narbonnensis, Legate of Augustus with Propraetorian Power of the Province of Arabia. To a Dear Father, in Accordance with his Will.

TB's energy took him to two other mountain tops around Petra, apart from the two ascents of Jabal Harun twice. The *Deir – Monastery* (Figure 9.12), so-called from the discovery of several Christian crosses cut into its walls - as TB observes and was apparently confirmed by Kennedy who visited separately. It is a Nabataean tomb, probably also of the 1st century AD, but is unusual. Partly its remoteness; partly the bareness of the interior; partly the evidence of a colonnaded structure in front and - even more, the ruins and remarkable circular structure opposite and on a higher level, known as Gethin's Circle (seen by David Roberts and Harriet Martineau).

Figure 9.12: The Deir – note human figures as scales (APAAMEG_20070409_DLK-0126)

TB again went on beyond the tomb to two successive high places from which he had a 'fine view'. This is probably the pair of outcrops from which one can indeed get a view both back towards Petra and west towards to Wadi Arabah. Today the locations are simple souvenir kiosks and tea shops, one bearing the apt name of the 'End of the World Café'.

Descending from a high place earlier in his visiting, TB had noticed an obelisk on the top of a hill above the theatre. Few people even today have visited it and it was not properly 'discovered' and recorded until a generation after TB's visit. This is the *High Place of Madhbah* from which, in additions to the remarkable ruins, wonderful views north over the Colonnaded Street and the other structures in the valley. TB approached it from the west and southwest along the Wadi Farasa with its important tombs. As TB writes, his companions wanted to turn back at the end of the valley, but he pressed on, spotted stairs leading up and after some struggles reached the plateau, the obelisk and the 'High Place of Sacrifice.'

On his way down TB discovered two more tombs, both of which are seldom visited by and are important examples. The Tomb of the Soldier (Figure 9.13) is so-called because of the statuary carved on the face which is sheathed in Graeco-Roman armour. As TB notes, the interior is remarkably ornate - especially when contrasted with the plainness of the Khazneh and Deir.

Immediately opposite is the second tomb TB visited, the so-called Triclinium Tomb. Once again, TB is rightly struck by the ornate interior (Figure 9.14).

Figure 9.13: Tomb of the Soldier (VHE_20171001_IMG_7730).

TB doubtless saw much else to which he gave no name or did not think to record. As noted, he did not see the Tunnel outside the Siq, so no surprise that he did not see the Crusader Castle of Wu'eira a short distance beyond it. Nor did he see any of the major structures excavated in the last generation. Most obvious as seen by tourists today, is the Great Temple just east of the Qasr el-Bint. Until excavation it was a mound of collapsed masonry and fallen column drums. Now it has been revealed as an astonishing structure, perhaps partly administrative; partly religious. The small 'theatre' inside it may have been either a Council Chamber or 'Ritual Theatre'. Also revealed have been the remarkable column capitals on which the volute scroll is an elephant's head. And, of course, he was unaware of the ruined churches, two now excavated, including one with the important cache or carbonized papyri.

TB was evidently hugely impressed by the place and walked and climbed considerable distances during his stay. Although he does not sum it up, we may be sure he would have shared the opinion of Jane Eames (1855: 283):

> *More than a hundred times since our arrival here, have we exclaimed, "well, really, this is the most remarkable place I ever saw," .... Petra is so utterly unlike any other place in the world, that there is nothing with which it can be compared.*

Figure 9.14: Triclinium Tomb – interior (Brünnow and von Domaszewski 1904: 1, 159, Fig. 181).

TB's walking took a heavy toll:

> I c(oul)d hardly get along. Yesterday & today I have been wearing the Sydney button shoes. All the rest being quite done for & they were in pieces.

He was soon having to put aside these 'Sydney shoes' and wear 'thin' ones he had had made in Galle. By the time he reached Hebron he was complaining he could not sight-see as he had no wearable shoes left and at Jerusalem almost his first act was to buy shoes.

Chapter 10

Jerusalem and Associated Trips

After two full days in Petra, the group left the following morning - Wednesday 29th March 1854, by the route they had entered. TB paused to copy some inscriptions as they left the ruins and, as they passed by, he was one of those who climbed Jabal Harun a second time. Soon after, the Fentons left the group (above) and the fourteen remaining began their march northwards up the Wadi Arabah to Jerusalem. They reached Hebron on Sunday 2nd April, after five days - a swifter journey than expected. Much of this section of the journey was a dull, hard grind. TB noted the increasing evidence of 'civilization' and farming as they got closer to Hebron. He also reported the remains of ruined villages, the small Nabataean-Roman-Byzantine town of Kurnub in particular. The pleasure of getting closer to their promised land was undermined by a bitter cold and a great deal of rain: *'it was cold & wet & very slippy for the Camels, who weren't fitted for anything but a dry sandy soil'*. They arrived at Hebron in a downpour, but their dragomen had been sent ahead and they had arranged for them to be accommodated in the Lazaretto, which was not in use, the quarantine having been called off (news of which TB had heard while at Aqaba). Horses had also been arranged for the next stage of the journey. After 31 days of riding camels, TB was glad make the transfer: *'I got off my camel & kissed his nose & said a very hearty goodbye.'*

Palestine

The next morning, regretting he had neither the opportunity nor the shoes to do any sight-seeing, they set off on the relatively short journey via Bethlehem to Jerusalem. TB noted the Pools of Solomon in passing - from which an aqueduct carried water to Jerusalem, 12km straight line distance to the northeast. The three large rectilinear reservoirs are still visible (Figure 10.1). They paused briefly to explore the convent and its church at Bethlehem. Curiously he did not return on a day visit from Jerusalem as one might have expected.

They reached Jerusalem on Monday 3rd April 1854 ... and TB was not impressed:

> *I was strangely disappointed. It was merely a wall & (a) few houses & minarets in the bare undulating stony country - hills appearing above & behind it, but all bare, without a tree. ... A filthy, dirty, narrow street, led us down to the Hotel.... (where) the landlord put us into a room in a house near wh(ich) he promised to furnish.*

TB was to remain in Jerusalem for two weeks, until Monday 17th April (Table 10.1). For the first two days he was in very bare lodgings but then Giovanni brought in some of their camping equipment, divided the room he was evidently sharing with Mr Kennedy, and set him up more comfortably and privately. During these two weeks he undertook extensive sight-seeing in and near the city, then a day-trip to Nebi Samwil and a longer outing to the Dead Sea. Happily, the period included Easter and its special festivals - many travellers to Petra aimed to reach Jerusalem in time for Easter. It was only later that TB named his hotel - the Mediterranean Hotel, *'one of the best, if not the best'* in the city and patronised by not just the famous - Herman

Table 10.1: Tourism around Jerusalem and associated trips (cf. Appendix 2).

Monday 3 April	Hebron to Jerusalem. Excursion to Pools of Solomon and Bethlehem: Church to see Holy Place	166-170
Tuesday 4 April	Jerusalem: City Walls, Pool of Siloam, Garden of Gethsemane, Church of the Holy Sepulchre	170-172
Wednesday 5 April	Jerusalem: Church of the Ascension, Bethany (Tomb of Lazarus)	173-177
Thursday 6 April	Jerusalem: Convent of the Knights of St Stephen, Tomb of the Kings, Tomb of the Judges, Jeremiah's cave	178-182
Friday 7 April	Jerusalem: Pool of Siloam, King's Gardens, Wailing Wall, Armenian Convent, Birket Es Sultan	182-186
Saturday 8 April	Jerusalem: Excursion to Village of Hanina, Nebi Samwil, Er Ram, Anata (Anathoth)	186-189
Sunday 9 April	Jerusalem: Valley of Hinnom, Hill of "Evil Counsel, Armenian & Latin Convents, Valley of Jehoshaphat	189-193
Monday 10 April	Jerusalem to Mar Sabba via Valley of Hinnom, Valley of Jehoshaphat, Valley of Kidron	193-198
Tuesday 11 April	Mar Sabba to Jericho via Dead Sea & River Jordan where Jesus was baptised	198-204
Wednesday 12 April	Jericho to Jerusalem via Spring of Elisha, Bethany, Mt of Olives, Valley of Jehoshaphat	205-206
Thursday 13 April	Jerusalem: Mt of Olives, Upper pool of Gideon	206-207
Friday 14 April (Good Friday)	Jerusalem: Mt. of Olives, Tombs in the Valley of Jehoshaphat, Church of the Holy Sepulchre	207-210
Saturday 15 April	Jerusalem: Greek Convent, Church of the Holy Sepulchre	210-214
Sunday 16 April (Easter Day)	Jerusalem: Church of the Holy Sepulchre	214-216
Monday 17 April (Easter Monday)	Jerusalem to Ain Es Yebrad via Shafat and Nebi Samwil	216-220

Figure 10.1: Pools of Solomon as seen in a photograph of 1890 - they would have been at least as isolated when seen by TB. Today they are in a built-up suburb of west Bethlehem (Library of Congress) (Public Domain).

Melville and Mark Twain, but – at that very moment, the important - Lord and Lady Falkland, returning from several years as Governor of Bombay Presidency (Gibson and Chapman 1995; Gibson *et al.* 2013: *passim*).

During their stay, TB seldom mentions his erstwhile travelling companions though it is clear much of his activity continued to be in the company of Mr Kennedy. It was while exploring with the latter that he encroached on the precinct of the Mosque of Omar and children stoned him. He was angry and indignant at the time but later wrote that:

> *I allowed myself to be very angry today when the boys & men threw stones at me - but I had no right to be. Though I would not know it, I was where I had no right to be. And I felt afterwards angry with myself instead at having allowed myself to feel any annoyance at an insult offered to me as a follower of him who was spit upon & buffeted & reviled for me.*

TB's tourism was the common items of western visitors keen to see the locations identified as those familiar from reading the Bible: Pools of Siloam, Mount of Olives, Garden of Gethsemane etc. etc. They could supplement the Bible references with what other visitors had published about the place though TB could be sharp in his judgment: '*Dr Robinson maintains that this is Siloam. He as usual think it necessary to disbelieve what others hold*' and when writing of Gethsemane, added '*this piece was walled-in to keep the memory of the place (& perhaps to make money)*'.

Much of TB's time in Jerusalem was spent in visiting churches and witnessing or participating in church services. He found them mixed, especially in the quality of the chanting. In Holy Seplchre on the day after Good Friday: '*The chanting & singing was execrable. I never heard some (- such?) thorough discord in my life.*' But then on Easter Sunday in the same place: '*The Latin Service ended & the Greek began. It was a gorgeous ceremony - multitudes of priests & 3 Bishops - one representing the Patriarch who is at Constantinople*'. He was shocked by the length and content of some sermons; others were just right. The highlight of his stay was, of course, being in Jerusalem over Easter and able to go to the Church of the Holy Sepulchre repeatedly. He was disappointed sometimes by the discordance and disorderliness. Two entries in his journal reflect his mixed feelings:

> *I am very doubtful of the good efforts of pilgrimmage. Try as I will to feel the sacredness of every spot about Jerusalem & to stir up holy thoughts. I find one goes to everything as a sight. I cannot feel the strong emotions others have described. But I have not been with a person whom I can sympathize with, or who sympathizes with me & that puts a restraint upon me. Certainly I felt the sacredness of our service yesterday as a great contrast to what I had seen since being in Jerusalem. It made me wish the Holy Land was in the hands of England.*

And later as he was leaving:

> *I have been nowhere more saddened with Church matters than I have been at Jerusalem. It is a concentrated picture of the world.*

The city was divided by the rivalry and quarrels of the various Christian denominations. It was also a city with growing numbers of Jews whom he contrasts with those he knew in England - possibly Jewish pilgrims from the Russian Empire:

I have been most struck with the fair hair & freshness of the Jews here. For the most part they are quite unlike the Jews we have seen in England. Very fair - bright red & white - with light hair. I never saw anything so fair.

Elsewhere in his journal he remarks on the efforts being made by Christian missionaries to convert Jews, and on Missionary Schools to provide employment and education.

The city was, of course, under Muslim government and control. At the highest level and supported by the consuls of the great powers, western visitors were commonly welcomed and treated with courtesy. Lower down the social scale there could be problems as TB recounts from two incidents. The first involved Col. and Mrs West discussed below, an affair that rumbled on before finally being settled. Amusingly the Turks discussed whether it was permissible to call a Russian an infidel but not an Englishman. Then there was the affair near Bethany where Mrs Heneage had been accosted threateningly with demands for bakshish (also discussed below). As TB observes, it made him cautious about his own solo trips in and around the city.

James Finn
(Clerkenwell, London, 13 July 1806 - Wimbledon, Surrey, 29 August 1872)
and
Elizabeth Finn (née McCaul)
(Warsaw, 14 March 1825 - Hammersmith, 18 January 1921).

As was common amongst British (or British Empire) visitors to Jerusalem, TB called on the British Consul and likely signed his name in the - the surviving document only begins in 1855. He met him briefly on the 6th April, was invited by him the next day to a meeting of the Jerusalem Literary Society which he had to decline as he was still too dishevelled after weeks of travel: '... I can't go in this wild man of the woods state with beard & black neck cloth.'

For seventeen years between 1846 and 1863, the British Consul in Jerusalem was James Finn (Figure 10.2) who provided support for British and other western travellers in the Holy Land: advice, guidance, contacts, intercession with the Ottoman authorities. Finn is a fascinating character some of which comes through in TB's report of the incident involving Col. and Mrs West (above, and cf. below) in which Finn interceded (Eliav 1997: 69):

> *Although his seventeen-year term in office was not the lengthiest of the British Consuls, James Finn became the most prominent and famous of all the Consuls sent by the Powers to serve in Jerusalem.*

Like his immediate predecessor, W. T. Young, the first British Consul at Jerusalem (1835-1846), Finn was not a career diplomat and, probably because of the troubles these two had, all their successors *were* diplomats. Finn's background was a very modest one (Abrahams 1978-80). Born in Clerkenwell, London to an Irish Catholic father and Protestant English mother, his evident potential brought him to the attention of Lord George Villiers, later 4th Earl of Clarendon, who paid for his education. He excelled in languages: he was evidently taught Greek and Latin early in life and he later claimed to *'read, write, and speak English, French, Italian, Spanish, "and a little German," vernacular Arabic, and "Hebrew more than Arabic."'* (Eliav 1997: 74, n. 191). In 1832, aged 26, he was appointed as a tutor in the household of another great aristocrat, Lord George Gordon, the 4th Earl of Aberdeen. During the next twelve years Finn developed a keen interest in Jewish history and published two books (on the Jews in Spain and Portugal and on those in China). Finn's two very powerful sponsors were crucial to his education and opportunities and, during the core years of his consulship, were respectively Foreign Secretary and Prime Minister.

Perhaps because of his poor origins, Finn was evidently very conscious of his status: a Consul and representing a great power in the Holy Land. He is characterised as arrogant and domineering, rigid and vengeful.

Like his predecessor, Finn was soon embroiled in quarrels with the senior Anglican bishops in Jerusalem and often with his fellow consuls. He insisted he was not allowing his strong religious beliefs to influence his decisions but was often accused of supporting proselytising efforts amongst the Jews - perhaps pushed by his stern and energetic wife and by the many independent missionaries in the city. Unlike other foreign consuls in Jerusalem, Finn travelled extensively throughout Palestine and far beyond (Finn 1867). Two of the latter trips are outstanding and directly relevant to this work. In April 1851 he travelled to Petra and – through discussion with the leaders of the beduin tribes through whose territory they were

Figure 10.2: Photograph of James Finn (Public Domain via Palestine Exploration Fund).

conducted, and of the *fellahin* at the ruins, dispelled some misunderstandings and negotiated terms on which future travellers might visit the ancient city without the fractious hostility so often encountered. His consular diary has numerous references to making arrangements on behalf of British visitors in the city (Blumberg 1980: passim). The arrangement survived at least as regards those travellers who went south from Jerusalem but by 1857 had collapsed as the chaos and violence that year reveals (Kennedy 2018). In May 1855, Finn undertook to lead an expedition *'Over the Jordan'* to the ruins at Amman and Jarash, comprising twelve British travellers. These included TB's friend and former travelling companion to Australia, Henry Stobart (above, Ch. 2), who had by then arrived from his diversion into India, and Lord Henry Scott.

Finn was very active in attempts to alleviate the considerable poverty amongst the Jewish residents in Jerusalem, often immigrants from distant lands. Some such attempts were resented by rabbis but the intended recipients were more receptive and were glad of the training and income. Through The Sarah Society founded by his wife, the Finns distributed fuel, clothing and food and some medical assistance to impoverished Russian Jews during the Crimean War, later they bought land near Jerusalem and established a farm to employ Jews at what they called Kerem Avraham, The Vineyard of Abraham (Polley 2019). Finally, they bought land at Artas south of Bethlehem and provided

Figure 10.3: Photograph of Elizabeth Anne Finn (Public Domain via Palestine Exploration Fund).

not just employment but training in farming. Important as these ventures were for the recipients, they led the Finns into debt which was part of their downfall and - together with the serious quarrels in which they became embroiled, led to Finn being removed in 1863. Officially he was to be transferred to be Consul of the Dardanelles but then, through an exchange with another new appointee, he was named at Consul at Erzurum in the interior of what is now north-eastern Turkey. However, he did not take up that post either. Instead, he retired to Britain where, in ill health, he died in 1872, still only 66.

Elizabeth Anne McCaul (Figure 10.3) was the second wife of James Finn. She was the daughter of the stern presbyterian missionary, Rev. Alexander McCaul and had been born in Warsaw where her parents were then working. When Finn was offered the position in Jerusalem, they married in January 1846 just before setting off. He was then almost 40 and she was 20. In Jerusalem she was a vigorous and forceful presence alongside her husband. She, too, was a polyglot - speaking also Hebrew and Yiddish, and a millenarian evangelical Christian. She was a painter and early photographer and supported the first resident Jewish photographer in Jerusalem, Mendel John Diness, who had been trained by James Graham (below).

Like her husband she wrote books which are a mine of fascinating information about their years in Palestine (E. A. Finn 1866; 1869, 1929). Neither of them left Palestine during their 17 years in the consulship and all their children were born there - and two died and were buried in Jerusalem. Of particular interest is Finn's multi-volume diary. Extensive extracts from it reveal the passion of Finn for the place and his deep sadness at leaving including this extract for 17 July 1863 the day they sailed from Jaffa for home (Abrahams 1978-80: 49):

And here ends my Syrian career of 17 years. I have done and suffered much in the time. May God forgive whatever mistakes I have made. I known (sic) of no crimes. I know that I have kept up well with a firm head and hand the honour of my country. Farewell ... to my fondly cherished plans of agriculture for Jews thwarted by the Bishopric and Jewish mission. Farewell to the old friends buried in the cemetery on Mount Zion. Farewell to my daughters, one buried there and another at the foot of Carmel. Farewell to my hopes of being myself buried in Jerusalem.

James and Elizabeth Finn were buried at St Mary's Churchyard in Wimbeldon (Figure 10.4). Another daughter, Constance Mary Finn, did survive, and, together with her mother, set up

> In Loving Memory of
> James Finn
> For Seventeen Years Her Majesty's
> Consul for Jerusalem and Palestine
> Born July 13 1806
> Died August 29 1872
>
> Also To The Memory Of
> Elizabeth Anne Finn
> His Dearly Beloved (?) Wife
> Who Was Born [March 14] 1825
> And Departed This Life [January 18] 1921

Figure 10.4: Tombstone of James and Elizabeth Finn (St Mary's Churchyard, Wimbledon) (Photo Courtesy: Tina Schofield).

and ran various organizations to support the poor, including the Society for Relief of Distressed Jews. She lived to age 99, unmarried. According to Abrahams (1978-80: 50):

> The Finns' Palestine interest was inherited by their daughter Constance Mary Finn who was born in Jerusalem in 1851. When I returned from Jerusalem after the victory of the Jewish army in 1948 and called on her, she, echoing what would have been her parents' sentiment, declared: 'I knew that the State of Israel would be resurrected, for Prophecy promises this!' It is pleasing to know that one of the first official acts of the newly established Israeli Embassy in London was to send their representatives to attend the funeral of Constance Mary Finn when she died at the venerable age of 99, as a mark of respectful honour in memory of James Finn, Her Britannic Majesty's Consul at Jerusalem.

***

Colonel and Mrs West

Most of what TB has to say about Finn is in the context of the incident involving the Wests recorded by him (above) in which the Consul had to play an official role.[1] TB did not witness the incident but did subsequently participate in the discussion. As recorded in the journal for Sunday 9th April 1854:

---

[1] Finn himself reports it quite briefly and without naming the Britons concerned in one of his books (Finn 1878: 2, 32). There may be a full account in one of Finn's numerous official reports now in The National Archives at Kew.

In consequence of this I missed a scene at the Consuls - going home from Church Col. & Mrs West met a file of soldiers. One took the trouble to step aside from the ranks & jostle Col. West - while another followed his example & did the same to Mrs West. Col. West recovered himself & struck at the soldier with a small stick he had in his hand - but missed him & hit the pouch on his back. The soldier dived into the line or ran in & out among the soldiers. Col. West following him. He complained to the Officer & the man said he had only called the Inglesi a "Giaour" (Infidel). The officer pooh poohed the whole affair. Col. West at once went to Mr Phinn (sic) - our Consul. Mr Phinn heard his story & sent to the Colonel of the Turkish Regiment - who came at last & made light of the whole affair. However it was insisted that an apology should be made & that in public & in presence of all the English that would be gathered together. The Col. refused - but at last yielded with a bad grace & the whole party (several English men had followed) adjourned to the barracks where the company or whatever it is was called out & the Col. made then a speech telling the men that they were sent here to keep order & <u>prevent</u> people being insulted. That they might only call Russians Giaours not English nor French & that the next time such a thing happen he would have a pair of scissors & slit the offender's tongue. Mr Heanage who was present said this was no figure of speech: he had been in Morroco (sic) & known this done.

The following day, Finn came to their hotel to report and discuss:

While at breakfast the Consul, Mr Phinn (sic), came in & discussed the affair with Col. West. The offending soldier had been sentenced to a fortnight's imprisonment & 2 men <u>who were thought to be near the man who had pushed Mrs West (!)</u> to a week each. I think I was the first to exclaim openly at this piece of injustice & then we all expressed our feelings & it ended by our all feeling that Col. West w(oul)d himself go & get all the sentences remitted. I also told the Consul what Giovanni had said - viz that it had been all accidental altog(ethe)r the soldier having slipped. This however was doubted. The Consul said the man had <u>not</u> made this excuse himself, as he would have done, had it been the case. I suspect the Consul came to feel rather whether there was much wish for severity on the part of the English visitors & finding that the feeling was all the other way, he said he would go to Col. West & suggest his interference on behalf of the offenders. A doubt was expressed about Col. West & I quietly said that probably the Consul's interference would be more influential than Col. West's & he looked across the table & said at once, he would speak to Col. West & if he was disinclined to move, he sh(oul)d feel that he had done his duty so far & would take the rest upon himself.

Finn's consular diary has several entries for Colonel West (Blumberg 1980: 162):

APRIL 9 - Sunday - Soldier assaulting Col. West - great uproar. The Col. struck him with his cane & followed him to the barracks with numerous English travellers- ·

The BimBashi invited to us - came from the other extremity of the city-

-Proceeded to Barracks near us - the men drawn out, then formed a circle round us & harangued against incivility & especially against the word Ghiaour - the travellers being present- these latter were then dismissed - and an investigation proceeded.

The man was sentenced to 15 days imprisonment - and two others near him were sentenced to a week each for having jostled Mrs. West. All our people satisfied.

APRIL 10 - Colonel West consented to mitigate the punishment of the soldiers.

Such incidents involving soldiery and opportunistic robbers were probably not uncommon (see 'Heneage' below) but the instruction of the Turkish colonel that - with the Crimean War now underway, only Russians could be referred to as Infidels and not the Turk's British and French allies, is interesting. Also of note is Finn's forceful treatment of the Turkish colonel.

Colonel and Mrs West cannot yet be identified. Finn's consular diary has Colonel West having a trip organized for him by Finn on 1 April then on 12th April Finn presides over a dispute between West and his dragoman involving a 'beating'. The Wests are noted in Bryce's list of boats on the Nile but just as *'Col. West and lady'* who he says *'passed at Mineh - met at Abou Simbel. Jny 18/54.'* Andrian, too, places them on the Nile at Abu Simbel on 21 January 1854 a few months before they appear in TB's journal in Jerusalem (Andrian 2011: 3, 26):

> *We found there the population of those 3 boats, viz. 3-4 English gentlemen and a rather pretty woman, Mrs. West, who has the courage to spend 5 weeks here, during which her husband wants to go on a hunting tour to and over Dongola[2], he later visited us on board and I was able to give him some good information with the help of my Berghaus map.*

Andrian was unimpressed by the English he was meeting as he went on:

> *On the whole these Englishmen, who are to be found everywhere, are a veritable nuisance, scribbling and defacing everything with their vulgar names, and running one after the other by the dozen to ascertain their Murray, from which they derive all their knowledge.*

Colonel West emerges as perhaps rather short-tempered, prone to striking out and keen on hunting. The insult to a young and pretty wife may also have been a factor.

***

Edward Heneage
(24 July 1802 – Hemel Hempstead, 25 June 1880)
and
Renee Elizabeth Levina (née Hoare) Heneage
(London, bapt. 26 October 1825 - Stag's End, Great Gaddesden, Herts, 13 March 1871).

TB being pelted with stones for straying into a religiously sensitive area and even the Wests jostled by soldiery were relatively minor matters compared to the actual robbery experienced by the Heneages:

> *Mr & Mrs Heneage had been out riding & beyond Bethany, 2 Arabs had demanded baksheesh, & seized the bridle of Mrs Heneage's horse. Mrs Heneage was frightened, she urged? him to give them something wh(ich) he did & they got away. It is unpleasant for I have been walking about, & like to do so, alone all about.*

Bethany - modern Al-Eizariya, lies just 4km east of the Temple Mount and the incident naturally made TB anxious for his own security while wandering alone around the outside of the city walls.

---

[2] In northern Sudan, *c.* 400km straight line distance to the south of Abu Simbel.

TB seems to have had close contact with this couple on several occasions while in Jerusalem. His reference to Heneage as the former owner of a yacht called the *'Sparrowhawk'* at Cowes enables us to identify him from amongst a number of members of this large family as Edward Heneage, son of a landed family from Hainton in Lincolnshire. After Eton and a B.A. (1824) and M.A. (1829) from Trinity College, Cambridge, he succeeded his older brother George as M.P for Grimsby from 1837-1852 and was later M.P. for Lincoln from 1865-68. In the 1861 and 1871 Censuses he is named as Deputy Lieutenant of Lincolnshire. When TB met him in Jerusalem he was on his honeymoon, his first wife (Charlotte Frances Ann Rolleston, daughter of Colonel Rolleston) having died in January 1853 leaving him with two small sons, the elder of whom - Frederick William Heneage (1843-1881), was to serve for over 19 years as a junior officer in the Royal Engineers (including eight in Canada and Bermuda), and on campaign in South Africa (Red River Crossing of 1870 and Zulu War of 1879) for which he was decorated. He died in New York (while stationed in Bermuda), apparently unmarried.

On 25 August 1853, about eight months after his bereavement, Edward Heneage had re-married. His bride, Renée Elizabeth Levina (sic) Hoare, daughter of Captain Richard Hoare, R.N., was 18 and he was 51. Their honeymoon seems to have included Morocco:

> *Dressed & went to dinner - where I had much talk with Mr Heneage - who has been out in Tanjiere (sic) ... Mr Heneage who was present said this was no figure of speech [cutting a man's tongue]: he had been in Morroco (sic) & known this done.*

When TB was leaving Jerusalem on 17th April he noted that *'All the party is broken up. The Heneages were gone this morning for Jaffa & Pagde [Page?] too.'* They seem not to have spent much more time in the Levant as implied in the published diary of Andrian when he was at Rhodes on 25th April 1854 (Andrian 2011: 3, 89):

> *We met the entire population of the French steamer that had just arrived, which had left Beyrut 48 hours after us, Tobins, Heneages, the whole pack of Americans, ...*

Edward and Renée Heneage were to have a daughter and five sons. Four of the latter were to have the unusual middle name of 'Fieschi', remembering one of Edward's great-grandmothers, Anna Maria Fieschi, daughter of Roboaldo Fieschi, Count de Lavagna, in Genoa who had married Thomas Heneage in 1728. The name was subsequently quite often found in the family including Edward's older brother George - who had been nicknamed 'Fish' because of it. The family had been Catholic but it is not clear if Edward and either of his wives were.

Their eldest child, Windsor Richard Heneage, was born just over a year after TB met the Heneages in Jerusalem (1855-1923). He was to study at Cambridge, but seems not to have pursued a career, simply recorded as *'living on private means'* in his census returns. One of his daughters – Margaret Renee Heneage, emigrated to Canada in 1906, married Hugh Kingston Llewllyn Statham ('rancher'), another immigrant there (who was killed in 1917) and had a son and daughter. They are recorded as living on Thetis Island just off the coast of Vancouver Island, where Margaret died in 1957.

Their second child, Hugh Edward Fieschi (1856-1878), Lieutenant of Artillery, died in India at Cawnpore (Kanpur), aged 22, unmarried. The third child, Alfred Rene Fieschi Heneage (1858-

1946), rose to the rank of Major and was awarded a DSO in the Boer War. He settled in British Columbia after retirement and is buried in the same grave plot as his sole sister, Eveline Mary Heneage (1862-1952), who died unmarried, on Thetis Island. Neither had married. Evidently Windsor's daughter Margaret (above) had moved to this remote place because she had an uncle and aunt there already. Finally came Everard Henry Fieschi Heneage (1860-1936), a graduate of Cambridge who became a Consulting Engineer. Little can be found of the next child, Thomas Fieschi Heneage (born 1864).

***

Although Kennedy was his principal companion throughout, TB met and sometimes spent time in Jerusalem with others he names.

James Graham
(Fereneze, Renfrewshire, 12 August 1808[3] - Paris, 12 November 1869)

TB encountered Mr Graham on 6th April 1854 at dinner in the Mediterranean Hotel soon after arriving in Jerusalem. He was unimpressed:

> *He talked both long & loud & made himself very conspicuous.... I don't like him, but I am prejudiced ...*

He learned then that Graham was connected to the 'Jews Miss(ionary) Soc(iety).' Despite his distaste, TB accepted Graham's invitation to visit a school for Jewish converts to Christianity run by a Miss Cooper (below) and the following day TB and Kennedy plus Mr Rodewald, were duly taken there by Graham. TB was to have gone with Graham on some tourism the next day but the latter went off without him. Graham appears a final time on 12th April drawing TBs attention on a religious matter with which he seemed to disagree.

James Graham was in fact a well-known resident of Jerusalem for several years (December 1853-July 1857)[4] in his mid-40s. He was a Scot, son of a minor landowner from near Paisley who went into banking which ultimately failed. He altered course to become a missionary and was then appointed as the Lay Superintendent of the London Society for Promoting Christianity Amongst the Jews and took up residence in Jerusalem in early December 1853, just three months before TB met him. He was evidently a devout Presbyterian and *'a man of rigid principles but also outspoken against injustice'* (Crawford 2008: 606). He seems also to have been 'difficult' - his photographic work (below) has been described as producing 'a brooding melancholy, as befitted this argumentative and principled Scot' (Crawford 2008: 605). Certainly, he seems to have fallen out with his employers and in particular the Anglo-Prussian Anglican Bishop, Samuel Gobat (1799-1879). A consequence was that he was effectively dismissed from his post, apparently for being too close to the converts and *'worldly'*. What that meant was that a year or two after TB met him, he quarrelled on several fronts with Gobat on matters of principal and morality on which he seems to have been in the right. After dismissal he left Jerusalem for several months to visit Egypt and then returned briefly before returning to Britain in July 1857. He was not the man to accept being treated badly and in 1858 published a

---

[3] The date is from his epitaph and corrects the year 1806 commonly given for his date of birth.
[4] Graham appears several times in Gerald Vesey's travel diary as a helpful source of introductions in Jerusalem (Vesey 2022: *passim*).

lengthy pamphlet explaining and documenting the quarrel with Gobat (Graham 1858). He had the support of a parallel pamphlet that same year by the artist William Holman Hunt who had been in Jerusalem, knew Graham and supported his case against Gobat (Hunt 1858). Although the hierarchy of the London Society supported Gobat, the two pamphlets were received well (Anon. 1858 review in *The Ecclesiastic and Theologian*). It is from Holman Hunt that we get a fuller picture of the man (Hunt 1905: 1, 425-6):

> He was a Churchman with a strong tendency to Presbyterianism; he was good-nature itself, but prosy, and an incorrigible procrastinator. Tall, fair, and brawny, riding beautifully, and having a deliberate polite gait and manner, he took rank at once as a person of distinction.

Graham is better-known today for what was a hobby developed while in Jerusalem: calotype photography. Indeed, he is regarded as the first photographer based in Jerusalem and important for his training of a local Jewish convert to Christianity, Mendel John Diness (c. 1827-1900) (Gavin and Rosovsky 1995) (cf. Elizabeth Anne Finn's role, above). Graham did not attempt to develop a commercial enterprise with this new art form but a significant number of his photographs survive in various archives - especially with the Palestine Exploration Fund in London (Wosford 2020) and in the Semitic Museum at Harvard University. Those from the Middle East bear dates from 1854 to 1857 and preserve views of places in Palestine and Egypt. Two of the people he got to know in Palestine were the Pre-Raphaelite artists William Holman Hunt (1827-1910) and Thomas Seddon (1821-1856). Both explicitly reported their indebtedness to Graham's photographs for their own work and it was a Graham photograph that underlay Hunt's famous painting, *The Scapegoat* (1855) (Lazard 1990; Crawford 2008).

After leaving Palestine in 1857, Graham travelled in the Levant including to Lebanon and Asia Minor, but seems then to have spent most of his time in Italy, and died in Paris in 1869 while returning to Naples. His surviving photographs include places in not just Naples but nearby at Pompeii and Pozzuoli.

***

Caroline Cooper
(Sulhampstead, Reading (?), 1806 - Jerusalem, 22 November 1859)

Matilda Creasey
(London, 13 August 1798 - Jerusalem, 3 September 1858).

It was James Graham who introduced TB to 'Miss Cooper' whom he describes briefly as:

> ... a lady who employs Jewesses sewing & mantua making & the clothes made are sold at a sort of bazaar & the percent applied to the support of the institution – as? they receive payment for their work.

TB says nothing more about Cooper who by then was 48. Caroline Cooper - described elsewhere as the daughter of a doctor (from near Reading perhaps) and of independent means, had arrived in Jerusalem in 1848. By then she was 42, never married but recently embraced an evangelical Chistianity. She had consulted her younger friend Elizabeth Anne Finn, wife of Consul James Finn (above), been encouraged to come and was initially accomodated by the Finns. She soon saw an opportunity: providing for the training of indigent Jewesses as seamstresses. Initially her means only stretched to supporting two or three of these women but she later received some financial support from organizations devoted to working with the growing Jewish population of the city. One source subsequently claims about 150 Jewesses were trained by Cooper and her assistant Matilda Creasey (below) (Blumberg 1980: 177 n. 11). Caroline Cooper receives several mentions in the publications of Mrs Finn but also extensively in the latter's semi-fictionalised book in which she is the forceful 'Miss Brandon' (1869). It is clear from Mrs Finn's accounts that there was tension not just with the rabbis who strongly disapproved of what temptations were placed before these Jewesses, but within the Christian community where some wanted the Finns and Coopers to be actively proselytising. In precisely 1854, the year of TB's visit, Mrs Finn noted the desperate plight of the Jewish population for any work and the careful line they trod in providing opportunities both in workshops in the city and on a farm ('the Plantation' at 'Kerem Avraham, or Abraham's Vineyard' (Blumberg 1980: 177 n. 11) they had bought outside Jerusalem at Artas (above) (1878: 2, 72-4):

Young Rabbis of the oldest and proudest families came to ask leave to join the Jews at work in the fields, and even Jewish women applied to us in their despair for field work; but the result of consultation on that subject was that such labour was neither expedient nor desirable for women, especially as there existed at that time in Jerusalem Miss Cooper's institution for teaching and employing Jewesses in needlework, of which needful art they had been for the most part deplorably ignorant. .... Others came back again and again and brought other people to intercede, that their petition might be granted. But this would have been imprudent on our part, and it was resolved to apply part of the money sent out from England for relief of distress, in opening a house in the city as an adjunct to Miss Cooper's School of Industry, where some more of the poor women might be employed in knitting and sewing. The difficulty of selection was very great. Ninety-five were soon at work. Miss Cooper already had eighty-five working in her own house, so that there were thus 160 women relieved, together with their families, making a total of 350. The number of men and boys at work at the same time was 203, with their families 450. ... On my return from a journey to Nazareth (July 22), I learned that the Rabbis had issued violent denunciations against both the Plantation and Miss Cooper's School of Industry. ... Strange to say, some of the Missionaries of the London Society were angered at the same time because we refused leave for them to come upon the ground expressly, for religious controversy, and this during the working hours. The object of the institution was to relieve distress by means of honest industry. But at the same time, the perfect freedom and religious liberty of the workpeople were respected. The Jews used to suspend their work for a few minutes at the hour when afternoon sacrifice was formerly offered in the Temple, which is now the time for prayer. But if the English Consul had authorised professed Missionaries of an English Society to come and preach and hold religious discussions, it might have savoured of attempting to convert needy people by taking undue advantage of their distress.

The young Lucy Matilda Cubley was in Palestine in 1854 and provided a drawing of a dozen Jewesses at work in Miss Cooper's 'Jewesses Institute' (1860: Plate V) (Figure 10.5). A lengthy caption describes the people and the work.

Another visitor describes the scene in June 1856 (Charles 1866: 65-6):

This morning we went with Miss Creasy, who kindly called for us, to see Miss Cooper's industrial school. ... The school interested us deeply. The Jewish women sat on divans around the walls of the various rooms busily engaged in sewing, knitting, & c. Miss Creasy told us Miss Cooper's Jewesses were known throughout Jerusalem by their superior neatness. Every day a chapter in

Figure 10.5: Young women at work in Miss Cooper's Jewesses Institute as seen in 1854 (Cubley 1860: Plate V).

the Old Testament is read to them, about which they are questioned, and a chapter in the New Testament, about which they cannot be questioned. The rabbis have more than once denounced and scattered the school; but after a time it has quietly gathered again.

Women missionaries like Miss Cooper were prominent in establishing such enterprises to support Jews in general and these were later usually incoporated into one of the formal organizations - in her case, the London Jews' Society (LJS), established in 1843 (Melman 1995: 178). Arrangements for incorporation were apparently made when Caroline Cooper made a visit home in late 1858/early 1859. She returned to Jerusalem on 18 March 1859 but died of 'Syrian fever' on 22 November 1859 and was buried on Mount Zion.

Before going to Jerusalem in 1848, Miss Cooper had sought assurances from Mrs Finn about the safety of the foreign community there. Of course, western travellers in Palestine frequently allude to the dangers of robbery and perhaps violence especially those who went down to the Jordan and Dead Sea. Although TB mentions only Miss Cooper from his visit, it is likely he would have encountered Miss Matilda Creasy who had joined Mrs Finn and Caroline Cooper in setting up 'The Sarah Society' to teach Sephardic Jewesses needlework. In 1858 there had been disturbances and robberies around Jerusalem and Finn's consular diary records under 7th September 1858 that Miss Creasey had left the city to visit the farm but never arrived. Searches were made and a few days later her body was found near the city, bludgeoned to death. Despite investigations by the Ottoman authorities and even arrests, no one was ever charged or punished (Stockdale 2006). She was 60.

***

Rev. John Wheeler Hayward
(Watlington, Oxon., 24 September (?), 1824 – Bingham, Notts., 2 August 1886).

TB encountered Hayward on 14th April and met up with him on each subsequent day till he left for Damascus on Monday 17th when Hayward came to say goodbye. They plainly knew each other:

> *After one Mr Hayward, whom I met to my astonishment in the street yesterday, came to walk with me.*

He never expands the name or explains the relationship but fortunately a reference in a book of Elizabeth Finn's (1929: 120) is helpful – 'Rev. J. W. Hayward'.

Rev. John Wheeler Hayward, son of a prosperous business family in South Oxfordshire/north Berkshire, had graduated from Trinity College, Cambridge: 'B.A. 1847; M.A. 1850. Ord. deacon (Oxford) 1849; priest, 1851'. Although TB was two years older, his late start at Oxford put him on an exact par with Hayward at Cambridge – namely B.A. 1847, M.A. 1850, though he was then ordained a little earlier than Hayward. Although the relationship may have stemmed from having grown up in the same limited part of Oxfordshire-Berkshire, their university background and choice of vocation, their closest relationship was when both were pupils at Rugby where Hayward was a younger contemporary and just a year or so older than TB's brother Samuel with whom he had entered in the same year (1839).

Hayward was apparently in Jerusalem at that time to support *'a young clergyman'* – identified by Mrs Finn as the 'Rev. W. Beamont'.[5] Hayward seems to have gone then or soon after to take up a post as a clergyman with the army in the Crimea – *'Chaplain to the Forces, in the Crimea, 1855'*, possibly in concert with his friend Beamont who was there at much the same time as *'chaplain in the camp hospitals of the British army in the Crimea'* (ODNB).

Hayward returned from the Crimea in 1855 to become Vicar of Granborough, Buckinghamshire (1855-1871), and on 5th November 1856, married Annette Jane Lane, daughter of Major General Ambrose Lane,[6] at Wood Eaton just outside Oxford, where her brother-in-law was vicar. They were to have at least one daughter and two sons.[7]

Hayward died quiet young (62), having spent the last 15 years of his life as Vicar of Flintham in Nottinghamshire.

***

---

[5] There is potential confusion. TB's young friend will have been William John Beamont (1828-1868), a near contemporary (but at Cambridge) who had made a trip to Sinai in 1853, returned home to be ordained in 1854 and immediately then gone out to Jerusalem where he taught at the English College and took a special interest in educating men to become missionaries to Abyssinia. His father, William Beamont (1797–1889), travelled to Jerusalem in 1854. The father published an account of his travels in Palestine and in Sinai (W. Beamont 1856) and both published accounts of a subsequent journey in Sinai (W. Beamont 1862; W. J. Beamont 1861).

[6] Often confused with a younger officer of the same name who served in Canada.

[7] Hayward's wife was descended from a Guernsey family (Le Mesurier), her elder sister, Louisa Clarke (née Lane) (1812-1883), was a prolific author as was *her* daughter, Theodora Louisa Lane-Clarke.

Figure 10.6: Mar Saba from the air (Copyright: Andrew Shiva).

Two days were devoted to an extensive trip away from Jerusalem. TB does not enumerate his party that set off on the morning of the 10th April but later refers to Kennedy and Yeatman being with him. As with many western travellers, especially around Easter, TB made a journey to the famous cliff-side convent of Mar Saba (Figure 10.6), 12km straight line distance southeast of Jerusalem. He gives a detailed account of the place and they camped nearby that first night.

Then on to the Dead Sea where he bathed but found the buoyancy unsettling. The salt water was painful in the eyes and left a coating he only managed to wash away that night in spring water at Jericho. In between, the party went to the supposed place on the Jordan where

Jesus was baptised. Until recently that was located on the west bank of the river but now an impressive complex of churches on the opposite, Jordanian bank, makes it more likely that was the important location of Bethany in early Christian times.[8] Interestingly, he noted that trees on the banks *'seemed eucalyptic, like the Australian though there was no aromatic smell like the gum trees'*.[9] They camped for a night near Jericho then the next morning he returned to Jerusalem via Bethany as Jesus had done on his final visit.

TB was keen to fulfil the plan he had made in Cairo to make a tour through Palestine and catch a steamer at Beirut to allow him to join another at Alexandria on 22nd June. Delays in Jerusalem meant he was now considering having to miss an intended detour via Damascus. However, during his last few days in Jerusalem he was able to resolve the issue and agreed with his dragoman John/Giovanni a plan to ride through northern Palestine to Damascus and then to Beirut to catch a steamer to Alexandria on 8th May. His last days in Jerusalem included buying souvenirs and getting a haircut - which almost ended in disaster due to a misunderstanding. One assumes that by this time he has long hair and beard (below, Beirut).

---

[8] 31°50'15.06"N; 35°33'0.10"E
[9] Australian trees were imported to Palestine about this time probably including fast-growing eucalypts. Cf. Mendel and Protasov 2019.

Chapter 11

Palestine, Syria and Lebanon and Return Home

The 'Desert Group' broke up at Jerusalem. When TB finally set off again it was in the company of Kennedy. Initially they were accompanied by the Rogerses but soon left them behind. Nevertheless, TB and Kennedy still occasionally encountered some of their erstwhile companions - especially the Eameses and Rodewald, then more of them again at Damascus then at Beirut.

Unlike Stobart who was in Jerusalem a few weeks later, TB did not take the opportunity to cross the Jordan to visit any of the cities of the Decapolis in northwest Jordan. An adventure too far when he had his sights set on getting home. This final month of his travels in the Levant took him though northern Palestine, to Damascus then on to Beirut to catch a steamer to Alexandria. In these last weeks of travel as 'home' got closer, TB was beginning to think about what was to come next:*'I must get to work as soon as I can after getting back. I sometimes wonder how I am to live.'*

Despite his keenness to get home, he continued to pursue visits to places of interest along the way (Table 11.1). Most were places of Christian religious interest, but Damascus had its own attractions as an ancient city. He passed up the opportunity while at Damascus to visit Palmyra, but he did divert between Damascus and Beirut to the immense ruins at Baalbek.

It is notable that while TB still seems full of energy, Kennedy is said at times to be tired and to have not accompanied him on his walks – as had sometimes been the case on the Long Desert Route (above). At Beirut Kennedy tried to persuade him to join him in travelling on to Constantinople but TB was determined now to return home. While they waited for their respective steamers the two nevertheless went to the mouth of the Nahr el-Kalb/Dog River estuary *c.* 12km straight line distance northeast of Beirut. They sailed there but TB returned three hours by donkey while Kennedy favoured the less wearing sea trip again. The objective was to view the famous inscriptions on the cliff face overlooking the narrow coastal road where it crosses the little estuary (Figure 11.1 (a)). There were almost twenty of these when TB visited, stretching from the Egyptian pharaoh Rameses II in the mid-13th century BC, through Neo-Assyrian and Neo-Babylonian six centuries later, Roman of the 3th and 4th centuries AD, a Mamluk Sultan of the 14th century then the French Emperor Napoleon III (1860-61). (Far in the future but something that would have been a matter of wonder to TB in view of his recent visit to the nascent colonies in Australia and New Zealand, is the inscription cut there in 1919 commemorating the advance of the largely ANZAC Desert Mounted Corps commanded by the Australian, Lt. General Sir Harry Chauvel). TB noted several of the more than twenty inscriptions and wrote down the first two lines of a long Latin one. The text is damaged and difficult to read but it can be dated to AD 215 and erected in the name of Caracalla (his nickname, as he was officially Marcus Aurelius Antoninus Pius Felix Augustus; Figure 11.1 (b)).

Palestine, Syria and Lebanon and Return Home

Table 11.1: The travels of Thomas Bowles in the Levant from 17th April till 8th May 1854 (cf. Appendix 2).

Monday 17 April 1854	Jerusalem to Ain Es Yebrad via Shafat and Nebi Samwil
Tuesday 18 April	Ain Es Yebrad to Nablous via Ain el Haramiyan, Sinjil, Khan Lubban, Jacob's Well, Gerizim, Samaritan Synagogue
Wednesday 19 April	Nablous to Jenin via Sebastiyah
Thursday 20 April	Jenin to Nazareth: Greek Church, Latin Convent
Friday 21 April	Nazareth to Tiberias via Erana, Cana, Plain of Esdraelon, Es Suk
Saturday 22 April	Tiberias to Wadi Mellahah via Mijdal, Ain et Tin, Tell Husn, Khan Minyeh, Khan Jubb Yusuf
Sunday 23 April	Wadi Mellahah to Banias (Paneas, Caesarea Philippi) via Ain Mallahad, Kalat el Hunin
Monday 24 April	Banias to Bejam
Tuesday 25 April	Bejam to Jedehda
Wednesday 26 April	Jedehda to Damascus: visit to Bazaar and Khan
Thursday 27 April	Damascus: Bazaar, Banker's house, Street called Straight, Khan
Friday 28 April	Damascus: Bazaar
Saturday 29 April	Damascus to Ez Zebedani via Valley of the Barada, Souk Wadi Barada (Abila)
Sunday 30 April	Ez Zebedani to Baalbec: Temple of Jupiter
Monday 1 May	Baalbec to Sahleh
Tuesday 2 May	Sahleh to Khan Hussein
Wednesday 3 May	Khan Hussein to Beirut
Thursday 4 May	Beirut
Friday 5 May	Beirut: Excursion to Dog River
Saturday 6 May	Beirut
Sunday 7 May	Beirut: Chapel of the American Mission, Maronite Church, Chapel of the Sisters of Charity
Monday 8 May	Leave Beirut on Steamer

Then Kennedy disappears from the narrative. After so long together it is surprising that TB says nothing about their final parting on 8th May.

In Damascus and Beirut, TB again encountered some of those with whom he had shared the hotel in Jerusalem, some of whom were to be companions again and again in the subsequent weeks.

Lucius Bentinck Cary
10th Viscount Falkland (5 November 1803 – Montpellier, France, 12 March 1884)
and
Amelia FitzClarence, Lady Falkland
(Teddington, 21 March 1807 – London, 2 July 1858)

Arriving at his hotel in Jerusalem, almost the first people TB encountered were the Falkland party:

Figure 11.1: (a) The earliest inscriptions as reproduced by the French artist Cassas in 1799; (b) a photograph of 1922 of the inscription of Caracalla recorded by TB.

Abb. 13. Römische Inschrift.

IMP · CAES · M · AVRELIVS
ANTONINVS · PIVS · FELIX · AVGVSTVS
PART · MAX · BRIT · MAX · GERM : MAXIMVS
PONTIFEX · MAXIMVS
MONTIBVS · INMINENTIBVS
LYCO · FLVMINI · CAESIS · VIAM · DELATAVIT
PER //
ANTONINIANAM · SVAM

It was quite full - Lord Falkland & his family being due on their way fr(om) Bombay. I went to the Salon in the Hotel, & sat by a stove & had a long talk with an old gentleman,[1] who turned out to be L(or)d Falkland. At dinner there were 11 men & 2 ladies beside ourselves & at a 2nd table by themselves L(or)d Falkland & his party consisting of 2 ladies & another man, Major Folay (sic).

TB encountered them again in the coming days in the course of sight-seeing and on one occasion - at Calvary on Easter Friday, he joined their party.

In April 1854, Lord and Lady Falkland were aged 50 and 47 respectively. They were a distinguished couple with a colourful background. Lord Falkland was the 10th holder of a viscountcy created in 1620 - and which survives still in the person of the 15th Viscount Falkland.[2] He had been Viscount Falkland since he was five (in 1809) when his father had been killed in a duel. The 10th Viscount had a broad and distinguished career: almost ten years as an army officer; in 1837 admitted to the Privy Council; Governor of Nova Scotia (1840-1846); Governor of Bombay Presidency (1848-1853). In 1830 he had married Lady Amelia Fitzclarence, an illegitimate daughter of King William IV (1830-37) and his long-standing mistress, the Anglo-Irish actress and courtesan, Dorothea Jordan - their tenth child.[3] She had been given away by her father who had just recently become king. From 1830-1837 she was the daughter of the reigning king then from 1837 cousin of the new monarch, Queen Victoria.

Lady Falkland died young - in 1858, just four years after TB met her, aged 51. In the intervening years she published a two-volume book based on her journals - *Chow-chow: Being Selections from a Journal Kept in India, Egypt, and Syria* (1857). It is a delightful read with extensive coverage of the period in which she and Lord Falkland were in contact with TB. Although she mentions or alludes to a number of people she encountered, none can be identified as TB. His reports on extensive conversations with the Falklands were more significant for him than them.

The Falklands and their military attaché, Major Foley (below), were on their way home from Bombay via Suez and Egypt when TB met them in Jerusalem. They made a trip on the Nile then sailed from Alexandria to Jaffa. They left Jerusalem on the 15th April - two days before TB; in view of their eminent status, they were accompanied by some Turkish soldiers provided by the Pasha. TB was to meet them again several times. When TB reached his hotel in Damascus the Falklands were already in residence and they overlapped for a few days. They all boarded the same steamer at Beirut and TB reports conversations with Lady Falkland both on the trip to Alexandria then on the next steamer to Malta then to Marseilles where they evidently separated on 20th May 1854.

---

[1] Lord Falkland was hardly old – just 50.
[2] Although created in the Peerage of Scotland and named for Falkland in Fife, the 1st Viscount was English. The Falkland Islands were named for the 5th Viscount and the current viscount - Lucius Edward William Plantagenet Cary, like all of the last five viscounts, includes the middle name 'Plantagenet' in recognition of a claim to be descended from that dynasty (1154-1485) which ended with the death of Richard III, *'in token of their descent from King Edward III and John of Gaunt through a female line of the Beaufort dukes of Somerset'* (Blumberg 1980: 162-63 n. 27).
[3] Lady Falkland had at least two of her family close by in India. Her brother, Lieut.-Gen. Lord Frederick FitzClarence, GCH, was Commander-in-Chief, Bombay (1852-54), the military counterpart to her husband. Her nephew, Capt. Hon Frederick FitzClarence, 10 Hussars, son of her eldest brother the 1st Earl of Munster, was one of Lord Falkland's two ADCs in Bombay and soon to see service in Crimea. On the other hand, the mortality rate in this otherwise large family of ten children, was high, not least precisely in the 1850s: her brother in Bombay died in July 1854 a few months after the Falklands had left for home and likely unknown to her; her youngest brother, Lord Augustus FitzClarence, had died the previous month (June 1854); another brother (Rear Admiral Lord Adolphus FitzClarence) and a sister (Elizabeth FitzClarence, Countess of Erroll) both died in 1856; and Amelia herself died in 1858.

After the initial conversation with Lord Falkland in Jerusalem, all of TB's later conversations were with Lady Falkland. He says they were extensive - *'a long talk'*, *'a good deal of talk'*, *'I sat talking to Lady Falkland till after Tea'*. All seem to have been on religious matters and she evidently had a keen interest in that.[4] In her own book, Lady Falkland regularly refers to making sketches (e.g. 1857: II, 193, 197, 199, 214, 243, 251, 254, 268, 276, 280). Near Marseilles, TB reports being shown some of them but offers no opinion. Lady Falkland seems to have compiled albums of sketches - mainly her own but including some by others including her husband. One such album is held in Library and Archives Canada and includes several items which may be from this period in the Middle East and four which certainly are and were likely amongst those she showed TB:[5]

- Djenin, Syria. 1854
- Village of Banias. 21 April 1854
- Tower of Rama Syria. 1854
- Ruins of Tiberias from the Plain. 1854

Unlike TB, Lady Falkland did enjoy her stay in Jerusalem (1857: 2, 223-4):

We left the city by the Damascus Gate [= 15 April 1854]. I never regretted leaving any place so much as I regretted leaving Jerusalem; and, as we lingered to take a farewell view of it, I thought how much I had left there unexplored. Owing to indisposition, and the rainy weather, I had lost some days of 'sightseeing.' It is not difficult to re-visit spots in Europe which have pleased one; but the Holy Land is not often visited twice in one's life. I therefore left 'the city of David ' with the sad reflection, I should never see it again.

Between Damascus and Baalbek, Lady Falkland's maid was injured and she was glad of assistance (1857: 2, 272-3):

We had not gone far, however, when an English physician, who had been staying at Damascus, rode by. He kindly gave all the assistance in his power, and promised to visit the poor sufferer at our encampment at the end of the day.

This may have been Dr Bryce.

***

General the Honourable Sir St. George Gerald Foley, K.C.B.
(London 10 July 1814 – London, 24 January 1897)

'Major Foley' is part of the Falkland party and TB, therefore encountered him first at their shared hotel in Jerusalem on 3rd April 1854 then would have parted from him at Marseilles along with the Falklands on 20th May 1854. In between they encountered one another in the course of sight-seeing in Jerusalem. At Damascus TB and Foley made a joint trip to the Bazaar

---

[4] Reinhart 2005: 28 describes her as 'a devout Anglican openly critical of the Catholic Church, occasionally conflating Catholicism with Hinduism and Islam which she considered heathen and rife with superstition.'
[5] Three sketches of people in Middle Eastern costume are in the album, dated 1837 and attributed to Lord Falkland. Presumably relating to an earlier visit to somewhere in the region.

to buy souvenirs. They surely had far more opportunity or conversation in the long voyages from Beirut to Alexandria then on via Malta to Marseilles. Although TB does not identify Foley other than as travelling with Lord and Lady Falkland and the latter's maid, the Army List for 1852 records him as: 'Military Secretary, Maj. Hon. G. St. G. Foley, Unatt(ached)'

At Malta TB mentions Foley for a final time and helps with his identity:

> *I met Major Foley on the way just coming back. He has found letters ordering him to join the Army in Turkey at Constantinople. It is a nuisance for him after 5 years absence not to get a peep at England & I think he is vexed now that he delayed in Egypt & Syria - but he is too lucky in getting this appointment to grumble.*

Evidently Foley was moving from 'Unattached' as the Military Secretary to Lord Falkland to an active post in the Crimean War then getting underway. It is not clear if he left the Falklands in Malta to travel direct to the war – as noted, below, at least part of the garrison of Malta was soon posted to the Crimea.

The Honourable St. George Gerald Foley was the brother of the 4th Baron Foley. His career can be traced from his enlistment in the infantry in 1832 (53rd Foot), service in Malta, then ADC to the Commander-in-Chief in Ireland, and Military Secretary to the Governor of Bombay (Lord Falkland). Now, in 1854, aged almost 40, he was posted to Crimea where he served as an Assistant Commissioner at French HQ and was awarded an Ottoman medal - the Order of the Medjidie, in 1854. He next major military service was in the Second Opium War (1856-60). He was awarded a knighthood (KCB), retired as a General and then served as Lieutenant Governor of Guernsey (1874-79). One of his two sons - Lt. Colonel Cyril Pelham Foley (1868-1936), also had a distinguished military career in both the Boer War and First World War, including service in the Royal Flying Corp. One of Foley's grandsons was to inherit the baronetcy as the 7th Baron Foley.

***

After Beirut TB's steamer touched briefly at Jaffa before continuing to Alexandria where he stayed for two days awaiting his next steamer (Table 11.2). Even now he had some of his Desert Group companions on board – Bryce, Ward and Freeman, and newer companions from Jerusalem onwards – especially the Falklands, Foley and Rycroft (below). At Alexandria he encountered the Fenton brothers whom he had not seen since they left Petra and had gone their own way into Sinai (above). While in Alexandria he learned of the recent passage through of companions from much earlier travels:

> *Tomorrow we start again towards Home. Fancy Stobart & his charge passed through here a fortnight ago on their way home - the 2 via Trieste. Schomberg Kerr by Gibraltar - and now to bed.*

At that stage he was not to know that Stobart and Lord Henry were soon to return to Egypt and then to Palestine and an adventure across the Jordan (above, Ch. 2).

After Alexandria when TB boarded the steamer for Malta he had the company of his Petra companions Ward and Bryce and his newer ones, the Falklands, Foley and Rycroft. He reports

Table 11.2: Travels of Thomas Bowles from Beirut to Home, 1854 (cf. Appendix 2).

Monday 8 May	Leave Beirut on Steamer	279-280
Tuesday 9 May	On steamer off Jaffa	280-281
Wednesday 10 May	On steamer between Jaffa and Alexandria	281-282
Thursday 11 May	Alexandria: English Church, Cleopatra's needle, Pompey's Pillar, Mahmoudie Canal, Pacha's Palace, Library, Roman Catholic Church	282-285
Friday 12 May	Leave Alexandria onboard Steamer	285-286
Saturday 13 May	Onboard Steamer "Mentor"	286
Sunday 14 May	Onboard Steamer "Mentor"	286-289
Wednesday 17 May	Valletta, Malta: St. John's Church, leave Malta onboard Steamer "Mentor"	289-291
Thursday 18 May	Onboard Steamer "Mentor"	291-292
Friday 19 May	Onboard Steamer "Mentor"	292-293
Saturday 20 May	Onboard Steamer "Mentor". Arrive Marseilles.	293
Sunday 21 May	Marseilles to Avignon	294-295
Monday 22 May	Avignon: Cathedral, Palace of the Popes, Museum	296-298
Tuesday 23 May	Avignon to Vienne	298-299
Wednesday 24 May	Vienne to Lyons and on train from Chalons to Paris	299-301
Thursday 25 May	Paris to Rouen: Cathedral	301-304
Friday 26 May	(On the Channel making for Southampton)	304
Saturday 27 May	(Home with family at Mill Hill)	304

conversations on religious matters with Lady Falkland but that he would not rise to the bait when Bryce – a 'Dissenter', attempted *'to get into an argument with me about Church matters - a state Church, the Sacraments - Form etc. - till I declined the discussion.'*

Nelson Rycroft
(Brighton, 11 March 1831 – Dummer, Basingstoke and Deane, 30 March 1894)

On the third occasion on which TB met up with Hayward (above, Ch. 10) – Easter Sunday, 16th April 1854, the latter was with a companion:

> *After luncheon I packed up & I went down to the Church - where I fell in with Hayward & Rycroft. We found nothing going on. So we perambulated the various parts of the Church & saw the sword of Godfrey de Bouillon etc.*

TB was to meet Rycroft again on 6th May at Beirut and a few days later Rycroft evidently was a fellow-passenger on the steamer from there to Alexandria. Then again on the final steamer to Malta – where he was helpful trying to arrange a cash advance for TB, and on to Marseilles. He appears for the last time on 21st May when *'our baggage went to the [Marseilles] Station & Rycroft's servant got our Tickets.'* They presumably travelled to Paris together on 24th and parted company there.

Before he turns up in Jerusalem with Hayward, Rycroft had been in Egypt, sailing on the Nile. Bryce records *'Messr Rycroft & Briggs'* on one of the European boats he knew of on the Nile in 1853 and says he passed them *'near Thebes Dec 29 [1853]'*. Two days later Andrian (2011: 3, 7-8), on the Nile between Kenneh and Thebes on 1st January 1854, recorded:

> *In the Temple of Venus we met 3 young Englishmen who came from Wadi Halfa and pleased me quite well, namely one of them, a M. Briggs of the Indian Army, ...*
>
> *... After I had made a brief visit on board the ship of Messrs. Briggs, Ryecroft (sic) and Elwin (sic), we left Kenneh at half past three,*

'Elwin' is presumably Valentine Dudley Henry Cary Elwes (1832-1908), at that time a Cornet with the 12th Lancers in India. The army is the likely connection between these three young men. Fortunately, the Register of Shepheard's Hotel records that *'Nelson Rycraft (sic) left for Upper Egypt, late Oct or early Nov.'* As Andrian refers to *'young Englishmen'*, we can be certain this is Nelson Rycroft, who inherited his father's title and became 4th Baronet Rycroft in 1864.

In 1854 when Bowles met him, he was just 23 years old. Census returns provide the framework of his life: in 1841 he was boarding as a 10 year old 'Pupil' at Berkhamsted School (founded 1541); in 1847 he was elected to Eton and in 1851 he is an Ensign stationed at Fulwood Barracks in Preston, Lancashire. As the eldest son, destined to inherit title and principal estates, there would have been no particular need to continue an army career. In 1861 he is a Justice of the Peace living at Chevening in Kent. In 1871 – having inherited after his father's death in 1864, he is Baronet and Landowner, farming 380 acres, at Kempshott House, near Basingstoke in Hampshire.[6] In 1881 he is Sherriff of Hampshire. In 1894, aged 63, he died quite young. He is buried at Dummer near his family seat.

Like many of the other young men TB encountered – though not TB himself, Rycroft married not long after his return from the East. On 27th July 1858 he married Juliana Ogilvy at Mains and Strathmartine, Angus, just north of Dundee - presumably the local church to her father's home at Baldovan House. Juliana was the eldest of six children. Her father, Sir John Ogilvy (1803-1890), was 9th Baronet Ogilvy of Inverquharity - a Baronetage, founded in 1626, not of Scotland but of Nova Scotia in Canada. Despite the family's deep roots in Scotland, her father - eldest child of Rear-Admiral Ogilvy, had been educated at Harrow and Christ Church, Oxford. He was later a Liberal MP for Dundee from 1857-1874. Juliana's mother was a daughter of Lord Henry Thomas Howard-Molyneux-Howard, younger brother of the 12th Duke of Norfolk. Juliana died 6 January 1917.

Nelson and Juliana Rycroft had five sons and a daughter. The second son became Major General Sir William Henry Rycroft (1861-1925), KCB, KCMG, Governor of British North Borneo 1922-1925 (where he died). Edmund Hugh Rycroft (1862-1932), was a Church of England clergyman. Margaret Charlotte Henrietta Rycroft (1871-1949), never married; likewise her twin brother, Evelyn George Rycroft (1871-1920). The youngest son, Charles Michael Richard Rycroft (1864-1897), was a Captain serving at Meean Mir, the Cantonment area of Lahore, when he died

---

[6] Kempshott House was once leased by the Prince Regent and used as a retreat for his life with his mistress, Mrs Fitzherbert. It housed German POWs in WW1 and 2 and was demolished in 1965 when the M3 motorway was built.

very young and unmarried. Nelson and Juliana's eldest son, Sir Richard Rycroft (1859-1925), inherited the baronetcy which is now held by a descendant, the 8th Baron Rycroft.

***

On the steamer between Alexandria and Malta, there was a request to TB to conduct a service but he observes that *'The applicant suggested that we were so many English - forgetting that though all English we are not all Churchmen [i.e. Anglican?]. I believe hardly the majority could be found to be churchmen.'* As he mentions Bryce in the next sentence it is likely he had in mind the presence of Dissenters.

Malta was an opportunity for some domestic pleasure: he was able to wash properly– in the very British hotel, then *'Delicious to get decent Tea & bread & butter & toast an <u>innocent</u> breakfast instead of the Early Dinner they give us on board the "Mentor".'* It was a chance to replenish his dwindling money as he was concerned he might have insufficient to get across France. Rycroft introduced him to a banking contact in Valletta. The latter asked for some security and TB was able to tell him he had a friend stationed there in the garrison – 'Arthur Coussemaker', a rare occasion when TB gives someone their personal name.

***

Arthur Lannay Coussmaker
(Farnborough?, 7 February 1833 – 20 December 1870)

TB's recording this man's first name may be because they were relatives of a sort. Arthur Coussmaker was the nephew of TB's brother-in-law Henry Woodyer,[7] the death of whose young wife in childbirth may have influenced TB to join Stobart on his Grand Tour just four months later (above, Ch. 1).

The Coussmakers were of Dutch heritage – perhaps Flemish from around Bailleul. John de Coussmaker probably arrived in Britain in 1688 with William of Orange and *c.* 1720 bought the Westwood Estate at Wyke-Normandy, near Farnborough in Surrey. The cemetery at St Mark's Church, has numerous tombstones and the church itself has several memorial plaques to this prominent landed gentry family.[8] There were still Coussmaker descendants until 1961 (Ashworth and Kinder 1998).

Arthur Lannoy[9] Coussmaker was the eldest of eight sons and six daughters of Lannoy Arthur Coussmaker. Two of the sons died in infancy, one became an engineer, one entered the church and four went into the army. Arthur Coussmaker was just 21 when he met TB on Malta in May 1854.

---

[7] Arthur Coussmaker's mother, Mary Anne Woodyer, was the sister of Henry Woodyer who was the widower of TB's sister Frances Martha.
[8] I am grateful to Terence Madigan for numerous photographs of tombstones, memorial plaques and documentary records from this church.
[9] Lannoy is often used as a first or middle name and – amongst mere Marys and Margarets of Arthur's thirteen siblings, are other unexpected names: Macclesfield, Lannette and Octavius as well as Archibald Lindsay Coussmaker. The origin of 'Lannoy' probably lies in the placename in northern France not far from Bailleul.

> *So I went to see for A. Coussemaker - I got a man to show me the way to the Barracks. I found him living at the Mess House.*

The 'Barracks' are presumably those at Florian on the same peninsula as Valletta and where the Badger family had spent many years (above, Ch. 3). No rank is given but he evidently left the service after a few years and at the time of his death was styled *'Captain Coussmaker'*. An entry for his father in *The County Families of the United Kingdom* (1869: 241) adds that 'issue' included:

> *Arthur Lannoy, educated at Sandhurst; late Capt., 3rd Buffs and 85th Light Infantry; b. 1833.*

The *'3rd Buffs'*, one of the oldest in the British army – 3rd in the line, and formerly known (appropriately for a Coussmaker), as 'the Holland Regiment'. It is now – through amalgamations, the Princess of Wales's Royal Regiment (Queen's and Royal Hampshires). The regiment was posted to Malta in April 1851, its near 600 men taking up position in the Floriana Barracks. It seems to have been in Greece in 1855, then at the siege of Sevastopol in the Crimean War. Whether Arthur Coussmaker was with the regiment is not known.

Arthur Coussmaker never married and died young – aged 37, and in the same year as his younger brother, Lieutenant Lannoy Richard Coussmaker, died aged 22 on service in South Africa.

***

TB had a pleasant final cruise from Malta, along the coast of Sardinia to reach Marseilles. His passport had not been returned to him at Malta and he had to visit the British legation for assistance. But he was in a familiar country with the amenities of home and no more of the *'dirt of the East'*. Although aiming for Paris he still had tourism in mind. First:

> *... a turn - round by the port & up above by a fort where we had a pretty view of the Entrance to the Harbour & saw a steamer start with troops for Constantinople - Poor fellows - they are sick enough by this time & wish themselves back I expect - though in the excitement of going "Glory" was everything I dare say & many of them perhaps will never see France again.*

They went first to Avignon by train: *'It was delicious to find oneself in a Train again. For 19 months I have gone back to the various modes of travelling in use amongst our forefathers.'* He was largely unimpressed by Avignon, the churches, Palace of the Popes and Museum; but he did like the scenery and was soon to embark on a steamer on the R. Rhône. That took him to Lyon where he crossed over to the R. Saône and continued to Chalons where he boarded a train for Paris - *'I determined to be a spend thrift & go first class',* and arrived early on 25th May 1854. Interestingly, he was subjected to a customs inspection at the station. Even then he was intent on further tourism but the closure of ticket offices prevented a trip to Rouen until the next day. As so often during his trip he visited churches and – this time, in Rouen, was delighted by the buildings and the choirs.

His journey was almost over and his entry on the final page concludes:

> *My train goes at 11.30 [for Dieppe presumably] & I must see what I can before Dinner at 10. This time tomorrow I hope please God I shall be on the English Channel making fast for Southampton - where this journal will come to an end. Fancy being with them again on Saturday Evening! I hope they will be all at Milton Hill.*

Part 2
The Journals

Chapter 12

The Journals: Character, writing, composition, survival

It was common for western travellers in Egypt and the Near East to keep a running account of their travels and activities. Most such travellers were attended by servants of some kind and it was a regular practice while the latter prepared the evening meal and/or set up camp, for the traveller to take time to write notes or even lengthy journals. Thomas Bowles frequently ends the entry for a day with a note that he retired and wrote what he calls his 'journal'.

Thomas Bowles evidently began writing his first Journal when he left home. The second Journal commences while in Australasia and the final one begins mid-sentence with his arrival in Cairo and then his account of his tourism in Egypt, his great expedition to Petra and then through the Levant before returning home.

Journal 1 is lost. Journal 2 is held in the National Library of Australia which bought it from a bookseller at Hay-on-Wye in 1967 for £120. Journal 3 was purchased from an antiquarian bookseller in New York City in 2019 (details for all three journals appear in Table 12.1). The two surviving journals are bound differently from one another and the third – which is the principal subject here, seems to have been re-bound and had the spine lettered: TRAVEL DIARY NEAR EAST 1854 (Figure 12.1a and Figure 12.1b). Its dimensions are 11 x 11 cm. It is written in ink throughout including the occasional sketches.

Thomas Bowles did not marry and his will shows him leaving an estate valued at £580 to his nephew Francis Wildman Selwood Bowles. There is no mention of what might be called his 'papers' and the circumstances in which the three diaries were split up with one even getting as far at New York, are unknown. Neither of the two sellers has a record of where their diary was acquired.

TB's journal was a personal record of people and activities on any given day. What he chose to include was no more than what he thought he might need to record for his own use. There was no intention the journals would be read by anyone else and – if he had followed the example of other travellers who subsequently published their writings, he would have likely explained a great deal more for the benefit of the reader. The difference can be seen in the

Table 12.1: Structure and coverage of the three Bowles Journals

Volume	Pages	Dates	Places
1	[1 - 291		Southampton – Cape Town – Sydney
2	292 – 753 (1 – 463)	28 October 1853 - 13 February 1854	Tanna Island, New Hebrides [= Vanuatu] – Hong Kong – Penang - Ceylon [= Sri Lanka] - Aden - Cairo
3	754 - 1057	13 February 1854 - 25 May 1854	Cairo – Mt Sinai – Aqaba – Petra – Jerusalem – Palestine – Damascus – Beirut – Alexandria – Malta – Marseilles – Rouen.

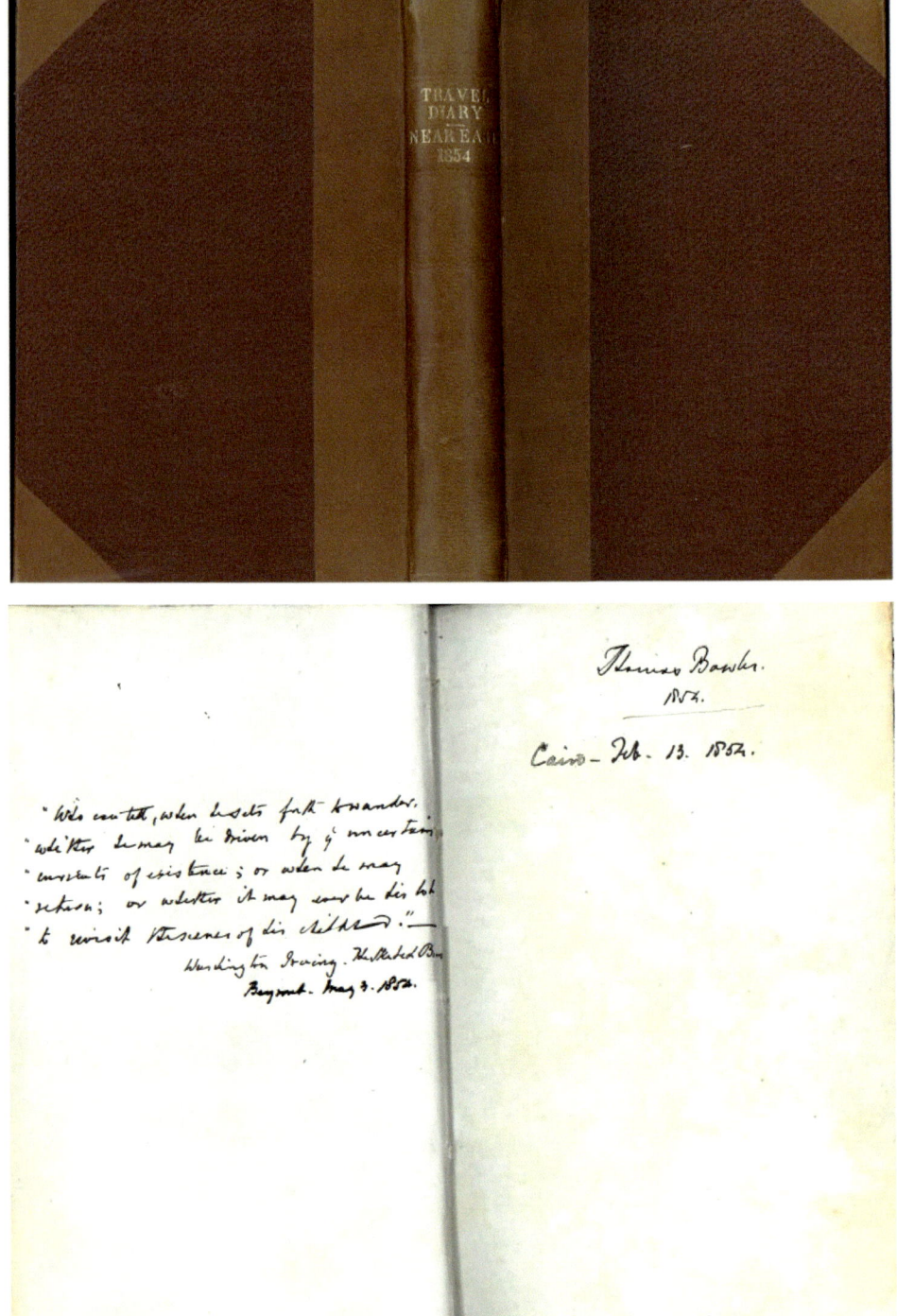

Figure 12.1: The third Travel Journal of Thomas Bowles: (Top) Cover; (Below) 'Title Page'.

parallel writings of his companion to Australasia, Henry Stobard (above, Chapter 2). The latter was writing letters home and recognised the need for something more than would have gone into a journal. This is part of one of his letters to his mother, a facsimile of which is also held in the National Library of Australia in Canberra. It was written while Stobart was at Brisbane, Moreton Bay in what is now New South Wales and dated 'Saturday Night July 9th 1853.' As the second item quoted from Stobart below shows, he was also keeping a journal though not always managing to keep it up-to-date (Letter 6, Page 1):

> *I herewith make a beginning of a Journal letter which I will try to keep regularly adding to whilst settled at one place, as we are here. There will be very little probably to jot down, but if I try to remember that I am writing to you at distance, to whom everything is new, I may probably take the trouble to put down many little things which may be interesting to myself hereafter, which when writing only for oneself one is apt to pass over, so gradually does one get accustomed to them, as not to be aware that there is any novelty to them. I will keep the journal daily just as I should have written it in my book, and as I sent off a letter to you yesterday morning I will commence with the after events of that day.*

Four weeks later on 4 August (Letter 7, Page 1):

> *Since Monday week until Tuesday we have been absent from Brisbane and I sit down with a very bad grace to write up my journal of the week which has been full of interest and amusement. For it is the most wearisome all occupations that I know to make up the back work of a journal, especially when you would like to put down a good deal – besides, so much is lost of the freshness as when written at the time. I took paper and pen down with me to the Bay where we have been, but with five in a small tent and no table and above all no spare time, it is impossible to keep it up. I must try and recall as well as I can.*

We need to differentiate between diaries/journals on the one hand and the extended letters sent home by travellers. In the former, authors are writing only for themselves and doubtless felt no need to record matters they would recollect. In letters, however, they were having to inform the recipient about things the latter would need to have explained. It need not surprise us, therefore, that Bowles can be exasperating in his vagueness about people he meets and travels with. It would be useful to have some of the letters he sent home for comparison. He certainly received and sent some, but none seems to have survived.

Chapter 13

Abbreviations and Symbols; Editing

Although we do not have the first volume of the Travel Journals we can infer from the second and third how Bowles evidently set out his pages. Each had a 'Header' in which he gave a page number, a date and a term to describe place or activity. In the second and third volumes he carried on the numbering from the first volume and added a new starting number for the new volume so that each has two page numbers. (He occasionally made a slip in the numbering which is noted in the transcription). For example, in Journal 3, we have the header (Figure 13.1):

      21   774   "Cairo"   17 Feb.

This is page 21 of the third journal and 774 in the cumulative counting. Following on from the bottom of the previous page, the first three lines may be transcribed as: *'There was a very handsome pulpit for the higher (21/774 "Cairo") priest to read the Koran from, with a staircase covered with Arabesques, the pulpit also & a canopy over the pulpit, with a sort of Egyptian Vase on the top of that.'* illustrating both abbreviations and handwriting.

In the transcript (Chapter 14), abbreviations have usually been expanded. Where there appear to be spelling errors these have usually been retained with the addition of '(sic)' or/and with the correct word in square brackets for clarity. Problems reading a word are indicated by the addition of '(?)' to the proposed reading or left as '****' where no viable reading was possible. Sometimes a word is entered in the text in square brackets usually where a personal or place name requires explanation – e.g. *'This was the Tomb of El Kaidbai [= Sultan Qaidbay] 1496 AD.'* Occasionally a footnote reference has been necessary. Elements in the text deserving more extensive commentary or explanation are dealt with in the relevant parts of Chapters 3 to 11 above.

In the transcription (Chapter 14) both page numbers and the place name/activity in the header have been retained for easy access to the original diary pages and inserted in parenthesis in the transcription at the correct point (as shown in the transcription of a few lines above).

Bowles used a number of abbreviations most of which are obvious contractions of the word followed by a colon or the first and last letters. E.g. wh: = which; c^d = could (Figure 13.2).

TBs punctuation has been retained except in one instance: He frequently uses a short dash (-), which can mean either a stop or a dash mid-sentence; I have substituted what seems best for the occasion. Underlining in the Journals has been retained.

TB includes a handful of simple sketches in his text. Only a few of the more important have been retained.

Abbreviations and Symbols; Editing

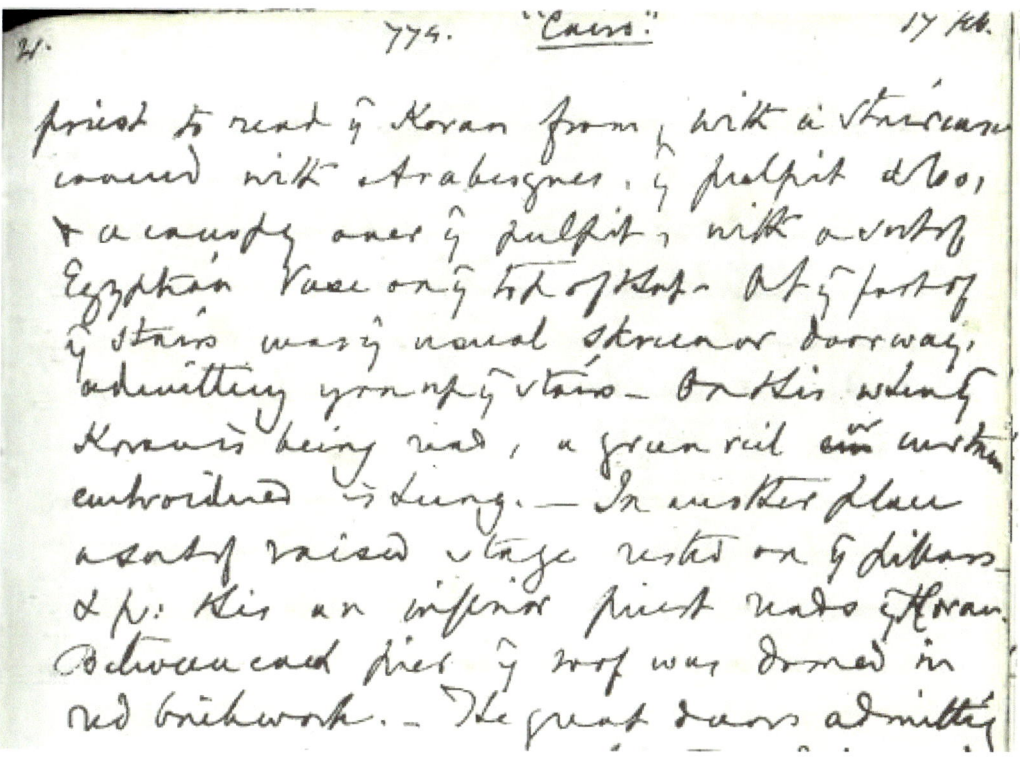

Figure 13.1: Page 21 of Travel Journal 3 illustrating the header, handwriting and some of the commonest abbreviations.

Abbreviation	Transcription	Abbreviation	Transcription
ĝ	the		by
	&		from
wd:	which		very
	could		everything
	should		O'clock
	would		ss e.g.: (passed)

Figure 13.2: Common abbreviations and frequent words as written in the Journals

Chapter 14

Transcription of the Travel Journals of Rev. Thomas Bowles

28 January (Galle, Sri Lanka) to 25 May 1854 (Rouen, France)

Transcribed by Alana Colbert
With edits and additions by David Kennedy

Travel Journal 2

(411/ 701 "Galle")

Sat. Jan(uar)y 28 [1854]

I got up and went out and found the steamer in. I was going down to the little jetty & met Dr {Garston}& Tupper. Dr Garston said his {sale} (412/ 702 Galle) was coming on that {morning} & I had better {sort} my bed & things to be sold. An offer I jumped at. I found I could do nothing more at present than just ascertain that the ship would go if possible again that day but not till the Evening. I went back to breakfast & found Mr Forbes had kindly set his 'Chef' to work on Breadfruit Curried & boiled & baked etc. I had just finished breakfast when in came Mr Talbot very kindly to bid me goodbye. I was a little sorry because I had ordered a Carriage in order to go & call both on Mrs Talbot & Mr Talbot at the {Cutjery}. Mr Forbes came & the session was continued some time. I then got a boat & went on board. Of course the Pursar (sic) was gone ashore. I went back & found him in the Verandah at the P & O. Office. I just gave him my Ticket & said good morning. I went back. Tupper came & went off to the ship. I went down the Library & consumed an hour or two. Then went back, called on Mrs Forbes- & {hired} some coolies & went down with (413/703 Leave Galle) my luggage. At the Office I fell in with the Pursar & Captain {Twinam} delighted me by saying that the Pursar had secured for me a Cabin to myself, the Starboard stern Cabin. I was very much charmed- it relieved me immensely & I went on board at once- got my few things into my Cabin- washed my hands & was just in time for Dinner. I can't say I like the Vessel. She is very narrow & the Cabins are poor. The height too between decks is considerably lower than in the Sincapore (sic). She has a flush deck, very narrow & not a seat or comfortable place to sit down upon. She rolls much & the throbbing & vibrating & noise of the screw are very great & annoying. About 7 we got off at last. Boats with torches were stationed all the way to show us the way out of the Harbour.[1] Heavy rain just before we started. I went up again & spent the Evening on deck, talking most of the time to the Pursar- who was a Chorister at Salisbury & was born (414/704) there. I went down & turned in.

<u>Sunday. January 29. 4 after Epiphany.</u>

I got up earlyish, after a most restless & uncomfortable night, feeling very poorly. After breakfast the Pursar came with the Captain's Compl(imen)ts to ask me if I would take the

---
[1] Contemporary reports describe the harbour at Galle as atrocious.

Service. We had Prayers. I think most of the people came. I found Adam's "Shaddow (sic) of the X" & "Distant Hills" in the Library & sat on deck reading. After dinner I went up on deck again & sat (most uncomfortably) on the Skylight talking to one of the Passengers, an Engineer who has been in the Chusan, going home. He is a Newcastle man & seems a nice fellow. He left a Wife behind, & now he has a daug(hter) whom he has never seen. I went down to Tea & sat next a Captain Hill, who has his wife, a little person like himself & little girl or girls & an orphan niece of his wife's. He knows Henry Tombs very well & says he is just gone home, with 3 Medals, likely to get his Company at once (415/ 705 S.S. Bombay) & his rank as Brevet Major at the same time.[2] He also talked a good deal about Sir C. Napier & told many anecdotes. At Dinner I had found that my opposite neighbour had been a good deal in Australia with his Reg(imen)t, Dr. Gammie & knew all the Coxes well. I was very glad to go to bed.

Monday. January 30.

I had a less painful night & got up earlyish. A perfectly smooth sea- the ship making 10 ½ under steam. After breakfast, Seton Karr offered me the "Edinburg(h)"[3] to read an article on the state of Parties in the Church, said to be by one of the Connybare (sic) family. I sat on deck & read it. It didn't do me much good. Sat lounging on my bed reading- wrote journal & got the Latitudes & Longitudes for today and yesterday. I find the observations are written out on a piece of paper & hung up in a small frame in the Companion - a very convenient place. I marked up my journal & read (416/706 S.S. Bombay) for some time in my Cabin again & then went on deck for an hour or so before dinner. One of the "Maldive" group was just off our beam about 5 miles. It was a low flat island covered with timber- near it on the west was a small low island with a few trees & between the 2 one or two native crafts. I couldn't learn anything about the Islands from the Captain beyond its being governed by a Native Rajah & being under English protection. We had made a good run of 250 miles. After dinner I went up on deck feeling wretchedly unwell. I read as long as I could see- very uncomfortable. There is no comfortable place to sit. Two only benches, shut against the bulwarks & so managed that the rail thrusts you forward. After Tea I went downstairs & made up my mind to see the Doctor. I went to his Cabin & found him out- but on going again he was there. Of course he made me up 2 pills & a draught, all to be taken (417/707 S.S. Bombay) at bedtime - what they were I don't know. I went to bed weary & ill.

Tuesday. January 31st.

An uncomfortable night. I got up & got a chair into my cabin & was only dressed when the breakfast was begun. I drank a cup of Tea & didn't want that. I sat still where I was & read till eleven, when I went to my Cabin for a while. Went on deck. I should enjoy it, if I were well. I fell in with the Doctor, who advised me to have some broth & ordered it for me. Sat & wrote journal & marked chart. We have made a run of 251 miles. I sat in my Cabin, where I found the Steward had got me a Chair, till near dinner, when I went up on deck, determining to have no dinner, but a plate of soup & a little pudding on deck. While the rest were at dinner I was

---

[2] Later Major-General Sir Henry Tombs, VC, KCB (1825-1874), son of Major-General John Tombs (1777-1848) of the Bengal Cavalry. The family was from Abingdon - where he and his six brothers followed their father into John Roysse's Free School (now Abingdon School), founded 1256, eight kilometres from where Bowles was born and 300m from where he is buried. These rural gentry families were surely well-known to one another. He won his Victoria Cross during the Indian Mutiny. (above, Ch. 4)
[3] Presumably Blackwood's *Edinburgh Magazine*.

able to get a Chair on deck. So as soon as ever they were gone down to dinner, I began to enjoy (418/708 S.S. Bombay) myself. One lady remained on deck. I had choice of all the other chairs. I picked out a good high straight backed armchair & sat down with my book, interrupted only by the arrival first of a plate of soup & then of pudding. After which I knew my comfort must soon come to an end. But dinner on board ship takes longer about than it does on land. It is a great means of killing time. Presently the diners came up & I vacated the Chair. I stood on deck still reading & finished my book, one of "Elliot Warburton's" out of the Ship's Library & then took it below & before it got dark hunted for another. It was darker than I thought however & I stumbled on a Volume of Shakespeare & carried it away to my cabin. I kept "Vespers" & said Evening Prayer & was able to read the Psalms & then the sun was quite gone & I could see no more. I went up on deck & walked for some little time & by (419/709 Scr(ew) S(teamer) Bombay) came down & read Shakespeare till just 10- when I got a mouthful of air on deck & turned in.

Wednesday. February 1st.

I awoke about 6 & got up - feeling rather better- but still in much pain. The heat has touched my liver I have no doubt. However I had less of the taste in my mouth & eat some breakfast. I sat reading Shakespeare afterwards & Seton Karr came & sat beside me with a book & of course we talked for an hour or more, of old Rugby fellows. I have often wondered what had become of so many who were there with me. They seem most of them to have gone to India. He says India seems to swarm with them. We found at noon that we had made a famous run, 277 miles. So that the ship has made over 11 knotts (sic) though I read & passed the time much in the usual way till dinner. I went (420/710) to dinner today, but eat very little. I couldn't get a sight of the Doctor all the morning. After dinner I went to my own Cabin & sat there till I could no longer see to read & taking advantage of the light to read the Evening Psalms & Lessons & other readings besides. I then went on deck & walked & sat about & idled time away till past 7- when I came down & read Shakespeare till bedtime. The lamps are put out exactly as the bell strikes 4. I have forgotten to mention that the men were suddenly piped to boat service soon after I went on deck in the Evening. Passengers thought someone was overboard. It seems they practise all these things in many of the P. & O. boats. At Suez they generally take the men by surprise & ring the <u>Fire bell</u> & set the Engines to work & then suppose the fire has gained on them & pipe all hands to the boats, when Captain, Officers & crew will all get into the boats, the Captain last & pull away from the ship, leaving not a soul on board & then on a given signal {recall} (421/711Scr.S. Bombay) back again. I tried to see the Doctor in the evening, but he was busy playing whist the whole Evening. He seems regularly to shirk seeing his patients. Any other man would have been round early in the morning to inquire after his patients. I was a little annoyed for I was very unwell - but determined finally to leave him alone, as he seemed so indifferent.

<u>Thursday February 2nd.</u> "Purification of the Virgin Mary"

I awoke early & got up. I am certainly something better. I dressed leisurely & staid in my Cabin reading. It is a great comfort having a Cabin to myself- the first time since I left England - except by the bye in the "Robert Syers", the misery of which I shall never forget. After breakfast I staid talking to Mr Lushington, about India & his work there. He has been Assistant Magistrate there in the Indigo country. Seton Karr came & joined us & presently Lushington finished his breakfast & went & Seton Karr & I sat on talking. I then went (422/712Scr.S.

Bombay) up on deck, and poked away forward, where I amused myself watching 2 Chinamen at work Carpenters. They are famous fellows. They want no bench. Anywhere does for them. They squat or sit or kneel or double themselves up into nothing till their head goes down almost to the ground between their knees, while their feet are flat on the ground all the while. Their tools are few & poor & you must give them their time - but their work is excellent. They were stretching a new rope too - an Indian rope. I thought it did not look very good. I got a comfortable seat on a bench & opened my Shakespeare book & read till past 12, "Love's Labour Lost". I fear I can't appreciate Shakespeare's Comedies at least some of them. Another good run 274 miles & we are only 300 odd miles from "Socotra" - so we shall pass that tomorrow. If this goes on we shall be at Aden on Sunday evening. After writing journal & marking up my chart I sat & (423/713 Scr. Str. Bombay) read & then went up on deck for a short bit before dinner. Fell in with Seton Karr & stood laughing & talking a great deal of nonsense. After dinner I went to my cabin & sat quietly reading Miss Martineau's Egypt & then the Evening Lessons & Psalms just as daylight served me. I can't tell the comfort of having a Cabin to myself- though it is but very small. I don't care a bit for that. I can be private & regular. I used to get into Stobart's Cabin in the 'Resolute' when he wasn't there - but always in fear of interruption. Then there was no privacy in the '{Nashemry?}'- and still less in the "Early Bird". I never could get by myself there. I must manage to get to Alexandria & secure the same Cabin in the steamer thence. I went on deck & walked for some time with Captain Hill - who seems a very nice person. I then sat & was unhappy enough to fall asleep & awoke coughing a great deal. I went down below & read about Cairo & the Pyramids etc. & turned in at 10.

Friday. February 3.

I awoke & found it very late. (424/714 Scr. Str. Bombay) I got dressed & all ready just as the breakfast gong was struck. After breakfast I sat talking with a man who sits opposite whose name is Kennedy. Then Seton Karr hove along side, and we two sat talking - & then I went to my Cabin & read the Morning Psalms & Lessons & went on deck, where I staid reading & talking to Captain Hill, till one o'clock. I went down & wrote journal & marked up chart etc. We have lost the breeze & so have made not quite so good a run, but very good, viz. 250 miles - we are therefore still 113 miles from Socotra. I expect we shall get to Aden on Monday. I hope so instead of Sunday. I am certainly better - I have lost my headache & can enjoy myself & certainly barring the disagreeable(nes)s of a crowded, throbbing, noisy steamer, it is very pleasant. The heat, 80°, is moderated to quite a pleasant temperature by the draft of the steamer (I suppose). The sky is blue & the sea is blue & today quite smooth. There is no motion, (425/715 Scr. Str. Bombay) except forwards at 10 knotts (sic) an hour, which I should be very sorry to dispense with. I read till near dinner time & then went on deck. After dinner I sat talking for some time & then went to my Cabin, where I read the Evening Psalms & Lessons as usual & said Evening Prayer. I then went on deck about ½ past 6 & walked. A school of porpoises afforded a few minutes of excitement. I went to Major Hill, who can't move from his chair with Rheumatism & had some talk with him, boldly, urging him to go straight to Malvern to Gully.[4] He had heard much of Gully & had ½ a mind to go there- but had lately had his confidence in him shaken by hearing of his mesmerising, or pretending to mesmerize some lady at Leamington, while living at a house near Malvern. I quietly asked him if he believed in Mesmerism. Oh. Dear yes! He had seen too much of it to disbelieve it & then told some stories, all of which he believed,

---

[4] Presumably James Manby Gully (1808-1883), famous as an advocate of hydrotherapy ('the water cure treatment'), which attracted such eminent people as Charles Darwin, Lord Lytton and Lord Tennyson.

much more wonderful than this affair of Gully's. I just suggested that this alarming (426/716 Scr. Str. Bombay) affair of Gulley's needed appear so wonderful after what he had told me, nor prevent his going to him. He led the way from Mesmerism to "Table turning" & we had a long talk. Then I continued my walk & Captain Hill joined me & we must have walked for an hour & ½. He gave me some fearful accounts of recklessness & extravagance in India & the way men run in debt at first. I went down & sat down to read, but I found myself next Captain Keane, who was showing tricks with Cards & he talked all the while & ended by sitting talking to me all the evening. I caught a glimpse of what I thought was land & out of a port & just went up to see before turning in; I found it was actually "Socotra" which we were abreast of & staid leaning over the bulwarks in the quiet moonlight looking at it & talking to the Pursar till near 11.

Saturday. February 4.

I awoke about 6 & got up - dressed & finished my reading etc. quietly & then sat on in my Cabin reading "The Far West" till breakfast time. (427/717 Scr. Str. Bombay) After breakfast I sat on talking to Mr Kennedy, till Seton Karr as usual came & sat down. After that I went on deck, went forward & sat down there after watching the men making a mat for some time & the Chinese Carpenters at their work. Another man was near in a very comfortable chair which I kept my eye on & when he vacated it & was quite gone, I put into it & sat enjoying the quiet & my book. The Captain came & had some talk & went away. At last I was interrupted by the Doctor coming to smoke a [**** (?)] after his Monday luncheon & after some conversation he mentioned Ilfracomb- where he said he was born. The Doctor's name is Vye. I mentioned the Buckles & he knew them. Then he went on to speak of Chanter & then to abuse him & Dr Pusey & to talk nonsense & show his utter ignorance, jumping from one wild assertion to another - one especially amused me. It was so wrong of Chanter to allow Pusey to preach when he was under censure. But I said it was censure protested against & merely (428/718 Scr. Str. Bombay) preventing his preaching in the University Pulpit. Still, he said, Chanter was so wrong in going against his superior - the Bishop of Oxford censured him. I urged that the Bishop had nothing to do with the University & that on the contrary he was one of Pusey's staunchest supporters. Then off he flew to what he called the similar presumptuous conduct of "H. of Exeter" (for which I pulled him short & said he was the Bishop) & the archbishop - in that horrid persecution of {Gordam}. He thought {Gordam} the best man that had ever lived - which of course I for one would not gainesay- but I suggested that his Theology was not very good. He asserted that it was & said he had read all the report of the trial & declared himself a profound Theologian & there I dropped the matter; just asking him to prove the Doctrine of "Prevenient Grace" ({Gordam}'s particular, from H. Scripture) & turned the subject & after talking for a few minutes on indifferent matters to blot out the theological con- (429/719 Scr. Str. Bombay) -versation, I came down to mark up Chart & write journal. We have made 245 miles quite as much as I expected & are 409 miles from Aden. So that we shall be there very early on Monday morning. I read & went on deck at 3- till just before dinner. After dinner we had an amusing discussion about Ceylon jewels- my right hand neighbour, I don't know his name had bought an opal ring. He fancied he had got a great catch. Captain Kean who has been in Ceylon for years is or fancied himself a judge of jewels. He said he would fetch 6 very pretty opals. I had been asking my neighbour where he had bought his opal ring & what he gave for it- he said, hardly anything- he had bought it at Galle. Captain Kean arrived with his opals which looked quite as good as that in the ring. After they had been looked at all round I offered him 6d for the lot- you're right he said- there (sic) only glass. My poor right hand neighbour was so disgusted (430/720 Scr. Str. Bombay) with his ring that he pitched it overboard. We began talking about Galle Cats Eyes', which were offered there for a very small sum. I had heard at Rundell's or some jewellers' in Town how scarce they were & consequently though not precious very dear & said so. When Mr Kennedy retired instantly as if he had been stung- It turned out that he had been let in like many others at Galle. After dinner I went & read quietly in my Cabin & before going on deck read Bible & Lessons & said Evening Prayer. I did not go up on deck indeed till very late & there walked a long time, avoiding other people & thinking my own thoughts. I then fell in with the Pursar & had a long talk with him & then was sitting down & Captain Hill came & sat down by my side & we sat talking till just 10 o'clock. Down to bed.

<u>Sunday 5th after Epiphany. February 5th.</u>

I got up & staid in my Cabin till breakfast time. I have always been in the habit of going (431/721 Scr. Str. Bombay) on deck before breakfast- but they wash decks so late here that they are swimming till breakfast time & having a Cabin, airy & fresh, to myself, I find it nicer to sit reading there. At 10 the crew were mustered & at ½ past we had prayers- I said prayers. Everyone was in the Saloon I believe. I went up & sat on deck reading one of the Library Books the "Rectory of Valehead", a nice book to find in a ship's Library & came down about 1 to mark up my chart & write journal. We have run 247 miles & are 174 miles from Aden. The rest of the day was spent much as usual. After dinner I went to my Cabin & read & said the Evening Service & sat sometime quietly in my Cabin & then late went up on deck. I walked sometime. All Sail was taken in & some of the awnings being taken down & preparations making for coaling. Land was seen, just some white point caught sight of. I couldn't see it. It was a glorious moonlight night & Captain Hill joined me & came & {haul} (432/722 "Aden" Scr. Str. Bombay) of the Bulwarks & we talked all the Evening. I went to bed tired out.

<u>Monday. February 6. Aden</u>

I awoke about ½ past 5 & at once got up - I thought we were to be at Aden about 5. I dressed quietly & just as I had done dressing some rocky land came in sight from my port. I went on deck & found we were coming up to Aden. About 7 we passed one of the H.M.C.S. (?)Steamers, big rigged & dropped anchor. We all staid quietly on board till after breakfast. Some little time after breakfast, I went quietly ashore & 2 other men in the same boat. A host of horses & donkeys instantly surrounded us & we were pestered to take this & that. I {escaped} the noise round me by jumping at once on a donkey & riding quietly away. At the point, a large Hotel has been built & the P. & O. Agent has a house & there are a number of Native huts in which the men who work at the Coal live. The other men also got animals, (433/723 "Aden") the other a horse & the other a donkey & followed. The road lay along the seaside. On the right rose a wall of hills & rocks & sand- the road took me up a hill & down again on the otherside (sic); in fact crossed a spur of the hills. On the left again was the sea- & on the right a level sandy bit of ground with here & there a plant something like a lupin, with a yellow flower beautifully pencilled & a very aromatic smell. There was also a low thorny bush- but very few. This road was quite straight for 2 miles I suppose. Arabs were at work collecting small stones which they brought in hay baskets on their heads & put upon the road. I met numbers of donkeys loaded with packs or ridden by Arabs or Turks & strings of Camels taking heavy cargoes on their backs. Their feet seemed to me to suffer from the stoney road. At last I went up a steepish hill up to the face of the rocks & a gate guarded by a few English soldiers, very solid & strong admitted me inside. (434/724 "Aden") The road had been cut through the solid rock for some way. The face of the cliff to the left has been scorched & a wall runs along the hills & is carried across above the road by rather a fine arch a great height above your head, a narrow road between the rocky walls takes you down hill into a sort of basin among the hills & you see the cantonments below you. The houses in the Town are mostly built of bamboo with mats; these perhaps are plaistered (sic) with chunam[5] perhaps not. Some parts of the Town are built of rough stone- with pillars & low arcades. I passed a rather pretty building entirely built of Chunam, with some nice carved work about it & colour. I think it was a mosque, but I couldn't make the boy understand me. I went to the Post Office & learnt where the English Church was building & that the Chaplain's name was Badger & that his house was close by.[6] So I (435/725 "Aden") started to the site. A spur runs out from the hills upon the flat on which the Town is built & close by the Barracks, (which are all built of bamboo & mat). The crown of this spur is being levelled by blasting & here the Church is to be built.[7] The English Burial ground is at the foot of the rock to the West of it & in it at present stands the Temporary <u>mat</u> Church. Close to where the Church is going to be built stands Mr. Badgers house. It is built of bamboo & mat work raised on a stone foundation. No windows, except Tatties [= tatamis] of plaited cane. I knocked sometime before getting admittance- at last when I was giving up in despair the door was opened by Mr Badger himself. I gave him my card & introduced myself & he welcomed me most kindly. I at once told him I was come to inquire about the Church & told him how I came to take an interest in it. It seems that while waiting for Woodyer's[8] plans a Government Engineer (436/726 "Aden") provided plans & they were sent to Bombay & approved of & now nothing can alter them. Then Woodyer's plans arrived & though they were much liked, they were thought too ornamental & that it would cost £7000 odd to build. I saw the plans & had a great mind to ask for them to take them back to Woodyer. He said he would send & see if he could get the plans from the Engineer. Meanwhile I sat & talked with him. I am sorry to say I was quite ignorant of Mr Badger's missions to the East & of his work "Nestorians & their Ritual" & could only fish for pegs to hang questions on. It seems that he went out there either by the wish or with the Consent of the late Archbishop. And all the good he was trying & likely to do was overthrown by the annoying interference of the American Independent Missionaries & Mr Layard & then by a member of the C.M.S. a Clergyman who it will hardly be believed (437/727 "Aden") associated himself with the Independent Missionaries & communicated with them. So much again for the C.M.S. He read me a long correspondence to show his difficulties in Aden. It was with the Brigadier who had written to him asking him to be president of a Branch of the Bible Society to supply China with some thousand Bibles, which he declined & thence arose a troublesome correspondence, charging him with having given money towards the R.C. Cathedral building & with having read R.C. books of devotion to R.C. soldiers in the Barrack Hospital. So the Puseyite cry flourishes even in a barren spot like Aden. It is a satisfaction to find a Puseyite Chaplain anywhere in India. He is evidently a clever & a learned man. He pressed me to have some food, but I had made up my mind to get back to the ship to dinner at 4 & have a quiet cool evening on deck. I mounted my donkey accordingly & rode away back (438/728 "Aden") towards "Steamer Point". By & bye Mr Badger pointed

---

[5] A type of plaster made from shell-lime and sand.
[6] See biography in Ch. 4
[7] The stone building - finally consecrated in 1863 as Christ Church, the Anglican garrison church, has had a chequered history since Yemen's independence in 1867 and especially recently in the current civil war.
[8] TB's brother-in-law, Henry Woodyer, a well-known church architect. See biographical note in Ch. 1.

out Charles Tombs's grave- with a hideous stone box tomb over it.[9] I met one or two parties of passengers on my way back. I got a boat & went off to the steamer & there found to my sorrow that dinner had been at 2- for what earthly reason I can't imagine. However the steward promised me some dinner & by the time I had washed my hands & made myself comfortable I found the end table laid out for 5. The first 2nd & 4th mates had had no dinner, the Engineer who is travelling home & another passenger whose name I didn't know & whom I had never spoken to before, a very nice gentlemanly young fellow. So that we were 6 at dinner. After that I sat down & wrote a bit of journal & then went to my Cabin & said Evening Prayer. Went up on deck- down to have a Cup of Tea at 7 & up again to see (439/729 "Aden") the start. The steamer for Bombay had come in during the afternoon & were going to coal all night long, so as to start the next morning after us. We were to start at 8 & as the clock pointed to 8, the screw began to revolve & we were off. I never saw greater punctuality. We have a new Passenger, said to be a French R.C. Bishop of what place I don't know- but he wears a long black Cassock & a cross round his neck & wears a large beard as well. The decks are crowded with 3 days coal- besides a great many sheep & stores. The Company have a depot of Coal at Suez- but it is so dear there, £8 a ton, that they never use it except on an emergency. I fell into conversation with Captain Hill as usual & we talked till past 10. Down to bed.

Tuesday. Feb. 7.

I awoke early & found we were just entering the Straits of Babel Man- (440/730 "Red Sea" Scr. Str. Bombay) deb [= Bab el Mandeb, the narrow entrance to the Red Sea]. Nothing to see, but barren rocky hills. After breakfast I went upon deck. The Arabian Coast was visible- rocky & sandy, without the appearance of a blade of grass. Later in the morning Mocha was distinctly visible with its' white houses- & white sails off the Town. All the coast is inhabited by a wild set of Arabs. Even between the Steamer point & Aden you have to take great care of the Arabs. There was a notice up in the Hotel signed by the Resident, advising people to be on their guard between the 2 places & after night fall not to think of going from one to the other without being well armed. The heat today is greater. The thermometer was at 87 early, but I think the sun had been <u>near</u> it;- at 12 it was 84¾. I sat in the Saloon writing journal etc. from ½ past 12 till past 2. I sat reading in my Cabin as it was late. Washed my hands for dinner, listening the (441/731 "Red Sea" Scr. Str. Bombay) while to a discussion just by my door with the Pursar as to where the R.C. Bishop should be put at dinner. To my grief I found on going out of my Cabin that he was to be put at our Table on a Camp stool between the 2 forms, i.e. between me & my right hand neighbour. However so it was & there was no help for it- but it was unkind of the pursar to put the French Bishop next to me of all persons. I'm sure, if he isn't told, the good man will never suspect that I am a Priest- which title by the bye he would not grant if he knew that I claimed it. I was as civil as I could be. Everybody else, English & Protestant like, treating him with the most utter contempt. I at once let him know I could not talk French; but, a regular Frenchman, of course he paid me the usual false compliments & insisted on talking- but he found at last that I could not understand him. After dinner I went to my Cabin & sat & read quietly & before going on deck said Evening Prayer. (442/732 "Red Sea" Scr. Str. Bombay) It was very hot & close. The wind had drawn round ahead & consequently draws all the heat of the Engines aft to us. I sat on the Taffrail a long time, where I got a draught of air & then

---

[9] Rev. Charles Tombs (1817-1846), Assistant Chaplain, died at Aden, survived by a wife and daughter. Son of Major-General John Tombs (1777-1848) of the Bengal Cavalry and older brother of Henry (above). See biographies in Ch. 4.

found a choir. The 1st mate, Pursar & a Passenger sang glees round the Capstan, lighting their book with a deck lantern- not very well. The Pursar was a Salisbury Chorister & the 1st Officer a Gloucester d(itt)o [i.e. a chorister]. Mr Miller, my neighbour at dinner, amused everybody by coming forward & volunteering a song or two & singing one or two half conversational songs, about Sophy or Polly, or some lovely creature. I fell in with Captain Hill later & as we were walking up & down, Lushington joined us & Captain Hill went down to bed. Mr. Lushington is nephew of Lord Cranworth & son of Dr Lushington. He lives near Guildford, or rather near Cobham, at a place, he told me, called Ockham Park. I asked him if he knew Henley Park & the Halse[10] & he knew them well & told me what I didn't (443/733 "Red Sea" Scr. Str. Bombay) know, that one of the Halseys had come out to India during the last 12 months & was already living on borrowed money to a great extent. He Lushington is going home after 2 years only in India for headache & inability consequently to work. Of course I tried to persuade him to go to Gulley. Down to bed.

February. 8. Wednesday.

After breakfast I sat talking sometime at the Table to Dr Gammie & Mr Kennedy. Dr Gammie was describing his being shipwrecked with troops in 2 transports in the one of the Andaman I(sland)s in the Bay of Bengal. Then the Pursar came for Miller who was sitting near to draw for cars across the desert & he came back successful, having drawn No. 3. I am of his party. So I am in luck- sure to be off in the first batch, within a few hours of getting to Suez. I went up on deck, after saying Morning Prayer & Litany in my Cabin & was fortunate in finding an empty (444/734 "Red Sea" Scr. Str. Bombay) chair, where I sat & read till past one. We have had a head wind & sea all night & have also managed to make 200 miles with the screw. We have passed several Islands- all much alike. Barren, cindery looking, rocks. While writing journal Dr Gammie came & sat down to ask me about the Coxes in Australia. It seems to be great pleasure to him to talk about them & no wonder for their hospitality of the past day seems only equalled by that of the present. After dinner I sat as usual in my Cabin & read. Then said Evening Prayer & read & went on deck. Captain Keane walked with me some time, talking nonsense & finally giving me an account of an outbreak at Sincapore (sic), quelled by the Ceylon Rifles. I went down at Tea Time & read the Rectory of Valehead till 9- when I went on deck. I walked sometime alone, shirking Miller, who altered his pace & tried to fall in with me, till at last I thought he would see I was avoiding him & so gave up my <u>solitary</u> & joined him & we walked together sometime. He (445/735 "Red Sea" Scr. Str. Bombay) went below & I was rejoicing in a solitary walk again, when it was again broken by Wood joining himself to me. And he staid walking with me till I went down to bed. It does not do to be selfish on board ship. One gets into the way of suiting oneself to any one & every one. It used to be torment to me to be with <u>some</u> people- there seemed to be a natural antipathy- but it wears off of necessity in travelling.

Thursday. Feb. 9.

I awoke at 6 & got up. After breakfast I went to my Cabin- but came out again & was sitting down to read, when Seton Karr came up to talk. Afterwards I went on deck. Where he produced "Whatley's Guide" which gave me some information I wanted & I sat reading till one o'clock-

---

[10] Henley Park can be traced as early as Domesday Book in the 11th century. The present house is largely the one bought in 1784 by the Halsey family which had made a fortune in India.

when I got the Latitude & Longitude from one of the officers & came below to mark up chart & write journal. We had a head wind & sea all yesterday & all night & have only done 199 miles. I sat in my Cabin (446/736 "Red Sea" Scr. Str. Bombay) as usual & kept 9th hour & then went on deck - where to my annoyance the French Bishop kept me hammering at [dog] French, till I got away to wash my hands. After dinner I sat reading in my Cabin, till decrease of light warned me to read the Psalms & Lessons & say Evening Prayer. When it was dark I went on deck intending to have a quiet walk by myself & enjoy the quiet. But the French Bishop joined himself to me & asked, a question which I made out with extreme difficulty, if I would play Chess. When I understood him I told him that I couldn't but offered to get him an Antagonist, which he declined & we walked & stood over the gangway a long time, he persisting in talking French & I trying to understand & answer. At last I went below to read & then got Mr Kennedy to play a game of chess with him- which he readily agreed to do. I went up & found him sitting against the Taffrail & came (sic) him below & sat him down to the Chess board- where I settled quietly down to my book. At 9 I went to put my book away & go on deck & found (447/737 "Red Sea" Scr. Str. Bombay) the same game still going on. But I had not had above 2 or 3 turns on the quarter deck when the Bishop came up & I got dragged into (or {not}) conversation again. However he relieved me by saying good night- after he had made an attempt to find out what I am- which I didn't tell him. I had a pleasant half hour by myself when he was gone & then went to bed.

Friday. February 10.

When I awoke I found it was very late & had to dress as fast as I could. After breakfast I went back to my Cabin to read what had been cut short by my laziness & said Litany & then went on deck- where I found a Chair & sat reading "Life at Grafenberg by a Convalescent" in 'Chambers's papers for the People' & then began an account of Ocean Steamers till one o'clock, when I got the Latitude & Longitude & came down to write journal. We have made a run of 217 miles. There is not a breath of wind- the sea as though oil had been poured on it. We must be about 500 miles from Suez. We hope to get in on Sunday (448/738 "Red Sea" Scr. Str. Bombay) Evening- in which case I should reach Cairo by midday on Monday. I sat in my Cabin till dinner. After dinner which was very uncomfortable for feeling that I ought to talk to my neighbour & couldn't, I sat in my Cabin reading & after reading Psalms & Lessons & saying Evening Prayer, I went up on deck forward, to avoid the Frenchman & the smoke. Here I found Seton Karr & we sat talking, till Tea time, when we both went down & I sat in the Saloon reading till near 9 when I went up on deck & walked in the Cool fresh air till 10. The wind had come in again from the North & it was deliciously fresh- the Indians called it cold.

Saturday. February. 11.

I got up soon after 6 & saw the sun rise. I sat in my Cabin reading till Breakfast time. After breakfast I went up on deck with a book & presently finding it very smoky & dusty on the ¼ deck, I went forward & on the forecastle found Mr Palmer & a chair & we sat (449/739 "Red Sea" Scr. Str. Bombay) talking from 11 to 1. He had been on leave from India to Australia & had only left Australia in September. He had been in Adelaide, just the country I wanted to hear about & had besides been one of the people on board the "Lady Augusta", a small steamer Captain Cadell had had built in Sydney for the Navigation of the Murray. I saw the vessel in Sydney & saw her stack on her hazardous voyage round the coast. I had heard of her safe arrival at Adelaide & of her getting across the bar into the "Murray". It seems they had a most successful voyage & there is now no doubt of the possibility of navigating the Murray. He bought me an account of the trip written by Kinloch, one of the passengers & having got the Latitude & Longitude I came down to mark chart & write journal. We have done 214 miles & shall not get in till Monday morning. After dinner I sat & read as usual & said Evening Prayer. I went on deck & was glad to fall in at once with (450/740 "Red Sea" Scr. Str. Bombay) Captain King with whom I walked for some time. I went down at Tea time & read till ½ past 8 & then went upon the deck again- where I walked till past 10.

Sunday. February 12. Septuagesima Sunday

I got up early. We were just entering the Straits of Jubal & the Gulf of Suez. The hills rose bare & rocky a little way from the shore with a sandy plain between them & the shore. After breakfast I packed up my things & put just a shirt etc. into a small bag to take with me in case we get to Cairo before the Camels with the baggage. At 10.30 we had prayers. And afterwards I went on deck & got the Captains Chart to see just where we were. We were off G(ebe)l Jehan & according to the map, over the first range of hills just opposite to us Mt Horeb & Sinai must have been- but we could not see them. The Pilot, an Arab, said they were there & pointed over the hills. I believe they are to be (451/741 Scr. Str. Bombay) seen, if you catch the right place & it is clear. They are about 30 miles from the shore & we were some 8 miles from the shore too. Opposite on the Egyptian side El Agribb a high hill rose with steep peaks into the clear air. The chart gives its height at 10 000 feet. The coast on each side was striking for its' ruggedness & barrenness. Yellow sand covered everything below & ran up the cliffs into the mountains. The light & shade on the mountains was very beautiful & the colours, bright reds & yellows. The sea perfectly smooth & the sky clear blue. We hope to reach Suez about midnight; when we shall be transferred to the vans at once.[11] They say in 4 hours after the anchor is down, every atom of cargo & luggage will be gone & the decks swept down & in 7 hours every camel & every passenger will be gone too & Suez will be perfectly quiet again. The (452/742) decks are covered with bales & boxes of freight & all our luggage. After dinner I went as once to my Cabin & said Evening Prayer & read & then went on deck again. It was very striking & I stood leaning over the port bulwarks with Seton Karr watching the sunset & the shifting lights. After the sun had set the western sky was all covered with the brick dusty red which Roberts' pictures give you. We were reported only 40 miles from Suez & had a fresh southerly breeze in our favour. I went down into the Saloon at Tea Time & sat quietly reading. It only made the time pass more slowly being on deck & was idle too. About ½ past 9 I went up on deck again & went on to the forecastle, where I found the chief officer & 2 or 3 of the passengers looking out for the lights of Suez- we burnt 2 or 3 blue lights & a couple of rockets were sent up. Presently the lights of Suez were made out & we were pronounced within 10 minutes. (453/743 "Suez" The Landing) We very soon found ourselves near- then we made out the hull of the "Zenobia", the Co's store ship, in the bright moonlight & that boat had been already posted with lights to show the way in;- then the Engines were 'eased' & next stopped & then reversed & at ¼ to 11 we dropped anchor. We were all ready to go ashore, but no boats came to take us. Presently the Master of the 'Zenobia' came along side- but he could tell us nothing. We didn't know whether to go to bed, or whether they would send for us & pack us off at once. In this state of uncertainty we staid, kicking our heels on deck, till the good eyes began

---

[11] Waghorn and Co. provided accommodation at Suez for passengers coming and going by sea and transport options they offered included 'spring vans', capable of carrying up to six passengers each across the desert between Cairo and Suez (see Ch. 4).

to see visions of a steamer making towards us for the shore- which were presently {naliged?} by a small iron steamer making her appearance towing a large baggage boat. She came along side & after a great deal of needless shouting & hollering on the part of the Arabs on board, she cast off the baggage boat (454/744 "Suez" The Landing) which was have (sic) round to the 'Bombays' starboard gangway & she herself was made fast on our port quarter. Then for the first time we became aware of the sea which was {roaring}. It is curious how soon the sea gets up here- the wind had only been blowing for an hour or less. We did not feel the motion in the "Bombay", but the small steamer pitched & rolled about so much that we wondered how anybody & especially women, were to get on board. At last they got all the passengers luggage on board & then we jumped down on to the paddle box & got down to her deck. Fortunately it was a bright moonlight night. If it had not been so I don't know how we should have been landed. We cast off & steamed away- as we looked back on the "Bombay" we all thought & said what a fine vessel she was. Old Captain {Baynton} sitting next to me said don't we give her a cheer. By all means I said & we both started up, saying (455/745 "Suez") 3 cheers for Captain {Treyear} & the Bombay- which were given with hearty goodwill. Of course the Comp(limen)t was returned & then we gave one cheer more & turned our looks toward the shore. This was about ½ past 12. We had some 3 or 4 miles to go to get to the wharf. A long nasty spit runs out, which we had to steam round, following a line of boats with Torches. We landed at the Co's wharf close to the Hotel & Transit Office amongst numbers of Arabs in their long dark dresses, carrying great flaring Torches. A long line of Camels were lying down waiting to receive their load & be off for Cairo. The passengers went into the Hotel. I, with some others, went at once to the Transit Office, where we got our Transit Tickets - having to sign our names to a paper stating that we had no Merchandise in our luggage. I then went upstairs into the Hotel. People had thrown themselves on the divans which ran all (456/746 "Suez") round the great room - while preparations were making at a long table for Supper, which soon arrived. I felt no inclination to eat at that time of night, but nevertheless had a Cup of Tea & some meat & capitol biscuit & butter, on the principle of taking in strength. In less than an hour a large bell rang, announcing the readiness of the first batch of Vans, 4 in number, each carrying 6. I had left the room & was out by the vans at the time - watching the curious scene. We were sheltered here & the sea was as smooth as possible. The bright moon & the numbers of Torches made it almost as light as day. Strings of Camels loaded were being led out of the inclosure round the wharf & others were being packed. Each camel is made to lie down- the Cargo is then made fast in strong net work bags on each side of the packsaddle & more is piled up in the top & made fast (457/747 "Suez" v(an) ac(ross) desert) with cords. The Camel moan while objects strongly to the whole business & growls loudly & snaps at everyone near (& the life would be no pleasant thing). Meanwhile horses & mules were being harnessed to the Vans. We all took our places & each van started in order. No.1- No.2. etc.- but before each started an official came to the door of the van with a paper & called over the names of the passengers. Then we started- much shouting & pulling & pushing etc. etc. before some of the Vans would start. The Vans are small omnibuses, on 2 wheels, carrying 3 people on each side the seats divided by arms. The driver sits on a small seat on the Top. There are double shafts & the leaders draw as in England. A horse boy sits on the shaft & each batch of vans is accompanied by a mounted courier, who has the charge & command of the whole. In case anything happens to a single Van all are made to (458/748 "Suez & the Desert") stop & keep together- in order that in case of any accident, all might assist.

<u>Monday Feb. 13 "Suez to Cairo".</u>

Our Van, No.3, went off with a rush about ½ past 2 or 3- round the corner, along the narrow street with the sea on one side, through the Gate & out at once upon the sandy desert. We passed the first Telegraph & not another thing was near you but sand & stones. The wind was blowing strong from the North & it was bitterly cold. We were glad to shut up the 2 little shutters that opened in front & wrap ourselves up well in coats & wrappers. I had put away my Coats etc. in my box at Ceylon, thinking in that heat that I should never want them again. I was told on board of the cold nights on the desert & had got out my plaid. That & my waterproof were not a bit too warm this night. We changed horses every 5 or 6 miles & at each change house got out & had a run. These stables, for excepting for No.'s 12, 8 & 4, (459/749 "The Desert Suez to Cairo") there are nothing else, are oblong yards with a well in front & the 3 other sides taken up with stalls. Nothing stands near them. All around is a sandy & stony undulating waste. Here & there a few scrabby thorns creeping on the ground & once or twice a single tree, of what sort I don't know; stood near or within sight. Along the front a long rope is pegged in the ground & to this the horses are fastened by a cord twisted round the fetlock. In the stables too the horses are fastened in the same way- the foreleg is fastened by the rope round the fetlock to a long rope running in front of the manger on the floor from end to end & another rope round from the fetlock of the forefoot & sometimes I think from one of the hind legs, to a ring or peg fastened into the ground behind. The start was generally a matter of difficult. One or other of the Vans was sure to refuse to start & then all had to wait- when once off the recusant van (460/750 "Suez to Cairo") was sure to go at full speed for some way. The driving was excellent. About 7 we reached Station No. 12 & here we found breakfast ready. Tough chickens & geese & bad Tea & bits of meat dressed with potatoes & onions & curries & biscuit. About 8 we were in the Vans again. Before breakfast I walked out over the sand at the back of the building by myself & got away & said Morning Prayers. It was strange & solemn & so I had felt it in the Gulf of Suez with Horeb & Sinai so near; & later in the Evening when I felt we must be somewhere near the spot where the {sea} stood on a {heap} & the Israelites had passed. I looked at the sandy shore & tried to fancy them encamped there & marching on & through the passage over the hills up to Horeb. We all dozed the first stage & I managed to get a little sleep, the first since 6 o'clock on Sunday morning. The next stage we all were wide awake- but I managed (461/751 "Desert Suez to Cairo") during the day to get some sleep in the stages. We had no books & if we had I couldn't have read- though the cars were far from being uneasy & the road the last part of the way a very fair macadamised road between banks of sand & stone. We had to keep all the wooden windows closed except those behind. The wind was strong & blew the sand about in clouds. As it was our eyes & everything got full of dust. There is nothing to say of the desert. It was a wide undulating mass of sand & thickly covered with stones & dark pebbles- not a blade of grass anywhere. We met & overtook camels & sometimes in strings of 4 or 5, sometimes singly & donkeys but most part of the way it was all desolate & dreary & the North East wind blew strongly & cold & mournful. About 11.30 we came to Station No. 8. Here we all declared that we didn't want to wait or eat & pressed the (462/752 "Desert Suez to Cairo") courier to let us get on. Opposite the station a perfectly straight road led up to a large white building, which is a palace of the Pasha's. He comes out here occasionally it seems to hunt antelopes etc.- a miserable place to stay. We soon got the horses & mules put to & the same thing went on again- short stages & constant changes & cold wind & sand etc. About 4 we came to Station No. 4 & here they had prepared a regular dinner. I was very sorry to see it- as I should like to have pushed on the remaining 20 miles

& got to Cairo to dinner. However the only thing was to make your dinner & look forward to Tea at Cairo. At last we got the (= to) the last station. Where we got out to stretch our cramped legs, there in the distance were the minarets & domes of Cairo & beyond them the Pyramids. We soon got over the first 2 or 3 miles & then found ourselves near the outskirts of Cairo. The desert ended & we (463/753 "Cairo" arrival) saw green fields & palms not far off. We came to a great white building, with a largeish inclosure & were told it was the Pasha's Palace & as we drove along the road, now with shaded trees on each side- we met 2 or 3 soldiers dashing along. Our Cars drew up at the side of the road & the Pasha himself came along, in a handsome Brougham & a bodyguard about the carriage. We took off our hats as he passed & he bowed courteously. He was dressed in a light blue dress:- but one couldn't see much of him as he passed. We next passed a large encampment of soldiers on the left. The number of Tents was very great. Numbers of people were passing along the road, in various costumes & ladies of soldiers carrying baskets of bread on their heads & camels without end carrying wood, which we guessed must be fire wood

Travel Journal 3

> "Who can tell, when he sets forth to wander, whither he may be drive by the uncertain currents of existence; or when he may return; or whether it may ever be his lot to revisit the scenes of childhood."
> Washington Irving. The Sketch Book.
> Beyrout. May 3. 1854.

Thomas Bowles
1854

Cairo - Feb – 13. 1854.

(1/754 "Cairo") going to the camp. We drove on & on & at last entered by a strong gate into Cairo & drove to the Sheph(e)ard's Hotel. Several Englishmen were sauntering at the door. "No room- They couldn't take anyone in". However I suggested that I meant to stay for a fortnight & they soon found me a room. I was taken upstairs at once, & a better room promised me in a day or two. The others went on to search at other Hotels. I washed & dressed but determined not to go to dinner wh(ich) was announced in a few minutes by the loud ringing of a gong. I went down & asked for a list of people in the Hotel & was shown into a reading room full of papers. It was very cold & no fire. Mr Kennedy & Mr Strachan came to try & get beds here & thinking of staying till the next packet. They said they did not care what sort of rooms they had if they c(oul)d be taken & were put into some wretched rooms, with a promise of better. Seaton (sic) Karr & Palmer came in- as soon as dinner was over we 3 had some Tea & then I went to bed & I believe they did the same. But on leaving the room I saw the reflection of a fire in the next room, & went in to (2/755 "Cairo") warm myself & was seized by the hand by the American, Mr Edwards, who came with us from Singapore. He is going on & on- he had not been able to get to Thebes. It is too late. But he has enjoyed Cairo very much & says he should like to be here longer. He has been somewhere every day & says there is enough to occupy you every day between 2 steamers. I went to bed about 10 & was asleep as soon as I laid down.

<u>Tuesday. Feb. 14</u>

I slept 2 nights in one & awoke about 7 feeling as if I had done good work during the last 9 hours. I tried hard to get a barber to come & cut my hair & Hassan said he would come presently but he never made his appearance till past 9 when the breakfast bell had already rung. I went down to breakfast. It was a lovely morning- fresh & cool with a bright cloudless sky. Before breakfast Mr Edwards the American came & asked me if I w(oul)d decide on engaging his Dragoman but I hesitated. The expense of 4°/ (sic) a day I didn't want to incur. I have nothing whatever for a servant to do & I thought the donkey boys would be sufficient guides. However (3/756 "Cairo") I said I would think about it. During breakfast I heard enough to convince me that unless you can talk it is absolutely necessary to have a Dragoman & as Mr Edwards spoke so warmly of this man, I determined to engage him & told Mr Edwards so after breakfast. I then started with Mr Kennedy & Strachan for Boulac- to see after my luggage. We went on a donkey a piece. It is a ride of about a mile or rather more down to the banks of the Nile. We passed a long string of camels going there carrying part of the "Bombay's" cargo. We found a large yard running along the river & in the middle a heap of luggage lying, being weighed. I soon picked out my things- some of them considerably the worse for the journey & my short bundle of sticks had been cut open, & the Kandy stick abstracted. The leather case of my desk was considerably cut too & one of my leather bags torn. We put all our things on a truck (sic) & sent them to Cairo & rode back on our donkeys. I went as soon as we got back to the British Consulate to see for Letters. My donkey boy took me to the French instead, & before I knew where I was I found myself in the presence of the French Consul- however I explained that it was a mistake & retired. At the English Consulate (4/757 "Cairo") I was told that Mr Bruce was ill. I explained that I had no wish to see him, but merely to get letters. I c(oul)d not get to see the Sec(retar)y or anyone, but only received a message that if I w(oul)d leave word where I was staying any letters should be sent to me. I then went Mr Walnes's the sub consul's. Here I put an order for the Shoobra Gardens & bespoke one another for the Mosque & anything to see wh(ich) an order was necessary & so back to the Hotel. I went down to have some luncheon & they proposed to go & see the Citadel. I went with them. We mounted our donkeys again & rode away- round the square & past the Hotel d'Europe where most our fellow passengers have got rooms & so on through the streets. It was amusing to ride through the streets. At first we went through a number of streets which all looked new & are so. I believe the Pacha is gradually getting the Town rebuilt, & introducing wider streets. We passed a handsome building, wh(ich) our guide said was a fountain & then got into older & narrower streets. On each sides the merchants or shopkeepers sat in what can only be called <u>holes</u>, places a few feet square in the walls, with their merchandise round them, & their pipes in their mouths. The (5/758 "Cairo") streets were crowded with people walking, women with their hideous nose bags or riding on donkeys, a mere lump of black silk, with perhaps one eye visible & men on donkeys each with his bag & camels loaded & taking up the whole street. Your own donkey boy continually calls out for people to get out of your way & strikes the donkeys of the inferior people, & the camels, on the head with his Kurbash to make them give way. We passed a great number of mosques & bits of curious architecture, minarets, old doorways with a very grand imitation of Norman zigzag & other work. At last we saw the minarets of the Citadel Mosque, & began to ascend towards it. The Citadel as you approach it does not give you the idea of strength- a mass of modern buildings & nothing else. You enter by a gate, in which there may be some strength as far as stone & mortar are concerned. Soldiers dressed shabbily in white, a mixed race of Egyptians & Nubians guarded the gate. We passed in & before a 2nd gate had to leave our donkeys. The 2nd gate led into the courtyard in front of the Mosque. The gate, the cloisters, the pavement are all of marble, rough & poor, but nice looking. The square is large- about the size of Queens Court quadrangle. In the (6/759 "Cairo") centre is a fountain covered with a sort of dome supported on pillars, wh(ich) belong to no order at all & are not handsome, save that the whole is of marble, or I am half inclined to think it is all alabaster. The Cloisters have pillars similar to those round the fountain, with round arches, the spaced {rills} filled in with shallow carving in alto relievo & the pattern nothing striking. The pillars, to my mind, are too high, & the arches too near the roof & the effect is general want of depth. Moreover the chord of each arch is joined by a thick square iron bar & the same runs from each capital to the wall behind- all which carries out the idea of want of solidity & strength. One side of the quadrangle is formed by the Mosque. The Exterior is to me unsatisfactory. It consists of first a high wall pierced with plain round headed windows. It has no moulding or lines or work of any sort to relieve it & ends abruptly at the top without cornice or anything to give depth or throw a shadow. Within this rises again a 2nd wall, supported on pillars & arches within, pierced with plain round headed windows; & broken by (7/760 "Cairo") a ½ dome in the Centre against it- & from the top rises the Central Dome. Each corner is filled in by a small dome, by a minaret & rising direct from the outer wall to a height of 300 foot nearly. I sh(oul)d think.

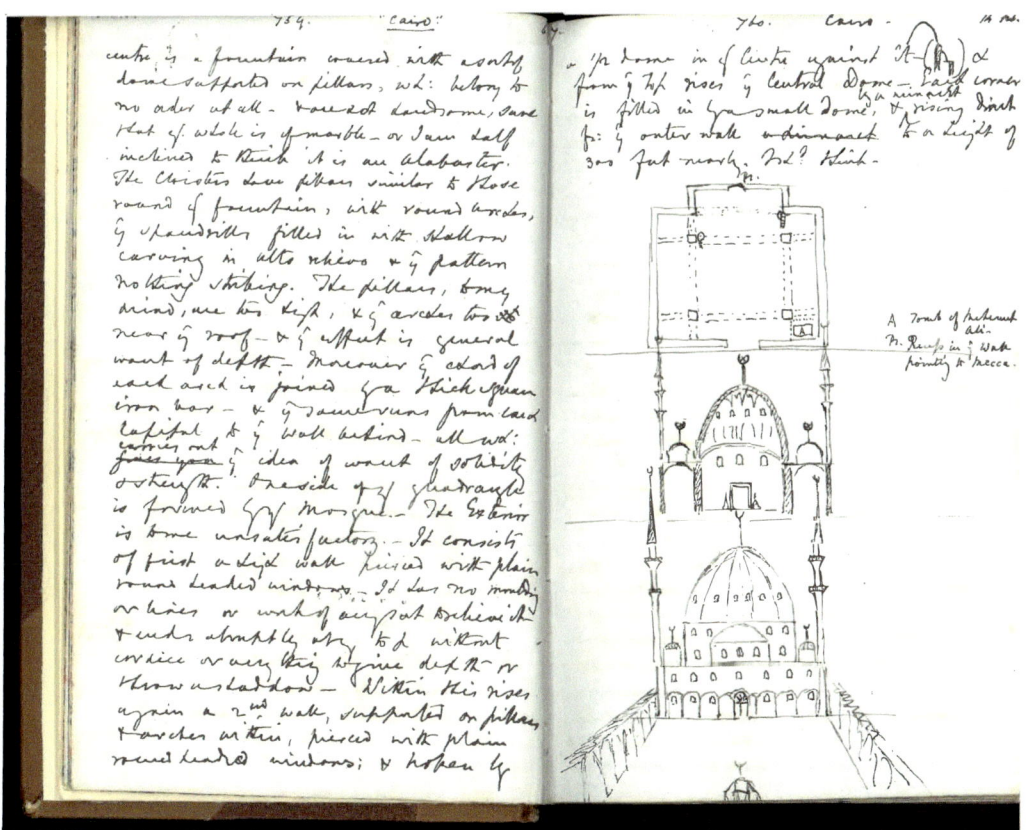

Figure 14.1: Sketch of Tomb of Mehmet Ali in Cairo.

(8/761 "Cairo") We had to take our shoes off at the door & go in in our socks. It seemed curious to take ones <u>shoes off</u> & keep one's <u>hat on</u>. The inside was very striking, from the excellent proportion & the colouring- although the prevalence of a dull green did not suit itself to my eye. There was too much green & yellow. They were the predominating colours. The Dome in the Centre is supported on 4 massive piers. I should think above 12 foot square. These rise plain from the floor, but have a sort of capital- indescribable. Our measurements did not agree- but the least made more than 65 feet from pier to pier. The windows were filled in with great square pieces of coloured glass. The walls wanted relief by carving or moulding- partly made up for by being altogether of alabaster & there is I think a general want of architectural detail- but yet from the proportions & masses of colour, the general effect inside was good. A tawdry cut glass chandelier hangs from the Centre & innumerable tin & glass lanterns hang from the ceiling everywhere. The floor was covered with matting- except towards the upper end; where Turkey Carpets covered the ground- at the further end, was a sort of recess or blank doorway & on (9/762 "Cairo") each side stood a heavy brass candlestick- a sort of ambo, low, was placed against the left hand pier- & a high pulpit against the right hand of the entrance to the Sanctuary, if so it may be called. Near the 2 doors 3 or 4 Arabs & Turks sat listening to a man chanting the Koran & in the right hand corner within a railing is the Tomb of Mehemet Ali. It was covered with a red cloth. I believe underneath it is an affair of French millinery black stuff, or velvet & spangles- but of this I am not sure. Several people were performing their devotions, & going through the 7 or 8 different postures. Here & there groups of 3 or 4 sat talking & laughing & probably discussing business. We went up a {spiral} staircase in the left hand minaret, & got into a gallery with an iron balustrade running round the inside of the mosque. Ascending higher we came out on the roof & higher still to the lower gallery of the minaret, where we found 2 muezzins just going to call the Faithful to prayers. We stood by them & it seemed as though the low chant never would be heard below. It was a sort of wailing plaintive chant, which I tried to carry away in my head, but couldn't & more words were repeated (10/763 "Cairo") than I had expected. When they had finished we mounted some 15 or 20 feet higher to an upper gallery- the highest point we would go to. From here we had a fine view of the city & country round. We first looked out to the West for the Pyramids- across the Nile, & over the green flat fields beyond, they stood. Directly opposite to us, the great Geizeh (sic) Pyramids- 3 in number. Then more again away to the left & further still the 3 more Pyramids of Sakkara. The thing which I believe strikes everyone, struck us. They didn't seem large. But one knew they were & when you learnt that they were 10 miles off, as far as from Milton Hill to Oxford (as the crow flies) & thought of how St. Mary's spire w(oul)d look if you c(oul)d see it, or the Radcliffe Library, then they began to grow. Below us lay the buildings of the Citadel. Just at our feet as we stood was the Quadrangle where Mehemet Ali slaughtered the Mamelukes & the Corner when Emir Bey jumped the wall & escaped. We went down afterwards to see the spot & it was a fearful leap indeed. To the south (I think) lay a sandy plain, covered with tombs & on one side of that high sandhills with a fort from which I fancy Cairo could be (11/764 "Cairo") quite commanded. On the other sides lay the Town. The numbers of mosques & minarets is quite marvellous. I believe they are numbered by hundreds. What struck me particularly was the general sandy colour of everything. Not only the ground, except along the Nile, where all was green as in England, but the whole Town was yellow. No trees relieve it & everything is one monotonous sandy colour. We came down- walked round the Quadrangle of the Mosque once again & passed out at the other side, past what used to be the Pasha's Palace- into the court where the Mamelukes were slaughtered & to Joseph's Well. This is a deep cut- about 260 feet deep & named after a Caliph of the name of Joseph, not the Patriarch. After this we mounted our donkeys & rode home the way we came. On the way, we stopped & went into the Turkish Bazaar. This is a very narrow street, full of Turkish shops- or rather holes in the walls in each of which a grave old Turk sits pipe in mouth, on the floor raised about 3 feet from the level of the street, (if it can be called a street, being about 5 feet wide). The floor is covered with a rug or carpet & in the front is generally a glass case, containing a few tempting wares, gold {lace}[smudge] (12/765 "Cairo") & round the walls of the <u>hole</u> are shelves with silks & stuffs etc. etc. A small Nubian slave nicely dressed is in attendance & hands down the different things, to show to the Customers or goes to the Warehouse, which is always somewhere just handy, but mysteriously invisible for others. You seat yourself on the Carpet X legged if you can or else use the floor of the <u>hole</u> as a chair & leave your legs in the street- the way I much prefer & examine the things set before you, and try by the aid of your guide interpreter, to beat down the unconscionable old fellow, who asks any exorbitant price that comes into his head. Meanwhile the slave comes from some unknown place, for the Merchants' house is somewhere in the Suburbs, with a newly filled long pipe & a small Tray of Coffee Cups in brass stands, steaming with rich Coffee- no sugar or milk. You smoke & drink your Coffee & say nothing, or occasionally repeat your offer & so the bargaining goes on. We went to 3 or 4 shops & then left without buying anything wh(ich) seems just as agreeable to the imperturbable old Turks, who salaam & bless you. We stopped to buy shoes in another bazaar. I mounted my donkey & went on, (13/766 "Cairo") expecting to find a packet of letters at the Hotel in which I was disappointed. Dinner & after dinner I went to my room & wrote journal, & went to bed early- at least as far as I could judge, for something has happened to my watch & it has taken to galloping so that I can't tell what time it is.

<u>Wednesday 15 February.</u>

I got up & read till breakfast time. When the gong sounded I went down. I decided on staying in & waiting for letters, which Mr Shep(he)ard promised to send or go for & while waiting for them I sent for a Barber, who came at his leisure, i.e. in about an hour & a half's time & clipped my hair, much as Richard w(oul)d cut one of the little Denton's. When he had finished of course we had a squabble about the paying- I said sixpence & he insisted on a shilling. I know that a Cairene would give him piastre 2 1/2, or perhaps ½ a piastre but he insisted on it that Mr Sheppard (sic) & everyone paid him a shilling. At last I produced half a crown, & asked him to change it, he produced one shilling & some piastres, & wanted to persuade me that a piastre was = to a shilling. This I would not have & I went to get change in (14/767 "Cairo") English money. He came back & produced some other coins, which I was not acquainted with. So I positively refused to pay him. He folded his arms & stood like Patience for some time while I laughed & went on with something I was doing. At last he got annoyed, & went to find someone. He came in with an impudent English boy in the house- who said the Turk complained of me & I must pay a shilling. The boy servants paid /6d. Well I said let him get me change & he shall have it. At last he got 3 shillings & I gave him a 5 franc piece & turned him out of the room. I went downstairs & to my joy found 3 letters- but not one from Home. I instantly started out with a guide reading Henry's charming letter as I went, to lunch again, at Briggs's & then other places but could find none. So I came back, had some luncheon & went up into my room to read my letters. Henry's gave me all sorts of home news, wh(ich) is the thing of all others to put into a letter. Everyday life & not strange news. Then a long letter from James {Rynik?} with a tail by Frederick & then one from Buckle, who has got a Son. I dare say he is very glad. I fear if I (15/768 "Cairo") were in his place I should not be. Henry's letter is next to one from Home, almost the same thing- but I have been building on letters from Anna & Alice & Sam & can't think how it is there are none. They must be at Alexandria. Later in the afternoon, after I had read & re-read my letters, & had sat a long time thinking about them & all in England, wishing I were going on the next morning with the rest of the passengers to Alexandria, I went out & found my way on a donkey to Mr [Alfred Septimus] Walne's the Vice Consuls', to get the admission I had asked for to the Mosques. After that I had strolled out in the dusk till dinner, which was at 5 today, that the room might be cleared for some performance in the Evening. After dinner I went to my room & read & wrote journal & wrote a note to go by Mr Sheppard (sic) in the morning to Mr Edwards at Alexandria. After which I took a turn under the trees in the square- the moon was light & people, mostly the English passengers were arriving on donkeys to attend the performance- which I found out consisted of a sort of Ballet, in aid of the principal performer a young Neapolitan I think. I went to my own room- read- & went to bed- thinking of England & Home & Grafham (16/769 "Cairo") a place which I have heard a great deal of, but have never seen. Fancy his having bought it after all. The letters have made me rather melancholy- but have quite determined me to get to work as soon as possible after getting home. I do wish I was settled in a living.

Thursday. Feb. 16.

I got up early. At breakfast we found the numbers diminished. Many had gone off by the steamer to Alexandria. It was proposed that we should do something, & finally settled that we should go to Shoobra after breakfast. I made enquiries for the Dragoman Mr Edwards had, got him hunted up, & found he had engaged himself to someone else. I was very glad, I shall go to most of the places with Mr Kennedy & Strachan & we have agreed to get a Dragoman between us. So I shall be saved expense. The wind was blowing & the air full of dust. We put blue veils on & started. The wind was in our back going & so we did not feel the full nuisance of it till we came back. We went out by the gate we had entered from Suez, but kept straight on instead of turning to the right towards Suez. The road was crowded with people, & camels & donkeys, bringing most of them, clover & other green food (17/770 "Cairo") for cattle into Cairo. The dust was dreadful. The road is a raised embankment with an avenue of trees the whole way to Shoobra. The trees are mostly acacia. Some of them are the Egyptian Sycamore. They have only been planted 30 years- but they have arched over & made a thick shade. On each side were irrigated fields, which looked very rich & luxuriant. We passed several houses, each standing within walls- not the sort of places that bear the stamp of Homes to me. We got to the river, & for a mile or more the road ran along the Nile's bank, which was here steep & high. At last we came to an ugly archway, & were told to dismount. We left our donkeys, and walked on about 100 yards or more through a very untidy bit of ground to a 2nd gate, where an old man demanded our "Permit", which I produced & got back again, thinking I should like to keep it- though why I hardly know. This led us into the gardens- I can't say much for them. It is not the season for gardens & nothing was in bloom but a few roses. But the whole thing is not what we would call a garden. It is more like a French Garden. Broad walks in all directions, with a low clipped hedge of myrtle,(18/771 "Cairo" Shoobra) & behind this an Espalier with roses etc., & inside lemon & orange trees & peach & other fruit trees, with flowers mixed. We came to a Kiosk, built of iron & glass & on our return got it opened. It has a balcony running round made of marble with a balustrade of iron & marble rail, cut & put together without care or finish. Along this were vases a semi-Egyptian pattern, made in France. Inside it is an octagonal room- floored with coloured marbles. The frame work of iron & the quarter part of the side consisting of glass. French & poor. A chandelier hung from the roof. We saw the celebrated fountain. It was undergoing repairs & some Frenchmen were painting the roofs etc. It consists of a large square basin, each side perhaps 200 foot long, with a marble island in the middle, with jets d'eau, vases & street lamps. This surrounded by a cloister. The walls & ceiling are being repainted in bad & paltry French style. At opposite corners are 2 rooms. One a billiard room with a table by some French maker, & at the opposite corner a sort of Drawing Room- pretty enough in its way. The floor & walls of inlaid panel- the {ground not} (19/772 "Cairo" Shoobra) being Walnut, with satin & other woods inlaid. A few poor French chairs- 2 handsome little tables & the ceiling white & silver. At one side of the garden stands the House. We could not see it, for the Haram was there. We returned to our donkeys, & rode home breathing dust. I never saw anything much more unpleasant. The whole atmosphere up some number of feet was a mass of ashes or dust. We got home for the tail of luncheon & afterwards I went upstairs & wrote journal etc. staying in the whole afternoon. After dinner I went into the smaller room & fell asleep in a chair. I went upstairs to bed & slept all night long.

Friday Feb 17.

At breakfast we settled to go at one to see the dancing & howling Dervishes & before that to ride to the Tomb of the Memlooks [= Mamluks]. We had an interesting ride through Cairo seeing many pretty bits, doorways & bits of pretty work. We went out at one of the gates & found ourselves at once amongst hills of sand & rubbish & amongst all this were Tombs, some used now & some old- much older. We passed by a great many & at last came to a mosque- the Dome was well (20/773 "Cairo") built & covered all over with fretwork. We dismounted & went up the steps to the door but here we were stopped & told to take off our shoes. We hesitated & were presently admitted with them on. This was the Tomb of El Kaidbai [= Sultan Qaidbay] 1496 AD. His Tomb was at one side of an inner room within the Mosque. Here also was a piece of granite stamped with the impression of the Prophet's naked feet. In another part to(o) was another stone, stamped with the Prophet's 2 feet in slippers. In the Mosque an Arab Priest sat reading the Koran in a sort of chant. We passed on to the 2nd tomb. That of a Sultan "El Barkouk" [= Barquq]. Here was a large quadrangle with originally a cloister made of high pillars & arches, some round & some pointed. Against most of these pillars clumsy buttresses had been built in consequence of their spreading & nearly letting down the vaulting. The mosque was at one side, an extension as it were of the cloister, by a 2nd vault supported on a 2nd row of pillars in place of the inner wall. There was a very handsome pulpit for the higher (21/774 "Cairo") priest to read the Koran from, with a staircase covered with Arabesques, the pulpit also & a canopy over the pulpit, with a sort of Egyptian Vase on the top of that. At the foot of the stairs the usual skreen (sic) or doorway, admitting you up the stairs. On this when the Koran is being read, a green veil or curtain embroidered is hung. In another place a sort of raised stage rested on the pillars & from this an inferior priest reads the Koran. Between each pier the roof was domed in red brickwork. The great doors admitting to all the mosques nearly that I have passed astonish me. Generally there is a high arch- pointed, filled in at the top above the door with a sort of Tabernacle work. Below this is a flat doorway- sometimes a large stone lies horizontally across from side to side. Most generally instead of one block you have 5 stones; they are put together much in the way we find a flat arch over a window in England. In very many cases these stones were dove tailed in to each other by carved work at the sides. Above this again you have a small round carrying arch, springing always from the stone on each side which rests on the jamb, if that is the name. Almost all these buildings (22/775 "Cairo") in Cairo are painted outside in alternate lines of red & white. Sometimes instead of red paint a red stone is used. In the Tomb of El Kaidbai two high arches in the mosque are pointed, horseshoed (sic) & built of alternate slabs of white stone, very like the Caen stone & black sandstone. It is very effective. After leaving the 2nd {mosque} we passed a small domed Tomb with a beautifully planned doorway, wh(ich) Mr Strachan drew for me. After this we had to go back to be in time at the College of Dervishes. We passed into Cairo by another gate, the Bab el Nasr & home as fast as our donkeys could get along the crowded streets. A mouthful of luncheon & then we started again out through the Suez gate, a short way along the {Khartoum} road, then to the left towards SW Cairo, wh(ich) was the Egyptian Babylon, or rather the Roman Babylon. We turned off this road again, to the right & in 5 minutes found ourselves at the College of Dervishes. This College originally belonged to another order of Dervishes, but Mahomet Ali or Ibrahim Pacha transferred it to the order who now hold it. We went into the courtyard & found a good many people assembled, many (23/776 "Cairo") of them English & German from the Hotel. In a few minutes time the Dervishes past (sic) into the Mosque & then we were admitted, having however to leave our boots at the door. The Mosque is a very poor whitewashed building with a Dome & windows in this, from wh(ich) women's heads carefully veiled looked down upon the Ceremony. The Dervishes were all seated in a ring, the sheikh sitting near the recess pointing towards Mecca. Two young men were chanting some account of the Prophet & his family. Presently this ceased, & all the Dervishes began pronouncing the word "Allah", in a deep voice, keeping time & swaying their bodies to & fro in time & bowing their heads. This ceased & they rose from their sitting postures & stood & then they began repeating "Allah" again & bowing in time. Their motions grew more & more violent- some 4 or 5 threw off their outer clothes & turbans, letting loose long hair (the Dervishes wear the hair long). A fife was introduced & a tom tom or two, which accompanied the chant of the 2 men. At intervals the motion became most {violent} & then relaxed & began again. All this time a Dervish (24/777 "Cairo" "Dervishes") in a high cap & dressed entirely in white girded round his waist, has been spinning round & round in the circle with his 2 arms extended, the right hand having the palm upwards & the left down. It was marvellous to see this man without stopping or relaxing spinning gracefully round & round in one spot for quite upwards of an hour. At the end of one of the paroxysms of excitement one of the 4 or 5 Dervishes mentioned above, staggered back from the ring shouting & tossing his head about, now against this wall, then head foremost against another wall. Then he fell on his knees & thumted (sic) his head on the floor & {arched} about apparently requiring 2 or 3 men to hold him. By & bye however he got calm, or left off his pretending if it was pretence & joined the circle again. It ended at last & we had to pay backsheesh which was put under a skin lying before the Sheikh, he pretending not to see it. We mounted our donkeys & rode into old Cairo, where we crossed to the river to the Island of Roda to see the "Nilometer". It was on the banks of this Island that Moses was said to have been found by Pharaoh's daughter. It was a large well 20 feet square, with a graduated pillar in the middle showing the height (25/778 "Cairo" Copts) of the water. It stands just at the side of the Nile, with which it has communication. There are some arches in the wall below, pointed, & with grand mouldings, of Saracenic building. Round the walls is a Coptic Inscription. To get to the Nilometer we had to pass through the garden of the Palace where Hassan Pasha lives. He is Prime minister. We crossed the branch of the River again & went into the Copts Quarter, to a Coptic Convent. We were admitted at once into the Church, a miserable dirty building, divided by several skreens (sic). A close skreen (sic) & doorway hid the altar but you could see through an opening in the screen (sic). The altar stood just inside.

It was apparently a mere wooden table, with nothing on it & behind the altar the circular apse had rows of steps or seats round {otherwise}. They bought candles & we went down a staircase into a low sort of crypt, & at one end was what they show as a cave where the blessed Virgin took refuge when she fled from Egypt. In the same place we also saw a font at which they said Coptic Children are regularly baptised. I got the Priest to show me the various vestments- a sort of Abba of (26/779 "Cairo") some white stuff, & embroidered in coloured floss silk on the front, back & the 2 cuffs. Over this they wear a girdle- then a strip of embroidered silk hanging from the neck down in front. The 11 Apostles & Virgin with our LORD above are embroidered on it. Over this a cape is worn, with the hood drawn over the head. We left & rode home, too tired to wish to see more on our way home. When we got home it was just time to dress for dinner & after dinner I came up to my room, helped to sermon from W°. 8 to W °. 20 (?) & wrote journal, falling asleep over it at last, when I gave it up & went to bed.

Saturday. Feb. 18.

I slept ill all night, as one generally does where one is to get up early to start for any place. I was continually waking through the night- the consequence of having determined vigorously the night before to get up at a certain time. At last it seemed to be about ½ past 5 & I got up- my watch had stopped & I am sadly at a loss. I dressed & went to Mr Kennedy's room & was sorry to find it empty. I went down & found him ready at the door, & Strachan drinking a cup of tea & (27/780 "Cairo" "Pyramids") there we mounted our donkeys & started for the Pyramids. We rode to Old Cairo- here we got some money changed to pay the Arabs with at the Pyramids. Got the donkeys on board a boat & ferried across & landed at the other side in the Village of Geezeh. This was originally a large suburb. It is said that the mamelukes had handsome palaces here, & after the fatigues of the day in Cairo, came out to their homes here. The palaces are gone now & you pass through miserable houses & masses of rubbish, mixed here & there with Tombs. A sort of fair or market was going on & a man buying a young horse trying to beat down the old Arab horse dealer through the street of which we passed. As soon as we had left the houses, we came out on a path along the banks of the Nile & branched off to the right at an Arab Village & through a large grove of Date Palms into the country. The Nile is low now, & you can get straight across country to the Pyramids. We followed a path leading through the fields, which were green with wheat, & clover, & Egyptian beans in full blossom. I thought & still think they are 'Lupins'. We passed one or 2 Arab Villages. These look at a distance like castles, or fortified places- with low towers & (28/781 "Cairo" "Pyramids") battlements. But as you get near you find that they are collections of huts, flat roofed, built of mud & sundried bricks & piled up on a mound of the same material so as to be above high water mark when the Nile overflows & fills the Valley. The fields were full of men, boys & women with herds of goats & a few cattle & camels & here & there a man ploughing. Before us on the rising sandy ground beyond the Valley stood the Great Pyramids. The Great Cheops & the smaller in which is Belzoni's Chamber. We all agreed that they did not look so very large. The usual impression I believe all people have. But there was nothing to compare them with. There were high Palms between us & them which looked no more than cabbages, but you are not helped by this, for the Pyramids stand out alone, on a sandy bank above the plain, without anything near that you can see from the valley. The only thing that showed their size was the time we took to get to them. Arabs here joined themselves to us. I take it they are always looking out for travellers & before we got to an Arab Village before us, we had six or 7 with us & several boys. However they behaved (29/782 "Cairo" "Pyramids") well. At last we began to leave the rich ground of the Nile & get into sand, mixed with some few green things & then to ascend up to the level of the Pyramids. It was only after emerging from the hills of sand & rubbish & finding yourself close under the Pyramids that you began to realise that they were huge masses of stonework. We rode straight up to the E. side of the Great Pyramid & round its N.E. angle & dismounted. On the top we saw a party of people. Frenchmen the Arabs said. We had started with just a cup of tea & a biscuit, intending to have a substantial breakfast of cold meat etc. here. It had gone on before us & was all right waiting for us. After riding 10 miles, I was all for eating breakfast & sitting quietly down & reading what Murray said, (I had brought his Handbook with me from the Hotel's Library) & then climbing the Pyramid at our leisure. But the other 2, who had taken care of themselves pretty well on tea & biscuits before we started, were both for going up the Pyramid first & so I gave way, though I was sure it would be hard work before breakfast. However having settled on this plan we set about starting at (30/783 "Cairo" "Pyramids") once. It always spoils my pleasure to have a host of officious people about, making a noise, & proposing their services. On the present occasion a host of Arabs were incessantly bothering you, two at least, if not more, insisting on helping you up. Besides a number of boys with bottles of water. It affronts one to have it supposed that you can't walk without help & a draught of water every 5 minutes. I confess I started with too slight ideas of the labour of reaching the top & at once shook off my 2 Arab friends & insisted on going alone. They hurried one along, the game being as we found afterwards to get you to the top, the first of the party. Well I went ½ way entirely alone, refusing help- but the labour of toiling up the great stones is more than I expected. The stones are at least 3 feet high & many considerably more, & its steepness is greater than it looks from below. Each stone recedes from the edge of the one below 8 or 10 inches. When I got ½ way I was quite done. No strength left in my legs & feeling faint from having eating (sic) nothing, it being now between 10 & 11. The rest of the way I had the 2 Arabs help & the way they pull you up by the hand is marvellous. My 2 Arabs were 2 brothers [space left] (31/784 "Cairo") & Mohammed Ali. We met the French party coming down, about 2/3 of the way up. At last we reached the top. If it were not for the labour, especially after having been at sea without exercise for some months almost continuously in the way of an unusual exertion of muscles, the ascent is no great thing. It can be accomplished in from 10 to 15 minutes. We sat down on the top & rested & then began gradually to get away from our noisy Arabs, good fellows as they were & wander & enjoy the view. It is very fine. To the East lies the Valley of the Nile, green & fresh as a bit of England, the fields full of people at work. Beyond this lay the Nile & Cairo, with clumps & belts of palms & a range of sand hills behind. To the West the unbroken desert, yellow & melancholy. At our feet to the north heaps of rubbish & huge stones & bits of walls etc. with the plain winding round following the course of the Nile towards Alexandria & lost behind the hills of sand. While to the South stood the other great Pyramid & at our feet the huge Sphinx gradually sinking deeper & deeper into the sand again, with remains of Tombs & smaller Pyramids & ruins (32/785 "Cairo" "Pyramids") about, & a waste of sand & stone & mounds covering tombs & relics of the world that lived & died ages ago, and beyond all the sky was cut by the sharp outlines of the Sakkara Pyramids & those of Memphis & some again beyond them further off almost lost in the distance at Dashur. Below, on this side, near the Sphinx & between that a high ruined wall, stood 3 trees, a Sycamore, a Palm & another tree. They looked like bushes down below, & I pitied them standing alone out of the sand the only green living things amongst the many ruins of the dead. I sat with Murray's plan on my knees & made out the different ruins & mounds below & so shifting my seat to the different sides of the Pyramid, I got a fair idea of the positions of the different ruins. I defy people to sit on the top of the Pyramids & compose the meditations which they put in the books they write when they get home again. It is easy enough to sit in your room & write about the Pyramid & say "I sat on a huge stone on the top of the Great Pyramid & thought how" etc. etc. etc.- but I am sure people don't do it when they are on the spot. The excitement is too great (33/786 "Cairo" "Pyramids") & worse still, you have a bevy of Arabs round you, asking for Baksheesh & suggesting that if you give them something <u>then</u> the sheikh won't see it & Englishmen very good & if you like one Arab run down this Pyramid & up to the top of the next in 10 minutes etc. etc. & that an old gentleman very good gave Ali 5 piastres to do this last week etc. & then they propose to dance an Arab dance then & there on the top of the Pyramid & all these things are against solemn meditations- the only thing is to insist on their keeping back while you enjoy the panorama & work out the positions of places & then take all their chattering good humouredly & when you are rested & have been there long enough go down again. These Arabs make a regular trade now of showing the Pyramids, as if they belong to them. The Sheikh receives his regular bit, wh(ich) goes to [*** (?)] a common fund for the good of the Village & something for himself as being Sheikh & pretends & I think does keep the men in order. Some of the Arabs talk fairish English, some few French & Italian. We got down more quickly than we went up- you jump from step to step. (34/787 "Cairo" "The Pyramids") but it could be awkward work but for the help of the Arabs. I believe if you fell you (would) most almost certainly be killed. It is like going down steep stairs, each step 3 foot high & very narrow, & no bannisters to catch hold of to save yourself. When we had got back to the N.E. angle we sat down & eat (sic) our breakfast of cold meat, had bread & Claret & water & then sauntered away while Hassan our Dragoman eat (sic) the remains. We strolled slowly along towards the Sphinx, passing under the E. Side of the Great Pyramid. It is when standing at the bottom near one corner, & looking up, that you realise the size of the Pyramids. You look along the base & see how far it is to the next angle & then your eye runs up line after line of great stones till it reaches the top (250 odd steps), & when it has got there, you wonder how such a building could ever have been put together. The stone is very like Caen stone- very fine- but full of shell. There are also masses of granite. The lesser Pyramid has several courses of black granite at the base, & the same white stone above. We passed (35/788 "Cairo" "The Pyramids") several tombs, & excavations, one in particular a deep square well, into which you look by holding the hand of an Arab behind you, & see that at the bottom a small opening on one side takes you into a horizontal passage & then we came to the Sphinx. A description of this may be seen in hundreds of books. It is a sad pity to allow the sand to bury it & it could be a fine work for Abbas Pasha to have it cleared away. The face is mutilated by the top of the nose. The sand is now high up the chest of the figure- we stood under its chin & it towered high above us. What must it have been when you c(oul)d stand between the fore paws & see the full {height} If you go a little way off not quite in front you get the full beauty of the face. It is that of perfect repose & dignity & as I looked at it & the perfection of the carving grew on me, as well as the calm repose about the whole figure, I quite felt what I remembered seeing noted in some book, the way in wh(ich) this & other gigantic figures in Egypt look out calmly over the world, while everything but themselves passes away. (36/789 "Cairo" "Pyramids") The Frenchmen came up just then to the Sphinx & one of them encouraged the destruction of the figure, by giving an Arab money to climb up the figure to the top of the head. Which of course was done by the help of small holes to put your feet in cut out of the neck & shoulders of the figure. We left then & went back to the Pyramid & prepared to go inside. I took off my coat & put on a very thin one I had brought on purpose. We settled that only 2 Arabs would go in with each of us & to one of each of these 2 we gave a candle. Then we climbed up to the entrance. The entrance is on the north side-some way up the side of the Pyramid. Huge masses of granite form the floor of the entrance & these oddly enough all slope down to the interior & large block vaulting from the arch above the entrance. When we were all ready we started I leading the way. First you have to stoop almost or quite to your hands & knees to get inside. (It gives you an uncomfortable feeling) & you go down, down, a long narrow low passage for some way. When you get to the bottom (37/790 "Cairo" "Pyramids") of this small passage, you are able to stand upright at a sort of corner. Before you is a small opening leading to the lower chamber called the Queen's chamber & a narrow passage, steep, leads you up to the larger one above. We climbed up on to the upper passage & went first up to the Upper Chamber. This was all up hill so low as to oblige you to stoop almost to hands & knees part of the way till you suddenly find yourself in the Chamber at the end. This is a large room; Murray gives the dimensions as 34 feet long, 17ft 7in wide & 19ft 2in high. At the right hand or west end of the apartment stands the great sarcophagus, empty & at the S. side you see a small hole, which is said to be an air hole communicating with the outside. The roof is flat, composed of great stones. My companions complained of the great heat & dust & oppression in the air, all which did not much trouble me but we shortened our stay & after the Arabs had shouted & screeched to their hearts content we began to beat our retreat. I heard of some man who struck up the appropriate song here the other day of "Down (38/791 "Cairo" "Pyramids") among the dead men". Just outside the entrance to the Chamber is a high narrow place & in the wall steps have been cut up to a small passage leading into a small chamber over the great chamber. This is I believe only about 3½ feet high. We passed on down the gallery again till we got to the place where the other gallery leads to the lower chamber. Mr Strachan had disappeared, but I fully meant going in & so in I went & Mr Kennedy followed. The passage was lower than the other, so much so that you had almost to go on hands & knees. We soon reached the Chamber. The stones of the roof are laid together like the arch over the entrance. On returning from this I stopped in the passage to examine the way in which it is built. I sat down & took the candle into my own hands. The great stones made parallel by the slope of the passage & are so beautifully put together that you can hardly make out the joints. The great stones wh(ich) form the roof of the passage seem to have cracked with the great weight above or else the stones met just in the (39/792 "Cairo") middle all down the passage. The walls are in most places covered with a crust of salt. It was very difficult going up & down these passages in shoes; the floor is very polished & slippery though notches have been cut all the way to keep you from slipping. If I went again I would take my shoes off. When we got out we found Strachan sitting in the entrance. I did not find the interior so tiring as they did. They came out wet through with perspiration. Thanks to Dr Gully I am more lightly clad than most men I meet. I have never taken to flannel again not withstanding all that people have said about hot climates & the extra need of flannel & I had put on a thin alpaca coat before going in. We sat down & rested a while & then went to see a tomb at the West side of the Great Pyramid. We entered by a small hole on our hands & knees & were well repaid- the walls are covered with painting of figures & bulls & boats. The figures are wonderfully done & the faces just like the faces of the Egyptians of the present day. I have been much struck several times with the great likeness between the donkey boys (40/793 "Cairo") & the engravings one sees of Old Egyptian figures. The countenance is just the same. After walking round & looking again & again at the Great Pyramid & seeing an Arab climb the smaller Pyramid, more difficult from the fact that near the top the outer coating which filled up the steps & made a smooth surface remains, we went to our donkeys & after satisfying the sheikh & giving each our own guides a trifle ourselves, we started back for Cairo. The Arabs came with us till we got near their village, when they shook hands & said goodbye & we jogged on. We came across a man ploughing with the old Egyptian plow. I got off my donkey (whose name by the bye is "Snooks") & stopped the plow. It had an Iron share- but no coulter. It was made like this- a flat beam at bottom with the iron share at the end (A). To this was joined with a pin or hinge of some sort the handle (B), which made a greater or less angle with the share beam according as it was pegged higher or lower to the upright post (C). The oxen drew from some (41/794 "Cairo") point below this I think, but am not sure. I turned it round & started the oxen down a new furrow for some 10 or 15 yards, much to the amusement of the man who was ploughing. The fault of it was that there was no share board, & so the earth was not turned over, you only made a channel & turned up the earth near the last furrow. At Gezeh we went out of our way to see the egg ovens, in which eggs have been hatched in the same way since the time of the Pharaohs. But after going a long way out of our way, & going on our hands & knees through a small hole into a place where the thermometer would stand I should think at 150 we found the process was not going on- but we saw the method. The eggs are wrapped up in mats & placed at the mouth of a chamber or furnace in which a great fire is kept up for 20 days, when the chickens make their appearance. We rode back to the ferry, got the donkeys with some difficulty into a small boat & thence into a larger, were carried in ourselves & then poled up the stream some way & took to oars & pulled across- by the time we had put across (42/795 "Cairo") the current had set us down the ½ mile to the landing place. But it was a long business We mounted our donkeys & pushed on, for it was getting towards dinner. Got back to Sheph(e)ards, washed & dressed & went down to dinner. An extremely forward young lady across the table opened the conversation. After dinner I went upstairs & wrote journal till I could no longer keep my eyes open.

Sunday Feb.19. Sexagesima S.

After breakfast I came upstairs & read the Xtian Year etc. & then went to Service. Mr Leader, a German, & I think a {Missionary} for the C(hurch) Miss(ion) to the Copts, has one service on the Sunday in a room fitted up decently as a Chapel. The Congregation consisted almost of some 30 or so from out hotel & 2 or 3 in native dresses. Dragoman or servants- perhaps Englishmen in Egyptian dress. After service we 3 took a turn in the shade in the square, discussing the possibility of carrying out what had been at first proposed in joke yesterday at the Pyramids & then taken up in earnest- viz. the making the (sic) journey by Sinai & Petra to Jerusalem & so through the Holy Land to Beyrout & (43/796 "Cairo") home (!) by Smyrna & Trieste & across the Continent. Hassan offered to take the contract- i.e. We pay down each so much & he stipulating to provide camels hence to Hebron, with tents, provisions & everything, & thence horses to Jerusalem. Keeping us at the Hotel in Jerusalem a week or fortnight, & conducting us all through Syria to Damascus & Beyrout. Of course I would give my ears to do it & if it is ever to be done, now is the time, when all expense of voyage from & back to England is saved. On the other hand, I don't like the idea of putting off my return to England. I have been dwelling on the thoughts of getting back in less than 3 weeks perhaps & besides this is the extra expense. I don't know what we should do. I got up this morning determined to think no more of it but go straight home & as far as present feelings go, that is what I would much rather do. I should not get home a bit before May if I went. However-----. After luncheon I came up to my room & wrote journal. Then Hassan came with a message to ask if I w(oul)d go out. I went down, & found Strachan, wanting to ride out to some Tomb or (44/797 "Cairo") about an hour & ½ away. I declined going, & kept him talking till it was too late to go. Then I came upstairs & said Evening Prayer- put on my hat & went out for a walk with him. We went down the Copts' Quarter & into the Coptic Church- a modern building- with nothing to see - & then on among the old streets & back to the garden, which we walked all round, & into dinner. The house was full of the outward bound passengers just arrived from Alexandria- ½ going on tonight & the rest tomorrow morning. After dinner we sat discussing Mt Sinai & the desert again & then I came up stairs & accomplished writing up to this point & now for bed.

Monday Feb. 20

Shortly after breakfast during wh(ich) I had had some talk to a Mr Ross who hopes to start on Thursday for Mt Sinai, Petra & Jerusalem about the (sic) crossing the Desert, we made a start for Heliopolis. Papers have come in by this mail- & war with Russia seems inevitable. The Russian minister has left London- & the English & French ministers recalled from St Petersburg. We rode through the part of Cairo to the East Gate discussing & regretting, & came to a halt (45/798 "Cairo" "Heliopolis") outside the Gate. There we were to mount 3 camels to see what the riding on a camel is like. We waited some 5 minutes, & then they came. By this time we had a small crowd round us, amused at our amusement. Mr Kennedy backed out of it at once, & said he w(oul)d not go on a camel for anything- but we both mounted, & persuaded him to do the same- in less than 20 yards, he cried out for his donkey, & w(oul)d not have a camel. We left him dismounting & getting on his donkey again, & rode slowly on. When he over took us he asked, "are you going at that pace all the {way}" – yes- Well then good bye, I'll go back- & without another word, he turned & cantered away back. However we shouted & made Hassan the Dragoman go after him, & at last he stopped & came back- & of course we had nothing for it but to insist on giving up the camels, & taking the donkeys, & then he w(oul)d not go back. I was sorry to give up the ride on the camel but there is nothing for it in travelling, but always to give way. We mounted our donkeys & rode away at a Canter- discussing again the Sinai & Syrian plans. Our road lay out towards the E. of Cairo & towards (46/799 "Cairo" "Heliopolis") the N.E. We crossed a sandy hill, & near the camp wh(ich) we had passed the day we came from Suez, & then must have crossed the Suez Road, but where I can't remember exactly, only we saw Abbas Pasha's palace on our right; after this we got into fields, & followed a path wh(ich) brought us to a sugar manufactory belonging to the Pasha- where camels burden with fresh sugar cane fr(om) the field was being unloaded. Not far beyond this we got into still richer fields, green & fresh, with green trees, & numbers of people about in the fields with flocks of sheep & goats & numbers of horses & camels. It was not unlike Lincolnshire here, barring the Palm trees. Further on again we came to a field in which were a row of she camels all fettered with their young ones running free- & in the distance many more. These we were told were the Pasha's. All about were tents with horses & flocks. These mostly belonged to Arabs. We passed close by the village of Mattareeh & here dismounted stopping to see 3 great masses of stone- 2 of wh(ich) were evidently parts of some obelisk- carved with inscriptions in hieroglyphic. We tried to make out the Dynasty but could not. We then (47/800 "Cairo" "Heliopolis") went on to see the obelisk, wh(ich) showed itself above the trees. We stopped at a garden close by some filthy, miserable Arab huts, over wh(ich) a number of vultures were hovering, attracted by the smell of the blood of some animal wh(ich) had just been killed. In the middle of the garden stood the obeslisk (sic). The ground has risen in the course of years so that the bottom is buried. The ground round has been dug away, but not sufficiently to show either of the 2 pedestals wh(ich) are still buried in the ground & commonly covered with water, though not now. It is about 62 feet above the level of the ground. 18 (?)[or 68 (?)] I believe above the upper pedestal. On 2 sides it has hieroglyphics. It is shattered at bottom & at the top of the pointed top the stone is cut as though it had had a stone or metal cap of some sort. We made out the name of Osirtasen I by the aid of the Table in Murray. This gives the date of B.C. 1740- 36 years before the arrival of Joseph in Egypt. The obelisk is of granite. This is supposed to have been 1 of 2 obelisks near the entrance to the great Temple of the Sun. Tradition says that an avenue of Sphinxes (or {serpents}) led up to the Temple. We went out in a N.W. direction & found a mass (48/801"Cairo" "Heliopolis") of yellow coarse marble, shattered, lying in a corn field, wh(ich) is said to be one of them. Two broken masses of stone had hieroglyphics on them- & one we fancied might have been the head. There were marks of the chisel on it- but we could not make it out distinctly. About 20 or 30 yards from this lay a piece of a granite pillar- perhaps 3½ foot long - & on this with Murray's instructions we made out distinctly ¾ of the {figure} of a man -& at his side part of the sign of Rameses the Great. The date is given B.C. 1355. From this we rode up on a great mound of rubbish & broken stone & pottery which ran round from N.W. to N. & thence to E. & so round to S.E. This is said to be the walls of the town, & the remains of houses wh(ich) all layto the N & NW of the Temple the site of which w(oul)dbe marked by the Obeslisk (sic). From this mound we looked N.E. over Birket El Hag, about 5 miles off, the rendezvous of the Mecca Caravan. Here we were then on the site of Heliopolis, once the seat of learning, where Eudorus & Plato had lived & studied- & was only surpassed & superseded by the Alexandrine schools (49/802 "Cairo") after the accession of the Ptolemies. This was [*** (?)] the daughter ("Asenath") of whose priest, "Potiphera", Pharoah had given in marrige to Joseph. It came across to me that it was the beginning of the Israelites in Egypt, & I longed to track their march from the Red Sea over Sinai & through the desert more & more. We turned homeward again, & turned aside at Matoréëh to see a Sycamore tree, under wh(ich)Tradition says that Joseph & the Virgin rested on their way fr(om) Palestine. It is an old & knotted tree, & may be of any age. Tradition says that its longevity arises fr(om) its having sheltered our lord, & that the salt "fountain of the sun" which the name of the village is said to mean, was made such by them. We madeustr on home, hurrying our donkeys as dinner was to be at 5, in consequence of a ball in the great room of the Hotel. We passed close by the Pasha's palace on our way. It is a miserable building. Plastered & painted yellow with bare square windows, all barred with iron bars & round the edge of the roof runs a sort of fancy railing of wood or iron, painted in bad taste with yellow & red &green. (50/803 "Cairo") The palace stands on a bare sandy plain a long wall with a part at the further end & contains a garden. Not a vestige of green looks over the wall. A courtyard is on one side with a shabby line of stables on one side & in it I saw some clothes hanging up to dry & yet this is the entrance. We got out on the Suez road which we followed & entered by the "Bab el[blank]"- something I can't remember & thundered our way through the narrow streets. We met the Pasha's only son- riding with several officers or servants riding both before & after him. People shrank out of the way, Hassan our dragoman tumbled off his donkey. We drew to the side- but I thought the man preceding the boy (he is only 17 or 18) was going to hit me with a long cane he held in his land which he lifted in to the air. It{roused} my {blood}& I nearly omitted to raise my hat. At dinner I had a long talk to a very nice young fellow who has walked a good deal in Switzerland. I sat in the reading room after- Kennedy, Strachan & I, with Mr Ross & another man. Much talk about (51/804 "Cairo") the Sinai journey. I am in a miserable state of indecision, & every other hour determines (sic) to cut short all discussion by settling to go on home as I had intended. Then the {feeling} rises, isn't it worth while to make a small sacrifice of time and money to trace out the scenes of God's dealings with the Israelites. The tales of our wandering in this world's wilderness & the scenes of our redemption fr(om) our fallen state from Blessed Lord & that I shall do it now at less expense than I can ever do it again. I came upstairs-{stressed} over expenses etc & wrote journal. People have been arriving to the ball in the house, & there is much noise of talking under my window, where they seem to have extemporised an entrance. Two waltzes have been played- & I must close this & go to bed. Tomorrow I must go to the agent of the Austrian Lloyds Comp(an)y- & make enquiries & perhaps decide. I may go to Sinai & Petra & Beyrout for £90 in addition. By steamer to Jaffa & through Syria to Constantinople & {home} for £40 in addition to what I have- or home at once. The Sinai route will get me home about the 1st of June. The Syrian [route] (52/805 "Cairo") alone about the 3rd of May. I went down & found Mr Kennedy & got upon the great subject again- & then we sat at the window near the large room looking at the Ball till past 12. A curious medley of people nearly all in tarbushes- a row of Levantine women sat on a sofa. Little things with trowsers (sic) & coloured hats & jackets & magnificent jewels

Tuesday Feb 21

After breakfast we went to Austrian Vice Consul to ask about the Steamers etc. He had been at the ball & was in bed. Then in to a church building which we found was for the Roman Catholics. Then to a German artist who takes photographs- Schrantz (sic). He had some beautiful things. I bought 2 drawings. After luncheon we went to the Vice-Consuls, where we saw a French man who gave us all the information we wanted- & from there we went through the town a long way to see a Mosque- that of Sultan Tayloon [= Ibn Tulun]- the oldest in Cairo, built AD 879 as proved by 2 Cufic inscriptions on the wall. It consists of a large square with minaret at each corner & a larger one attached to a {brick} centre of the west side. There was originally an arcade all round of Painted horseshoe (53/806 "Cairo") arches, but the place has been turned into a {farmhouse}& 3 sides have had the arches blocked up. The part used for prayer is still untouched. It consists of 5 long aisles separated by piers made of a piece of wall 4 ft. long with pilasters supporting painted horse shoe arches, with {rough} border of tracing or arabesque round them- above the pier the wall is pierced with a horseshoe arch. The pulpit is very handsome & the recess towards Mecca beautifully inlaid & worked. There is also <u>one</u> semicircular horseshoe arch in a wall connecting the cloister with the great minaret. The place is now disused except as above, falling to decay & filthy to the last degree. From here we went to another mosque Sultan el [**** (?)] also disused but one of the most beautiful things I have seen. It was worked in coloured stones & marbles- but it is impossible to describe it. Home- very wretched about the Sinai trip & wishing I had not been led into it. I am very sorry to be away longer & I am frightened about money matters. It was not quite dinner time, & I took a turn with Mr Strachan. After dinner Mr Kennedy & I had a talk to Mr Shephard (sic) about Hassan, whom he pronounced too young & inexperienced for the trip. (54/807 "Cairo") He also said it w(oul)d be difficult to raise [**** (?)]. We told Hassan, who was in great trouble. I sat in the reading room talking to the American & others. Much disquiet.

Wednesday Feb. 22

I awoke all night long & expecting it was 5 o'cl(ock) at last I got up & went to Mr Kennedy's room & found it was later. We dressed & had a cup of tea & biscuit & at 7 instead of 6, we mounted donkeys & started for Sakkara & Memphis. 16 miles there & more if we returned to Memphis. We cantered along to Old Cairo, & then crossed in a boat. After that we followed the river & went straight on at the 1st arab village instead of turning up to the Great Pyramids. It was a lovely day & the country rich {eventful} with multitudes of people & cattle in the fields. We struck the causeway & left it. It was being repaired & a few men managing & directing large parties of boys & girls who carried mould in baskets on their heads from the field adjoining. Each village is obliged to provide so many workmen a month to keep the causeway in {repair}. At last we got near the line of sand hills ({on}) the further edge of the valley & ascended them. We stopped at a great hole into (55/808 "Cairo") a deep place & lifting our candles we went in- the whole place was full of pottery- long narrow pots- in these have been Ibis mummies- & on searching we found a great quantity quite perfect. I carried away 2 long [**** (?)] one, & found the bones & beak quite perfect. From here we walked across the sand among mounds wh(ich) perhaps contain great treasures. To Mr Mariette's house, a Frenchman who is excavating.[12] We saw his assistant who readily gave us permission to visit excavations & sent a man. Before inspecting the Tomb however we sat down to eat our Breakfasts. Then we went into the tomb. The vaults consist of 5 sides a quadrangle under ground with one side longer than the others. In the entrance nearly of the right hand passage is the granite cover of one of the sarcophagi. This is being removed. All along the corridor are branches to the right & left. In each there is a great granite sarcophagus. They were empty I believe.. The sarcophagi were of granite one block- 18ft by 23. & nearly 10ft 6in high. How they were ever put into the vaults is the extraordinary (56/809 "Cairo") thing. From here we walked across the sand desert to another tomb wh(ich) has been excavated last {quarter} Thence we walked across the sand again- where I found part of the [***** (?)]. From Sakkara we rode on about an hour to 'Memphis'- here the statue lies in a great hole. All the winter he is buried. At this time the pit was nearly dry. He lies with his ear & side on the ground. The face is extremely beautiful. There are over 2 broken remains near. We staid (sic) some time here & then hurried home. It was just 6 when we got home. After dinner I sat at table & had some talk with a young engineer working on the Railway here- & then came upstairs & wrote journal.

Thursday Feb 23.

I was miserable all the morning while dressing about this journey through the desert- I believe an undecided man pays dearly for his weakness or whatever the deficiency is. Whenever I have to make a decision on any point, I suffer real agonies- I become feverish, & got quite ill. The question haunts me by night as well as by day. I begin by (57/810 "Cairo") by (sic) being very eager about a thing: then the wish for it quite passes away & reaction comes on & then I become miserable & ill & am unable to make up my mind & torment myself by a hundred different feelings & changes of mind in a few hours. After breakfast Kennedy asked me to come to see the Dragoman & then I had to decide & did decide to go. But since then I have been at ease & long decided I will not allow myself to discuss the matter again. Mr Kennedy & I went out to make some inquiries & arrangments. After luncheon Mr Strachan joined us, & we went to the Vice Consuls & to the Turkish bazaar & to a watchmakers shop. Then on my way home I stopped at the Hotel d'Europe & sat some time with Mr Palmer- William Palmer of Magd(alen) College. I did not know him before- he half proposes going with us. I confess I am afraid of him- he seems the "don" quite but it would be a great thing to have with us a man who knows so much. After dinner I came to my room & read the "Guardian" of Jan 25- wh(ich) I had found & written (sic) to Tupper at the Vice Consuls. Journal & to bed. By the bye Mr. Palmer enlightened me about the Sakkara Tombs. He says (58/811 "Cairo") they contain are the cemetary of the sacred bulls & each of these huge sarcophagi contained a bull or cow mummy. They must probably have been previously opened.

---

[12] François Auguste Ferdinand Mariette (1821-1881).

Friday. Feb. 24. F(eas)t of St. Matthias

I got up this morning really glad that things were settled. I would have made myself miserable about the delay in reaching home & the expense etc etc that I had determination when I made the decision yesterday, that having so decided I w(oul)d not allow my mind to dwell on the question, but enjoy the journey thoroughly, & get what good I can for body & mind out of it & I hope for much for both. After breakfast we went to Mr & Mrs Briggs- where oddly enough I found a letter from Alice dated August. Then we did some shopping- & finally went to the P. & O. Comp(an)y's office, to ask about transmission of our baggage to England. Mr Kennedy bought a revolver, with wh(ich) he must defend me as well as himself. After luncheon Mr Strachan & I went to Mr Schrantz's about the photographs. We looked over some drawings of Petra, Mt Sinai, Baalbec & Upper Egypt & then Mr Kennedy joined us- & we went into some of the Bazaars to see (59/812 "Cairo") some jewellery etc. but were unsuccessful after that we went to Sultan Hassan's Mosque which I had not seen tho' they had. It certainly is fine though I don't think it deserves all that Murray says of it. Outside the porch is magnificent from its height &general character. The rest is too like a factory. The mosque consists of an open space- large & with a handsome fountain in the middle- on each side is a square recess joined to the open square by a fine arch each. The East recess has a magnificent pointed arch- 69 foot- & in span. In the midst of the recess stands, a marble platform for the junior priest & a handsome pulpit & beyond is another square surmounted by a dome, & containing the Sultan's Tomb on which lies a Koran. It was built about 1860 AD. I have forgotten to write that in the morning in the New Street. We met a marriage procession. A number of people followed by the Bride walking, covered with jewels under a red veil or {canopy}. She had also a red veil on. In the evening we met a funeral. The body was carried in an open coffin covered with a cloth & was followed by a number of women (60/813 "Cairo") wailing & tossing or waving hankerchiefs about. We walked back part of the way& then got on our donkeys & rode briskly home- when Strachan & I walked as far as Boulac & home to dinner. After dinner I came up to my room & read & wrote.

Saturday Feb 25.

After breakfast I came upstairs & was arranging some packing when Mr Palmer came to call. He sat with me for an hour I should think & gave me a good many hints about our visit to Syria, all of which I made note of at the time. He thinks of going to Suez in one of the Vans & on from there by Camel to Sinai only & back again. After that I found Mr Kennedy & got him to come to the Police Office to have our passports noted. He went home from thence & I onto the New Street & back to luncheon. After luncheon I went to call on Mr {Lieder}, to see if I could get a small pocket Bible but without success. However I picked up Bohn's "Early travels in Palestine" containing all the great travellers accounts from the time of Alfreddown to Maundrell in 1600 ad. Thence I went to the watch makers where I got my watch going again all right- much to my delight & then (61/814 "Cairo") went to a hatter, & succeeded in getting a very nice black "wide-awake". I must kick the crown of my tall black hat in. I can't wear it all through the desert & I have no hat box. After that Mr. Strachan & I rode by slave market, where we saw some five or six Abysinian & Nubian girls or young women. It was a painful sight. I did not care after that to go in to the large square to see the boys. We went to a bazaar after we bought some attar of roses. Here were three kinds of attar of roses.1. What the man called the best, from Mecca- 2; from Constantiople- far the best to my mind- & 3- from Tunis which had a vinegary smell about it. After that we walked to we home together to the

Hotel. Where Mr Kennedy joined us & we walked out into the County for nearly an hour. At dinner we sat opposite (I think) to Captain Butler & what seemed to be the same party, 3 other men. After dinner I sat in my own room. I cut part of the "Early travels in Palestine", wrote journal & finished my letter to Alice, besides writing a few lines to Tupper. Our camels are to start on Monday afternoon & camp outside Cairo & we shall sleep here & go early on Tuesday. (62/815 "Cairo").

Sunday Quinquagesima. Feb. 26.

We went to service at ½ past 10 & afterwards Mr Kennedy & I walked in the garden in the square till one o'clock. We fell upon Church & Doctrinal matters. He fell foul of the "Puseyites". After luncheon I went to my room & Hassan came asking me to make up his accounts. I proposed deferring it to the next morning but he begged me to do it then & we set to work. He can't write & has been carrying them in his head. This took me till past 6. Then I said Evening Prayers & went out for a turn in the garden, where I took a regular constitutional till just dinner time. Giovanni says, we are to have our luggage ready by 12. So we should really start ourselves on Tuesday. After dinner I came to my room- wrote journal- made up my wideawake which I have washed & packed.

Monday. Feb. 27.

Today has been a busy day. After breakfast we had to make arrangement & by 12 our luggage was all packed on the camels & they started. Since then I have been finishing letters, doing accounts, & seeing about the tranmission of my box & canes to Southampton, a much more difficult business than I (63/816 "Cairo") expected. Late in the afternoon I fell in with the other 2 & we went for a walk & back to dinner at 5 as preparations were making for the ball which is now going on- a very fair band is playing below & I am writing to the tune of a waltz. Most unpleasant news has come from Syria. Fighting between the Latins & Greeks & report says that the Latin Bishop has been killed & the French & "English" Consuls have left Jersualem for Jaffa- & additional troops have been sent for to Jerusalem. People are frightened & changing their places &some giving up their arranged journey across the desert. We however are fairly started & must go. I [***(?)]that we can [*** (?)] as far as Hebron & when we get there, if it is not safe to go on, we must strike across to Jaffa. We have only made our arrangement to Hebron. This is the last night in the civilised world for some time & the last night I shall spend indoors- I can't say in bed, for I shall have as good a bed in the tent I have little doubt as I have here. But it is anxious being on the eve of starting for the desert. God preserve us & bless this journey to me. It is as if we're treading in his steps. I am very glad. I asked (64/817 "Cairo" & Start) to go. I best now go to bed. To be ready for our first days camel-ing tomorrow.

Shrove Tuesday Feb. 28

I got up about ½ past 6 & partly dressed & bathed & then read & waited for Hassan who was to bring a barber to cut my hair short. He came presently & I had my hair cut- cropped I might say. At 9 to my surprise the breakfast gong sounded. After the {former} ball it was past 10 before we got breakfast. After breakfast I paid my bill & took my letters to the clerk to be posted & left for Tupper & Henry Scott etc. I was just asking him about my letter to Mr Holton the P. & O. Agent at Alexandria, when he said "Why Mr Holton has arrived here just now- oh, there he is". I at once went to him & explained my wants & gave him the letter. He was very civil & said he would see all about my luggage & it would be quite safe. We then put our saddlebags etc.

down & got on our donkeys. Mr Strachan came with us part of the way intending to turn off & go visit the Petrified Wood. We rode out by a new way, down past the slave market, & out at the gate close by & so away from Cairo & among the tombs outside. We came out in the Suez road opposite Abbas Pacha's palace & so jogged along our donkeys past the (65/818 Start for the Desert) 1st rest house. Then we saw some camels lying down away to the right & sent to know whose they were. They were waiting for some other party. Away we went for some distance & then Hassan who was with us said Mr Strachan must turn off for the Petrified Forest & we said "Goodbye". Past the 2nd station & just by it came up with a number of camels which turned out not to be ours, but the caravan; of the 2 men one of whom has been in Spain, but they said ours were just in front & we soon overtook them. We made the Sheikh's acquaintance & put saddlebags etc. on our 2 riding camels & then as it was cold rather (sic) we did not mount but walked on. We passed all our eleven camels & looked to see about our luggage & then went on ahead of them. A most disagreeable day- the wind high- fortunately in our backs, the dust & sand, {aye} even small pebbles, flying in clouds. We wore our goggles & found them No.10 (sic). We came presently to the 3rd station. It was about one & not having made much of a breakfast began to feel that the sun was "over the main yard". We waited there till our camels came up. When Giovanni produced his bags. We made our riding camels kneel down while we sat down (66/818 (sic) "Start from Cairo – 1st Day") on the sand & eat some bread (a luxury for the 1st few days) &cold sausage & meat. To say nothing of sand. We had just finished discussing this & were mounting our camels, when the caravan of the 2 brothers [= Fentons] came up, however I was on my camel & away before they overtook me. We had barely gone a mile, perhaps ¾, when to our surprise we found our caravan had turned aside off the road & were unloading the camels. One felt inclined to rebel- but you must do as you're bid all the world over & so we said nothing but stood & watched the pitching of the tent- no easy matter in the high winds. It was a very pretty sight as & I wished I could sketch. Our camels were all kneeling & being unloaded. Luggage was lying about the tent erecting & the other party streaming down towards us. We took a turn across the undulating sand until our tent would be ready- we found a little gully, where we were sheltered under the hill & lay down on the sand out of the way of the wind. It was rather annoying being stopped at 2:30 in the afternoon, with nothing to do. However one must make the best of it & we got up & went back to speak to the other travellers. They too were standing by their tent, (67/819) evidently rather not knowing what to do with themselves. We went back to our tent, where we found everything in nice order- we unpacked our things & made ourselves comfortable & then went out with Kennedy's pistol, to try it & to take a good walk. We got away <u>down</u> the wind so as to know our way back again readily. We had some trouble in loading the new revolver- the bullets would not fit. Then one barrel was fired at a mark & we started for a good walk. By & bye I wanted my knife & found it was gone. I had lent Mr Kennedy where he loaded the pistol & determined to {strike} across till I found our trail & trace it back to the place. After some trouble, owing to our having tried a short cut, I came on the trail- followed it to the spot & searched in vain for the knife. We went up over a small rise in the ground which I had run over after a piece of paper we had put up as a mark & the wind had carried away & coming back I chanced upon my friend the knife & glad I was to recover it. A knife & a watch are 2 indispensables when you are travelling. We got back to the tent where we found a fourth party arrived- their tent sporting the stars & stripes.[13]

---

[13] Western travellers in the Ottoman Empire were required to advertise their nationality to ensure the privileges to which they were entitled by treaty arrangements with the imperial authorities (Nance 2009).

Giovanni (68/820 "Desert – 1st Day") proposed dinner as it was getting dark. I was amazed to find silver forks & clean napkins, soup, chickens & cauliflower- another dish. Some very nice pudding- a good cheese- oranges figs & biscuits. Followed by coffee, journal writing & a cup of tea. Kennedy began the {process} of journal writing- we read some of "Wilson" & turned in. So end of the first day in the desert.

<u>Ash Wednesday. March 1.</u>

I had not a very uncomfortable night the first in the tent. I had just gone off to sleep when Mr Kennedy went out of the tent & coming in again awoke me hammering in the peg. It was rather cold towards the morning too. At about a quarter past 5 Giovanni called us & I turned out & lit the candles. No shaving & no bath & so dressing (only) was quickly performed. We had breakfast in the other tent while they took down our tent & when Mr Kennedy had finished his breakfast we started & walked on. It was fine then, but the wind was getting up & clouds with it. The wind was piercing- quite as cold as that cold day we crossed from Suez in the vans. We walked as far as to the next station, about 4 miles on- No. 4- at which we had dinner in {crossing} in the vans. Here we stopped & went in & tried to talk to the Arabs in (69/819 (sic) Cairo to Suez – 2nd Day) the stables. By & bye the camels came up. Ours first having made the earliest start. I still walked on- Mr Kennedy rode by & bye. I had a long walk of some miles by myself trying to think of the day & its duties &then got on my camel& got my Bible out of my saddlebags, & read the service for the day. We halted for 10 minutes at station No. 6 & then rode on. After which I got down & went with Giovanni across the sand after some desert partridges. He is quite a keen sportsman but he missed. They got up {wind}. However he followed them for 2 or 3 miles. This afternoon we saw the "mirage" twice. The first time I should have been satisfied there was a large lake in the distance. One saw the hills reflected. The 2nd time it took the form of groups of trees, with small pools of water. In one place there was a rolling motion like a row of clothes fluttering in the wind. I rode my camel about over the sand, whenever I could see a bit of dry grass or thorn & so he got something at all events to eat. Rain came on once or twice & it was very cold. We came to station No. 8 & then under the lee of the building Giovanni who is a "whale" at his work secured us the best place & pitched our tent. (70/822 (sic) Cairo to Suez – 2nd Day) at once but as it is generally some time before one's tent is fit to receive one, we together with the 2 brothers whose names we don't yet know started off to walk to a hunting Palace Abbas Pacha has built on a sand ridge about 2 1/2 miles off. We had got about 2/3 of the way when we saw the clouds rolling along in an extraordinary way along the hills & presently the palace was completely hidden & before we knew anything about it, a sandstorm came upon us in a gale of wind with a thick cloud of sand. There was nothing for it but to stand with one's back to it. I got into a hole first, but it was much worse there. We turned back & walked back to the tents where we made ourselves comfortable & then I sat down to write (my) journal. It is <u>not</u> the weather one would <u>choose</u> for being out of doors- but it will mend tomorrow I dare say. I expect I shall turn in v(er)y soon after dinner.

<u>Thursday March 2.</u>

It was bitterly cold all the evening & if we had not wrapped ourselves up in coats & plaids we should have been half frozen. Every now & then a {heavy} gust came down shaking the tent as if it would shake it down & once or twice we had a shower of rain. But as (71/823 Cairo to Suez – 3rd Day) the evening went on it got finer & when I went out about ½ past 9 to run & warm myself, the wind was going down & the stars were bright. I turned in to bed with delight. About 4 I was awaken by great barking of the dogs & I got up & took my watch outside to see the time. My companions would have it that it was late. I got another glorious sleep & was awoke by Giovanni about 5:15. By 6 we were at breakfast. A quiet still morning with not a cloud to be seen & as the sun rose above the desert the day broke out most lovely. By ½ past 6 I was off- Kennedy wisely determined to rest his foot & wait for the camels. I walked away alone, but allowed myself to be overtaken by the 2 brothers Fenton. About 9 we met 4 of the van's carrying the passengers to Cairo. About 10 we allowed the camels to overtake us & I mounted. But before this I had been out alone on the sandy plain away from the road. I sat down & enjoyed the quiet. I never heard much quieter silence & stillness. The sky was cloudless & the shadows on the sand hills very soft & pretty. We saw a fine mirage early. I started on my camel intending to read, but was saluted by Mr Rogers & then found Mrs Rogers was close to me- & I could only ride along talking to her (73 (no page 72)/824 3rd Day) At the next station we met 4 more vans & I heard that Tupper was in one of the next. We rode on together again & at the next station about 1/2 past 12 we all dismounted for Tiffin & then finding the vans were expected, I sent our Dragoman on & determined to wait till they came. In about 1/2 an hour they came. Mr Kennedy fell in with some man connected with his house in China who was not a little surprised to meet him in the Desert. Here again I heard that Tupper was in the next (the last) batch of vans. I walked most of the next 10 miles & at last as we got near the 12th station we met large caravans of camels carrying the baggage & goods from the steamer. One caravan was very large. They were in 3 rows & perhaps from 20 to 30 in each string. On getting close to the station I saw the vans standing near the door & walked on quickly & found Tupper in the room eating dinner. He was much astonished to see me. Poor fellow he is [****(?)]. He has not recovered his voice & he says his throat in suffering. I left him eating his food & went out & ran after my camel which had gone on with the rest & brought her track to the (74/825 3rd Day) rest house. I succeeded in making her kneel down but she was very uneasy at being left by the others & with a stranger. Tupper came out & we had some talk & then the vans went on their way. I mounted my camel & strolled away along the track. About 2 miles from the station I found one of our Arabs in the road. He had waited to show me where we were to camp & in 2 miles he called my attention & pointed to the tents away on the right- ½ a mile from the road. I found the tents pitched & Kennedy making himself comfortable. I went out just to see the sort of place we had camped in & then came in & put out my things & washed myself & was going to write journal- but Giovanni came to lay the table for dinner. After dinner I set to work at my journal & read. Early to bed. I have enjoyed today very much. Tupper commended my change of plans. I am to meet him at Jerusalem. I am very glad that I came. I am sure it will do me good. I feel as I did in Australia. I have quite got into the {thing} today & the time has passed very quickly. I got a great deal of exercise. I suppose I have walked 15 miles today & early getting up & early going to bed (75/826 4th Day - Suez) always suit me especially.

Friday March 3 Into Suez

The same routine as yesterday. We have quite settled down to the regular desert life. At a few minutes after 5 we were called & at 6 precisely we had our cup of tea & at half past I started alone to walk on- Kennedy having {lamed} himself. - I walked away, finding myself the first away. I walked on past 2 stations, about 2 miles, glad to be alone & meditating. Past the 2nd station I lay down on a heap of stones just by the track waiting for the camels. It was more than 4 hours {since} I had started. I believe I had a knap (sic). Presently I discovered my camel coming along with Kennedy & Giovanni. I was very glad to mount. The Red Sea lay before us & I had made a sketch, about a mile back. At 3 we camped just outside Suez & after walking along the beach waiting for Kennedy who was bathing, we walked into Suez. Kennedy posted a letter & we strolled into the bazaar. Home to the tent (home is quite the word I feel regularly at home in our tent now). In the evening I wrote journal & read. It is the pleasantest day we (76//827 5th Day. Ayun Musa) have had- mild & pleasant- before it has been very cold.

Saturday. March 4 Suez to "Ayun Musa".

I got up at 5- after a cup of tea at 6- great discussion whether we should go round with the camels or across & down to Ayun Musa, or only just across, meeting our riding camels there. I had proposed to Kennedy that we should do much as our Dragoman suggested & when asked what we were going to do, we said cross & ride on from there. And so it ended. We walked into Suez- spent 4 or 5 hours there- during which I made a sketch from the flat top of the hotel & bought a "Penang Lawyer" or walking stick & some braid in the bazaar for my glass. Most of them had breakfast at the hotel- we however were quite satisfied with what we had had. We crossed & mounted our camels & followed the track across the plain of sand inwards & down southward. I let my camels saunter along browsing on the dryish bits of plants here & there- which I sat comfortably on his back reading the day's Psalms and Lessons as usual & then I pushed on to overtake the rest. Our caravan had gone on straight to this spot, where we expected to water the camels & then push on for 2 or 3 more hours. About 3 (77/828 5th Day. Ayun Musa) we reached a green spot which we had seen some time before us. It was the "Ayun Musa". The wells are 7 in number with green bushes & a few date palms round them. They have been enclosed with fences & a few Arabs live here- cultivating a small piece of ground about the wells. I was sorry to find our tents pitched- it seems (a) waste of time. However the only thing in the East is to take every thing as it comes, & not fash. There seems no reason to doubt that this is the well where the Israelites found water after crossing the Red Sea. The exact spot where they crossed the sea it is impossible to ascertain. Just southwest of Suez is a high hill called "Jebel Atakah" the end of the range which we have had on our right as we came across to Suez. Behind from us then, that is on the S.W. of them, one, perhaps the most prevalent theory, places the Israelites route- a valley runs down between 2 ranges. Jebel Atakah on the north & Jebel Abu Deraj on the south. There the sea w(oul)d be about 6 miles wide- quite distance enough for a nights' hurried march of of 1 000 000 people with cattle etc. This too would lead them across exactly to the "Ayun Musa" & that the removal of the water was miraculous & that the Israelites were not enabled to cross by reason of a very low ebb tide as rationalists (78/829 5th Day. Ayun Musa) suggest is satisfactorily proved by fact that here is deep water above which the steamer proceed to anchor- although it is sufficently plain I think that the sea has retreated, leaving the position of the Town of Suez much nearer the head of the sea than the same spot must have been at the time of the Exodus. Our camels only took 4 hours going round to the spot opposite the town. On our left after crossing, i.e. on the east of the sea, about 5 miles inland, was a range of hills, bare & grey, called "Jebel er-Radah". It was between these & the sea, the sandy plain which we have partly crossed toay, that the wilderness of Etham is thought to be. We have had a glorious day. The sky blue & nice soft breeze cooling the air, (though about 1 it was very hot, the sun scorching you from behind by reflection from the sand) & the lights & shades on the hills, which are precipitous & bare & melancholy looking, very [bra*** (?)]. It was a day to be alone & think. It seems the

Vice Consul who is also Agent of the P. & O. Company has a house here at the wells as being the only place where any thing green grows near Suez. While sitting writing this in my tent, he has just passed by with 2 ladies & some the Officers of the P. & O. Steamer "Hindostan". They seem annoyed (79/830 5th Day. Ayun Musa) to find an encampment & stared with extra power in their eyes. The younger Fenton came in with Foster's book on the Wadi Mocatteb [= Wadi Mukatteb][14] & we all went out together to see the wells- the older brother joining us at the tent door& following Palmer's map of the wanderings of the children of Israel which provided more discussion. The first well we went to was higher upon a mound of sand. It was just a puddle of dirty water 3 foot by 1 in the sand & the curious thing to me was that it did not soak through the sand all round & run away. I suspect if it was dug & cleared out a regularly built well might be found. From here we went down to the garden or inclosed (sic) ground near which our tents stood & there found a large stone built tank, with the water bubbling up fairly in the middle. From this the whole garden is irrigated. We walked through the garden. It is surrounded by what I think were tamarisk trees & a few date palms were growing in it. By means of irrigation this bit of sand, formerly quite bare, now produces famous cabbages, carrots, onions, parsley, etc. We passed through & out at the other side passing a 2nd well in the garden. We went then to 3 other regularly built wells (80/831 6th Day. Ayun Musa on to near Wadi Wardan) & so round home to the tents where we found dinner just going in. After dinner I read & about 7 went out for a turn over the sandy plan at the back of our tents & up the sand hills where the well we first saw is with its its one Palm tree, originally the only one. From here I had a view down over the 2 miles of plain between us & the Red Sea, on which the moon was shining bright & so to the Jebel Atakah &the lower range on the other side. It was very pretty. Below were our tents & the Arabs & camels all lying together in parties round their fires. I went in & to bed.

Sunday 1st in Lent. March 5th.

I awoke & got up & lighted a candle & found it was too soon to get up, 20 minutes to 5 but I would not go to bed again & was glad to have extra time- it being a Sunday morning. I dressed & had some time to myself & then had time for a walk for 20 minutes before Kennedy was ready for breakfast. We started early. The 2 Fentons & I walking on ahead. There is no road near so that it is less easy to find the way. We passed over unlevel sandy & stony ground for four or 5 miles & then {even} pulled up by the slight tracks dividing with a ridge of stones laid between them to mark the (81/832 6th Day. Ayun Musa on to near Wadi Wardan) 2 ways. Here we stopped & waited till the first batch of camels, ours, came up. I still walked on alone for a long time at one side of the Caravans & by & bye went on & joined the younger Fenton & we had a very pleasant walk together for some miles more till near 12 when our 2 parties made a halt for luncheon. We had come some 12 miles about by this time. After luncheon I mounted my camel & Kennedy & I went on- the other 2 staying behind some time longer. I read the Psalms & Lessons & Epistles & Gospel & there the sun was so hot &glaring that I could not look at a book any longer &put on my blue goggles & we rode on talking together for some time. I read him the Christian Year for the day. Our road lay all day over a vast plain partly sandy,

---

[14] The Rev. Charles Forster (1787-1871) had a keen interest in the Bible, Arab lands, and early Aramaic inscriptions. The book referred to here is probably Forster 1853; cf. 1844. His better-known book - notable for including photographs of inscriptions and casts of inscriptions (Forster 1862), owed a great deal to material collected for him by the Butler brothers whom we find travelling in Sinai with Bowles at this same time (above Ch. 5).

but mostly hardish gravel with loose stones. The rock freestanding here & there- sandstone- with pieces of coral rock & a great deal of {'Falk'}. We passed several Wadis. They were but just depressions in the sand, down which the rain water runs from the hills to the sea. The names of some seemed doubtful & not very worth mentioning. Wilson[15] speaks of a "Wadi A {Ahide}, or {Ahtha}, mentioned also by L(or)d Lindsay - & suggests some connection between its name & <u>Etham</u> (82/833 6th Day) I questioned all the Arabs but not one knew the name. From this we passed over 10 miles of dead level burning plain to Wadi {Sadz}. Mirage continually. It was very hot to our left was Jebel Rahah rather drew in & a conspicuous hill stood out well, called Jebel Sadr. We hoped to have got on to camp at "Wadi Wardan" but our Arabs turned off before & we camped about ½ a mile from the track on our left. I went out & walked up a highish hill where I knelt down & said Evening Prayer. I was glad afterward to going - home for the wind put up suddenly &drove the loose sand in clouds. When I got back to the tent, I found everything, deep in sand. Bed covered & everything. I sat down & wrote journal before dinner. I have felt all day as if we had been travelling over a great dry steep field, or plain trampled down by multitudes of people, where nothing had ever grown again. It has here been a lovely day- bright & clear- a very hot sun- but wind enough behind us most of the day to keep us from feeling the heat - except for 2 or 3 hours in the afternoon, when it was very hot. Camels & men felt it. The men from time to time would suddenly lie down just where they were on the plain & appeared to be asleep in a minute. They would sleep for ½ an hour or (83/834 7th Day) more & then woke up, shake themselves &trot on to overtake the caravan or party to which they belonged. The wind dropped again while we were at dinner & we both sat reading. I went out for my usual turn before going to bed & then turned in.

<u>Monday March 6th</u>

Everybody was extra early this morning. So much so that when Kennedy & I walked off as we did just at ½ past 6. After walking a cup of tea & some biscuit etc.- we found the American Mr Rogers & his wife as well as the 2 other Americans following us & were soon overtaken by the younger Fenton. I have omitted to mention a disturbance I had in the early morning. About 3 I was awoken by the side of the tent coming down on me. I thought at first that it was merely the wind blowing it on me, but when thoroughly awake found the side was all down. I got up & put on some clothes & went out & found one of our camels had torn up 2 of the tent pegs & was lying down among the lines. I was just hammering the pegs in again, when our cook came out of the other tent & did it for me. Kennedy soon fell behind & waited for his camel. I walked on with Mr Fenton but found (84/835 7th Day) our road lay still over the level plain, the stones on which shone in the morning sun. They are all polished, I fancy, by the sand which is constantly travelling over them. Gradually the hills on the east began to draw in a little & the ground to break. Presently we came to the "Wadi Wardan", a broad riverbed & in some places with a high perpendicular bank of gravel just like a river. Here were signs of this having been water- a surface of caked sand dry. We saw traces of gazelles & of some animal with a small foot which we supposed might be a rabbit, or perhaps a Coney, which seems not to be the rabbit as I always supposed, but some animal something like. We had glimpses of the Red Sea all day at times & the colouring on the hills to the East was very fine. We passed the

---

[15] The Rev. John Wilson is frequently cited by Bowles. His published account of this journey is easily downloaded on the internet (free) and very readable (Wilson 1847). His own story is a remarkable one including founding what is now the University of Mumbai and later serving as Moderator of the Church of Scotland.

Wadi Amarah & sand stone hills that us in from the Sea. About half past one we passed a very curious stone which Dr Wilson calls Hajar-er- {Rabbat}- though our Arabs give that name to another which we had passed before- Dr. Wilson however is right. It is a great square mass of stratified stone on a sand mound- how it came there is another question. Half a mile beyond this we came to a sand hill with 2 palm trees or bushes & near them a well. This is said to be <u>Marah</u> & I think (85/836 7th Day "Marah") it most likely is so. It is about 30 miles down the coast from the probable places of the passage of the Red Sea. The water is not brackish, but is bitter enough to make your mouth a little rough & on pronouncing the word <u>Murrah</u> to the Arab who was with me, he said it was not 'Taib' [= good] it was <u>Murrah</u>. I saw in Wilson's book that his Arabs on seeing him about to drink some of the water tried to stop him crying "Murrah, Murrah" which is the Arabic of the present day for "<u>bitter</u>". A few miles beyond this we came to another broad winter water course called Wadi Gharandel. In the bed for some distance are shrubs & a few palms & people have fancied it to be <u>Elim</u> but it is too near <u>Marah</u>. The camels rushed about to gnaw the bushes & there was no holding them. I hoped we should have pushed on a few miles further to what is more likely Elim, but it was ½ past 3 & 2 of the sheikhs decided on halting here & in a few minutes the tents were erecting & in a few minutes more the Wadi was full of camels straying about. I took my book & walked away intending to sit & read but I caught what I thought would make a pretty sketch & made a mess of that & then went across the Wadi to a large bit (86/837 7th Day) of harden ground, where the sand stone rock had been washed bare & apparently crushed & broken down. Then I climbed a steep gravel bank or hill & went down into another very striking spur of the Wadi. There must be a great rush of water here at times from the appearance of the land. It was a quiet private place & I knelt down & said Evening Prayers. I have forgotten to say that during the day we met a small caravan coming from Sinai. An old monk, with white hair, was going to Cairo to appeal to the Pacha in behalf of the Arabs who live about & serve the Convent. It seems that the small Pacha of their district has ordered the arabs to build a wall or some building for him & as they know he will never pay them, they refuse to [*** (?)] their work for the Convent. He has insisted & the old monk was going to appeal. It was pleasant to see the affectionate greetings between them &our arabs. They were all of the same tribe. They shook hands all & talked for a few minutes. Our Dragoman gave the new comers some tobacco & we rode on. Giovanni gives these arabs a very high character. He says we might drop anything however valuable here & we should be {sure} to recover it. They are known to live longer & are generally a (87/838 8th Day. "W. Gharandal to the Sea") very moral people. Drunkeness is unknown. Our's are very attentive & civil. They have always a smile on their face & run to do anything for you. After dinner I wrote journal & read.

<u>Tuesday March 7th</u>

We were all away early in a lovely morning. Mr Kennedy & I started first- up the Wadi & then to the left up another gully, where I was last entering which led us by a tortuous course, amongst walls of sandstone up upon higher ground again. The flat ground is all past & we are among the hills & rocks: white, yellow, black, grey, red & the hills & peaks a little distance off rose & violet colours & softened down into a beautiful picture. We passed over some broken ground & down into a hollow, where stood a circular mound of stones- another like it stood on a high bank above. The arabs said they were to commemorate the leap & consequent death of someone pursued in battle who took the leap & fell dead & there he was buried. Three or four miles after having our camp ground in the Wadi Gharandel we came by a winding up & down hill road, into the Wadi {Nsert} or {Weisat} Dr Wilson calls it. There we found several springs under some palm trees. Dr Wilson call this Elim & it seems like it. Palms grow freely & though (88/839 "to encampment by the Sea") there are now no wells but only small holes in the sand with water in them, most likely if you dug you would find remains of stone work. Poking about after the whole caravan had passed I found what I am sure was another spring. I tried to make a couple of sketches of the place but I find sketching beyond me. On leaving the Wadi we met a caravan with several arabs & there was the usual embracing between them & our arabs. I was walking above them, on the hill above when I saw everyone drawn to one spot & surrounding something. I made my way down & ran to the place & found them gathered round two sick men. One was being held up & was drinking eagerly some water out of the hands of our Sheikh. The other was lying on the ground looking half dead. It turned out that they were 2 Egyptian soldiers, who had run away from Sinai or somewhere there. They had meant to get across the desert to Gaza & so away from the Country but had been wandering about among the hills for 19 days probably lost. The Arabs carried them off prisoners to restore them at Sinai. We now saw Jebel Serbal- a fine high mountain but a little too flat on the top. Nevertheless it looked well this morning, a hazy pink light floating over it. It is thought by some (89/840 "to the Sea") to be Mount Sinai. It being very hard to ascertain which of the many peaks are the one on which Moses received the law. It is a considerable distance from what is called <u>the</u> Mount Sinai or Djebel Musa. The Fentons with whom I was walked stopped to have Tiffin & I walked on ahead alone, enjoying it. The hills soft & pink closing in & withdrawing alternately. We passed through the "Wadi Thall" in one part of which were a few (2 or 3) palms. The White Broom we found in several place, in the Wadis, in some of which there must be apparently a great rush of water at some time in the year. The W. Thall ramifies very much & it was a long time before we got out of it. About ½ past 11 Kennedy & I had some Tiffin & then I mounted my "Kemel taib" [= good camel] & rode the rest of the way to our night's quarters. We passed through the "Wadi Shahka", with high cliffs of sandstone. This took us about an hour & then passing by a barrier of rock & sand in the middle into the Wadi Teiyibah (pronounced Ti-ye-bah). The strata in the rocks were very much distorted in this valley & there are some grand seams of red & yellow. Here we found a palm or two, some Tamarisks, which the camels are especially fond of & some (90/841 8th Day – 9th Day. Seaside to W. Makabbeb (sic)) springs of water. We came round several sharp elbows & suddenly out upon the sea. Here we turned to the south & kept along the plain between the sea & the hills. The hills were of all colours- pure white (of chalk I suppose), grey, black & red. We paced slowly on for an hour or more & camped in a small valley, where the hills approach close almost to the sea. After seeing the tents pitched I strolled away along the beach thinking to get a sketch & ended by having a bathe & home to dinner. After which journal writing, reading & to bed.

<u>Wednesday March 8th</u>

We have had a most interesting journey today & so far from time hanging heavily as I dare say it must be sometimes to everybody on camelback, it has gone too quickly. On first starting we tracked along the shore for a short distance- rounded a point & then came out on a large sandy & stoney plain between the sea & the hills. We saw traces of gazelles everywhere & of some smaller animal which turned out to be a hare. One was startled by some of the party. Stunted shrubs grew about- among which it was very difficult to drive the camels. I was walking as usual. Indeed I have been on my feet the greater part of the day. The hills bounding this plain on the east, (91/842 9th Day to W. Macatteb (sic)) or inland side, were very striking. Sandstone, chalk (perfectly white & ghost like). {Porpherite} & some hills quite black. Wilson speaks of these as tufa- I enjoyed the colouring & lines made by fissures etc. & the lights & dark more. It took us 3 hours to cross this plain- after which we found the entrance of a Wadi & struck in to a Wadi called W. Lagam (or Lajam Wilson calls it). We were in the midst of savage country all the rest of the day. A fine sandstone hill rose before us the name I {can't}[**** (?)]. In it was striking looking strata of this sandstone. Twenty minutes took us throught this Wadi when we turned sharp to the right & entered the W. Shellal. The Wadi is just like our Road between high steep mountains. One c(oul)d hardly persuade oneself that it was not a road. The hills were red sandstone & granite & yellow sandstone. It was a time & place to study Geology- for in this barren place, the strata are {unearthed} & free from vegetation & soil & the strata particularly visible as we wound up the Wadi, we were now in front of the strata & then in the rear & then saw them laterally with the great masses of granite & greenstone that in the upheaval rose out of the north & (92/843 9th Day to W. Macatteb) bore the other formations about them. We ascended for sometime & then came to what seemed to be the end but a narrow way zig zagged up among & behind the exposed strata which made capital walls & so over a ridge, too narrow as I thought for the loaded camels. This is called the <u>Nakb {Badra}</u>. Then some of our baggage camels, not of ours but of some of the party, had accidents, falling under their loads, which had to be recovered. I don't think they were hurt. We continued sometime then entered the W. Badra on descending instead of rising. It is impossible to describe the scenery. It seemed to be one wreck & ruin of the hills. The forms the rocks took when displaced from their horizontal position are wild & strange & with it all is the deadness & desolateness of the wilderness. The sandstone hills too crumble & {break} away which makes it look still more miserable. Next we came into the Wadi <u>Sedr</u> & here our Sheikh asked me if I would go to see some caves in a Wadi near- called the W. <u>Maghara.</u> Kennedy & I mounted our camels & trotted on with the sheikh for about half an hour, where we entered the <u>W. Maghara to the left.</u> Here we dismounted & leaving our camels (93/844 9th Day to W. Macatteb) began behind the sandstone mountain on the left. It was hard work. The hill was steep & the whole face of the rocks broken & smashed so that we had to climb up over great lumps of stone, broken from their place & many moveable to by no small I anger of those who followed. Our Sheikh seemed to be a little uncertain of the spot for he led us a long dance up & along the face of the hill. At last he bought us to a cave but hurried us on, saying as we gathered from his Arabic not a word of which could we understand & signs that the great thing was further on. So on we clambered till after a great deal of complaining on Mr. Kennedy's part he stopped & sat down before another small cave, with some Egyptian Inscriptions, at the side cut in the sandstone & still as sharp, the greater part, as the day they were cut. Who & when were these hieroglyphics cut {by} Wilkinson says "they contain the names of Rimai, who appears to have been the same as Papi of a name adopted at a later time by Sobaco II of Shofo, Suphis or Cheops. The founder of the great pyramid & another Memphis Pharoah found in the great pyramid". (94/845 9th Day to W. Macatteb) And so these mines, as they seem to be, if these names were written at the time of the Egyptian kings to whom they apply, must have been worked before the time of Abraham. After looking at these places for some time, we climbed down again, & as we went down the valley [*** (?)] our route again, we discovered inscriptions in several places, with Egyptian hieroglyphics, in one of which was the oval of a King of the 17th Dynasty & besides these, several the first we had seen, written in the Sineitic (sic) character, the same as the mysterious inscriptions in the Valley Mucatteb or Valley of "Inscriptions". The 2 Fentons had joined me, & Mr Kennedy had gone on. We staid some time looking about & then got our camels & rode on after the Caravan. After half an hour's riding, our Arabs proclaimed the <u>Wadi</u>

Mucatteb. We soon discovered the walls or sandstone sides of the valley to be covered with inscriptions. They seem to have been cut in with some pointed instrument, not scratched or chiselled. Among them we found a greek (sic) inscription. I got off my camel & climbed up to it. It was of 5 lines in the top line– MHNA- in the 2nd KOCMA & the 3rd I could not make out but Kosmas is (95/846 9th Day to W. Macatteb) said to have incised these inscriptions about 4 or 500 AD. We saw also another Greek inscription in which I made out the {word} *thagaser*. Camels, men, donkeys & lines upon lines of inscriptions everywhere. We rode on & found our camp in a hollow Wadi to the right. I dismounted & then went out again to have another look at the inscriptions, but it grew dark & was very cold & dinner was nearly ready so I went back. Dinner, wrote journal, read & went to bed. The 2 Egyptians, our prisoners, cried & lamented so much last night, that the Dragoman took compassion & this morning released them from the Arabs & let them go with a small supply of biscuit. They warned them they would perish but they said they would rather be starved & lost than be taken back to the Pacha & so they went away. They have no clothes hardly- only a few biscuits & such directions as the Dragoman gave them about the places to find wells in. They say that the poor fellows have just enough to keep life in them for 4 or 5 days & no more. I heard nothing of this till this afternoon & if I had, I (96/847 9th Day to W. Macatteb) would have done nothing. All provisions & water belong to Giovanni & not to us & we have no right to do anything but eat & drink what we want ourselves. I have omitted to chronicle a narrow escape from a fall today. I was getting on my camel who was kneeling on the ground, with a bark stick in my hand, when just as I was raising my leg to put it over the camel's back, he started up suddenly as they constantly will do. I was carried up into the air, hanging by one hand & a knee against his side, the leg partly over & so pitched down on to his head which was near the ground- as I was falling he rose from his knees upon his fore feet, thereby throwing me back. However I caught hold of his head & as he straightened his hind legs & became erect, I was thrown forward again & they said made a complete somersault over his head but still [***ked(?)] on & finally landed on my legs when he dropped his head again- without being at all hurt if I had not had well hold of his head, I must have had a nasty fall, probably on my head. (97/848 10th Day to W. Feiran)

Thursday. March 9.

After breakfast the question rose to see how to manage the day. I for one didn't mean hurrying on without seeing more of the Inscriptions & Kennedy said he should content himself with what he would see from the camel. I started him off to walk on & then went back to the camp- where the 2 Fentons joined me & asked what I was going to do. I desired my camel to be left & we 3 started back down the valley to examine the Inscriptions we had passed yesterday. I had looked into Foster's book & therefore knew something of his theory & I had deteremined 2 or 3 points to search for. First he says that it is quite evident the inscriptions were <u>all </u>those of <u>one age </u>& done by one people & next he decides the idea of crosses being found about the inscriptions. After spending more than 3 hours at one part of the valley & another hour further south, we all three agreed that some of the Sinaitic from other inscriptions are of much earlier date than others & also that the cross is found repeatedly above & before Sinaitic inscriptions punctuated as they are & indirectly of the same date. There are other crosses upon the rocks- generally the Greek cross- sometimes alone, sometimes at the beginning of or near to Greek inscriptions (98/849 10th Day to W. Feiran) but I don't think that one of these Greek Inscriptions or crosses were punctuated as the Sinetic are, but all are <u>cut.</u> <u>I also saw the </u>+ placed above Sinetic Inscriptions. We were fortunate enough to stumble upon 2 or 3 of the instances given & translated in Foster's book & in those we compared with his book, his versions of the inscriptions was inaccurate & such inaccuracy must interfer with the right translation. It was bitterly cold & about eleven we started upon the valley, walking on before our camels to warm ourselves We repeatedly had to turn aside & stop to see inscriptions. To describe them is impossible. The W. Mocatteb runs 6 or 7 miles- sometimes wider, sometimes narrower. The sides are sandstone, on which the Inscriptions & drawings are cut. Huge masses have rolled down & on these inscriptions are also cut- in many instances, we judged, since the mass fell- as inscriptions are cut on the fresh side of the rock, where it split off from the cliff. We rode & walked 2 hours down the valley & then came apon the Fenton's servant with their luncheon. They offered me some but I knew they had no more than they wanted & so though I had had breakfast at 6, I (99/850 10th Day to W. Feiran) declined. However they insisted on my having a biscuit at least & brought to me on the camel a couple of small biscuits & an orange. All of which I shared with my Arab & went on, having lest I should be late for dinner. It was very cold except- when sheltered when the sun was very hot. I walked most of the way. A short distance before where I left the Fentons we had entered the W. Feiran. The valley was very striking. If one painted the rocks in their genuine colours, people would laugh. High granite hills of red, grey, black & other colours with purple porphyry & red, yellow & white sandstone. Add to this the effects of time & veins of green stone with the lights & shades & you have a really brilliant picture. The valley opened in one place, & I found thorns growing there. Then besides thorns a few palms & gradually I got into the richer part of the valley- where the village of Feiran is, & the valley is filled with palms. It just gave what you wanted something green & rich, as a contrast to the wild barrenness of the rocks. We passed behind Djebel Serbal. It glowed with colours in the later afternoon. Further on I found a sort of (100/85110th Day to W. Feiran. 11th Day to W. Sheikh) {mole} embankment to keep the flood within bounds & here were gardens with vegetables. The rocks were covered with ruinous remains of houses built of[**** (?)] stone & at one corner a larger ruin which is said to be the remains of an Episcopal palace. It was the seat of a bishop in the time I think of Constantine. I soon found the tents. Earlier by ½ an hour's march than I had expected & was glad to have dinner- wrote journal & go to bed.

Friday. March 10.

Before going to bed last night I was so cold that I went out to get a run to warm myself. It was a bright cold frosty night with the moon making it quite light. The Wadi looked well. We were encamped near some of the native Arabs in their native black tents. Our tents were standing on an open space with large thorns; & tamarisks & palms surrounded it. Further off the high granite cliffs rose many feet into the air & in one place Djebel Serbal looked over them. The Arabs had made large fires & were sitting round them. Some asleep & some talking & laughing. They were wonderfully cheerful & merry. It was bitterly cold but I rugged myself up snug & slept sound. This morning we were up by time & Kennedy & I walked off in the clear sharp morning by (101/852 11th Day to W. Sheikh) the valley. We soon got beyond the rich part. We greeted the Americans as we passed. They complained bitterly of cold. The thermometer went down to 3 degrees on being held in the air a few minutes. I have little doubt it would have been lower if left. I hardly know what to notice in our day's march. It was much the same the whole day. A long march up the Wadi, passing from the W.Feiran into the W. Shekh. High granite rocks on each side- grey, black & red & others which people who pretended is a little science called basalt & porphyry. There were some curious strata among the grey granite- what I don't know called a "dyke" in geology. A stratum of some red rock, the grains much finer than those of granite, ran perpendically or sloping through the granite, sometimes a good many [****** (?)] & where it was perpendicular, it would stand out at me. The granite like a redwall. The granite being apparently decayed while that had endured. It was hot & cold- very cold in the wind. All around about. Sometimes walking & the riding again. At last we came near a great wall of red porphyry or some rock & about a ¼ to 4 camped in a pleasant sheltered nook- catching the afternoon sun. I thought this red range was not far off & persuaded Kennedy to come & try to climb (102/853 11th Day to W. Sheikh) it. We mounted a hill of granite at the side of the Wadi & then found a valley seperating us still from the mountain. I pushed on & then found a rise which when I had mounted revealed a 2nd valley. However I went on & got to the red rock which I proceeded to climb- making for a gap which I thought would show me Sinai & all the country beyond. I climbed with some difficulty at this but when at the top found myself still far from the ridge of the chain & that my view was still shut in by a shoulder behind. I had however a fine view N eastward & towards Djebel Serbal. I then made haste to go down again. Suspecting that it would be more difficult than going up & so it proved. However I got down safely & was glad to get back to our tent & rest. The evening was spent much in the usual way- writing journal, reading & a run before bed to get warm up & down a camel track in the valley. The cold was two or three degrees greater than the night before & we were obliged to sit in great coats & with our plaids round our legs.

Saturday. March 11. To Mt. Sinai

A bitterly cold night. Our cook called us at 5 as he said, but I found my watch said 20 minutes to. I jumped up & lit my candle & found the dirty water my basin frozen hard & the water in the tin can frozen (103/854 12th Day to the Convent of Mt Sinai) hard. No time was lost in dressing & before 6 sometime (sic) by me we had breakfast. The Fentons also were early. Giovanni [= John hereafter] our Dragoman we found had started with their Dragoman at 3 to get to the Convent & secure us beds & a room. We were soon a foot & overtook the Fentons & walked away up the valley. The thermometer in their tent had been 29½ in the morning when they got up. We passed up the valley in order to get through the chain of red mountains I had tried to climb last night by a fine gap. It looks like a gigantic "cutting". The walls rising perpendicularly . We passed through & entered the Wadi Ed Deir which we were some 3 hours in passing. We all 4 walked on together- enjoying the lovely morning & getting unfrozen. At last our Arabs began to exclaim "Djebel Musa" (Sinai) & point round into a Wadi we were coming to. We entered this Wadi on our right & in about ½ an hour rounded a point to the left. Here we found some loaded camels & a little further on came upon the encampment just packing up & recognized the American & his wife [= Eameses] with one of the two other Americans [= either Yeatman or Ward] with him & Mr Ross who left the Thursday before us. They gave a bad account (104/855 12th Day Convent of S. Catherine - Mt Sinai) of the monks who, they said, imposed on them in {lazy} way insisting on their having guides whom they didn't want & then sending in a bill for food for every day, although they have lived in their own tents & on their own food, & had nothing from the Convent. Of course they refused to pay. We said goodbye & went on to the Convent – under whose walls we dismounted. We were directed from the top of the walls to go round to the rear of the building, by a way between the Convent & a garden wall, & here we found a low square hole in the wall which we had to go nearly on our hands & knees to get through. This led us into a small court, a modern addition to the Convent & so through a small door, leading into a winding passage, apparently in the wall of the building, guarded by 3 heavy iron doors, admitting us into the interior. This it is impossible to describe: it is so spoilt & choked with modern additions & shabby buildings. We were taken up some rickety stairs to another passage opening into four or 5 small rooms, with a divan round each. The apartments for strangers. Here we found 2 or 3 monks who received us very courteously. One an old man, much bent, who has been here forty years. Here too we found Mr (105/856 12th Day. Mt Sinai) Freeman & his companions whose names I don't know. We had our coats & saddlebags etc brought up & determined to ascend Djebel Musa that afternoon. Some trouble about Mrs Rogers. She seemed much upset by the cold & we found there was some difficulty about rooms for all the party. Fenton, Kennedy & I at once offered our room for Mrs Rogers & ordered our things to be removed- assuring them that we had no idea we were keeping them from a room. This they would not hear of. By and by it was all arranged & we each kept our room. Then the monks made objection to our going to Djebel Musa as it was too late etc but I put it to one of them, Petros that 4½ hours was quite time enough which he at last was obliged to admit. We discussed the question with the Fentons who strongly wished to go & I went to Petros again. He began by telling me how many piastres for the Arab guards & how many Franks for the monk whom we <u>must</u> take as a guide & that we would want luncheon this & that. All which I cut short by saying we had had luncheon & should be home to dinner. Disgusted to see the mercenary spirit w(hic)h had [***** (?)] him to dissuade us from going this day. (106/857 12th Day. Mt Sinai) I asked for a guide & we went down to start. Again we were disgusted. 'Were we 2 parties or {one} As each party must have a guide.' I said one would be enough. However we found 2 monks sent with us & 2 arabs carrying water & provisions whom we sent back, but they insisted on going. So we said no more. So much for diagreeables & disappointment at finding the monks of Saint Catherine not a bit better than the monks generally are reported & believed by the world to be. <u>Djebel Musa</u> is one of the highest peaks among this mass of mountains which most people agree in thinking to be the actual Mt Sinai. The mount where Moses received the 2 tablets of stone twice which was all on a flame when god came down upon it & talked to Moses. I believe myself that it is quite impossible to ascertain which is the mountain mentioned & therefore I am content to take this which tradition has almost marked out as the spot & for in honour of which there are most satisfactory reasons. Other people have adopted Djebel Catherine & others again take Sak-Safeh (= Ras Safsafa?), just to the north of Djebel Musa. Djebel Catherine is the highest of all- but further from the probable camping ground (107/858 12th Day. Mt Sinai) of the children of Israel. Sak-Safeh is almost impossible. The arguments pro & con I can say nothing about. We started up the steep rock behind the convent. The path has been so much used, that it has been made into a rude flight of stairs. The Empress Helena is said to have caused this to be made. About 20 minutes after we had started we came to a well of delicious water in a rock & higher up a ruined the Chapel of the B. Virgin. Here pilgrims had to cross originally & then received a ticket with which they would pass through a narrow gateway where in order to proceed with their donations in the chapel on the summit.We passed a 2nd gateway for the same purpose & came out upon a more level bit, where we found a fine Cyprus growing & a chapel of Elijah. It is here that Elijah is thought to have been when he went out with his head wrapped in his mantle & god spoke to him in the still small voice "What doesn't thou have Elijah." He fled from Jezebel to Beersheba & threw himself down under a juniper tree, which {tree} Mr Wilson makes out from the Hebrew to be the White Broom, which we have found in blossom (108/859 12th Day. Mt Sinai) repeatedly. Here the Angel awoke him & he was fed with the food prepared for him & went on the strength of that food 40 days & nights to Horeb. Horeb seem undoubtedly to be the general name for all this block of mountains, though most travellers have given that name particularly to one mountain rising out of the W. Er Rahah. W. of Dejebel thus at Sak Saleh. It is not unlikely therefore that Elijah may have come to take refuge among these mountains & even to the very spot pointed out. Outside the cave a small chapel & outer rooms for pilgrims have been built. All now neglected & in a ruinous state, with names scribbled in many languages. The cave itself is to all appearances a natural cave. It was dark & I could not see in but I groped in some way & felt the rock with my hands & it seems just in its natural state. The altar has been built just on the right hand of the entrance against the bare rock. We were now more than ½ way to the top. The rest was very steep but the whole way a regular stairs of rock has been made, so that there is hardly more trouble than in going up a very long staircase. We came upon the edge of the precipice in one place which runs sheer down upon W. El Shu' eib? or Valley of Jethro Wilson calls it. Just before this the Arabs with us took a few steps aside to see the mark of Mahomet's camels foot. (109/860 12th Day. Mt Sinai) What he was doing I don't know. While they were loitering I went on alone & had some time on the summit alone before they came. The view is very striking, only spoilt by 2 ruined chapels, one belongs to the Greeks, the other to the Mohammadam- both equally desecrated. To the south we saw the Red Sea & one of the islands at the mouth of Gulf of Akaba [= Aqaba], probably Tiran. The mountains of Arabia & between them & us we caught glimpses of the Gulf of Akaba, among the tumbled masses of wild desolate rock on that side. Westward of us was "Djebel Katherin". And more north we just saw the Red Sea in the blue distance near Suez & the line of 'Djebel {Attak}' beyond, with 'Abu Derag'. North of us was a wild scene of desolation but very grand- peak after peak of mountain, all in confusion. Bright red & grey granite & black porphyry & yellow sandstone with the peaks of J.Serbal further off. Below us at our feet lay the valley of Elijah, westward of the flat place where his cave was shown us. One side, the east [*** (?)] to the "Djebel Sak Safeh" which Wilson denominated Horeb. Others have thought it to be Sinai, "the Mount of God" itself. Below to the N.E. but hidden by the rocks between was the Convent, in the Djebel Shu' Eib or Valley of Jethro & J. Er Rahah north & N.W. from that. Djebel Shu'Eib runs further south under the {E.} precipitious side of Djebel Musa. Tradition makes this the site (110/861 Mt Sinai) the battle with the Amalekites & that Moses sat where we were standing on the summit of Sinai, with his hand upraised till the Amalekites were destroyed. But that battle was fought in Rephidim which can hardly be the valley below & it must have been under one of the outer flanks of the chain of Horeb that that battle took place I think. <u>Er Rahah</u> is probably, I think very probably the place where the Israelites were encamped when Moses went up into the mount to receive the law. Above & under Jebel Shu Eib a rock is shown with a hollow in it, where the golden calf is said to have been molten. It may or may not be. Djebel Musa has many circumstances in favour of its' being Sinai. It is easy of access. The place where Elijah's cave is might be easily enough the place where Aaron Joshua Nadab & Abihu & the elders were left while Moses went up into the cloud. From it you cannot see Er Rahah so that Moses & Joshua may have been there without hearing the noise of the Israelites & their murmurings nor seen the making of the molten calf. When sent down by god they would have passed down naturally to the Shu Eib where the Convent stands & up the little valley to Er Rahah & as they rose up into the great Wadi Er Rahah, where the Israelites were encamped they would first hear the noise & see the calf under (111/862 Mt Sinai) Sah Safeh, or Horeb, where it is said Moses break (sic) the Tables. Er Rahah is the only place where so large a multitude could have encamped. Far away to the East & North East the mountains lay one beyond another & in the distance a wide sandy plain, with a high table land or flat range of mountains beyond, with at [*** (?)] the chalk or sandstone sides. We spent about an hour & a half on the top, reading & comparing maps etc. & reading the parts of the

Bible relating the account & then thought of entering a great crevice was shown us close to the top where tradition says Moses was placed by God, when he passed by & Moses saw his back parts as God passed by. Again it might or might not be. We all agreed that of all mountain scenery we had ever seen (& among us 4 we had seen the Alps, Apennines, Spanish mountains, Chinese mountains of the interior, Australian, English, Scotch & Irish & Welsh) we had never seen anything so strange as these are. Wild & grand & desolate. <u>Horeb</u> D. Wilson remarks means <u>Desolation</u> without a particle of anything green anywhere to be seen although there is a great deal more of prickly shrub & grass about upon the mountains than you imagine at first sight. The mountains are just masses of rock of different kinds &(112/863 Mt Sinai) nothing more. The colouring as the day wore on towards sunset & the shifting lights upon the red, grey & black mountains is indescribable. The sky without a cloud, clear blue though not deeper as far as I could see than usual. It was very solemn to get away by oneself & listen to the silence.Very solemn to feel oneself on in all probability or if not near the spot which was all in a flame with God's presence. We came down rather a longer way. For some time very steep & rough & wearing to the shoes & then passed through a great crevice which Abbas Pacha has had blasted through one of the buttresses of the mountain, where in the rock I found what was either a green lichen growing <u>in</u> the rock seen whenever it splits off mostly, or else fossil fern. I examined a great many pieces & broke off other pieces & thought it very like some of the smaller & more delicate ferns I found in Australia & about. From here we descended by a hard path or road which the Pacha has had made lately, in order to <u>drive</u> up the mountain. 500 men who pressed for the work, soldiers & others & it was done in about 6 months. We dined & wrote journal.

<u>Sunday. 2nd in Lent.</u>

I was very unwell in the night. I have found the dry desert air affect me like the air of the sea. I was very unwell all day. (113/864 13th Day. Mt Sinai) We inspected the church. The Sanctuary is ½ domed with old mosaics. The church doors are the original. 1300 years old put up by Justinian. Behind the Sanctuary is the Chapel of the "Burning Bush". However I managed to get to part of their service between 5 & 6 & again went to see the Celebration of Mass about 8. It began with a procession. The chanting was careless, nasal & bad altogether The Celebrant & all the priests generally went through the service in a perfunctional & very irreverant manner. Great part of it I could not make out. The Consecration took place with closed doors & the doors were opened & closed several times besides during the service. Also solemnity was disturbed too by the bad behaviour of some of the dirty lay brothers or attendants talking in the aisle & only going forward & then to light or carry a candle etc. etc. One or two old men seemed more devout than the others. One can't tell. I was sadly disappointed. Thank god for the blessings we have in England. We took a walk in the garden with 2 of the priests. One of whom spoke Italian & a little French. It consists of terraces. All completely irrigated & is green & thriving. Magnificent (the word is not too superlatives they me like great patterns). [*** (?)], olives, almonds, oranges, lemons, figs & vegetables beside some [*** (?)] with vines. The rest then went over the {journey}. I was too poorly & went & sat & saw (114/865 13th Day. 14th Day Mt Sinai) part of the morning service. I had suggested a regular service, but it was not taken up by the rest & I said nothing more. The whole Convent seemed to neglect the Sunday. A priest complained to one of the travellers that the Sunday Service was too long. Father Petros escorted Mr & Mrs Rogers up to Djebel Musa a whole day's work & so was absent for Vespers at 3- as were 2 others who went just before 3 with Kennedy & the Fentons to visit the "Smitten Rock" etc. Work was going on in the Convent all the day. Taking in & measuring wheat etc. I staid at home & later when the Convent was in the shade & cold, sauntered out down W. Shu Eib & sat one on the rocks in the sun. Reading the Christian Year & saying prayers there by myself. I went early to bed. I hope I shall be better tomorrow after the physick the Fentons supplied me with.

Monday. March 13.

I am thankful to say that I am much better today & only feeling the effects of the medicine I took in no small quantities yesterday. After breakfast I wrote & read. Then we, i.e. Mr Kennedy, 2 Fentons & 2 Butlers went to see some manuscripts kept in the Bishop's lodgings here. They used always to have a Bishop here. I expect a sort of mitred superior. But have & not had one for 40 years. There are his apartments & a niceish chapel next- opening out of the (115/866 14th Day Mt Sinai) sitting room. Here are some curious pictures. We were also shown the head of the Crosier. It is of silver gilt with jewels & subjects- very handsome & the workmanship good. Not gothic work. A chalice was bought out too- but not handsome. One of the manuscripts was in gilt letters on vellum. The 4 gospels- said to have been written by Emperor Theodosius in Greek. The 2nd was in a woman's hand apparently- in microscopic writing- a Psalter. We also saw the Bishop's stole- very handsomely embroided. We went from there into the garden to see the Charnel House. A priest went in with a censer of incense first. The bodies are buried ontside. The monks by themselves & the priests by themselves. When the flesh is decayed, the bones are removed into a vaulted building. All the skulls of the monks are piled in one place & all the legs & arms in another. The priests are stacked away in the same way but in an nicer chamber- in which are also the skulls & bones of the Bishops in boxes- with an account on parchment in each box, what Bishop they belonged to. We saw the remains of one Bishop in part of his vestments & in a corner of the vault as you enter, are the remains of an old monk many years {Master} to the Convent, still in excellent preservation. It sits in the corner on a raised platform. The head erect on the neck. The skin covering (116/867 14th Day Mt Sinai) the skull except behind. The face as when he died. The rest of the body was clothed in a dress given by some Russian Lady who visited the Convent. After this we some of us went out. I wanted to see the "Smitten Rock" which they visited yesterday & the Wadi Er Rahah- the supposed encampment of the Israelites. Mr Kennedy & the 2 Fentons went with me. We first went up into Er Rahah & then I turned back, & went along the foot of Sak Safer to the Wadi El Hagar or Hejar with an Arab guide. I passed several Sinaitic Inscriptions. This {taking} is very hard & rugged- more broken & terrible than any I have seen. Someway up the valley we came to a large granite boulder I should think from 10 to 15 foot high. Perhaps higher- on this are some horizontal {figures} one below the other all down the stone & running down the stone with them a sort of water mark. Behind however the explaination is more evident than in front. There you see distinictly that a vein of porphery or red granite runs down the rock- which has split out in parts in front. This looks like the fact but it is not impossible that it may be the rock & by God's Will preserved as evidence of the miracle. I tried to make a sketch & {returned}. When I got back to the W. Er Rahah. I spent some (117/868 14th Day Mt Sinai. 15th Day Start for Aqaba) time there, making a sort of plan & putting down the names given me by my Arab guide & sketching 3 or 4 bits. I always regret that I can't sketch. I wish I had a little instruction. If I had I think I could manage it in a rough way. Home just before the gate was closed. Dinner & wrote a letter to Anna to go by an express which Capt. Butler who came on Saturday[16] is going to send to Suez- journal- packing as we start early tomorrow & to bed. Much better.

---

[16] 11 March 1854, the day Bowles also arrived.

Tuesday, March 14

We were called at 5 & got up to the usual 6 o'clock breakfast but we did not get away till about 8. The Fentons have settled to go on straight & they were rather late. We left Giovanni to settle the Convent account. They manage to make up a long bill for servants, guides etc. & leave you to make them a present for use of rooms. I confess I thought the payments we had been <u>obliged</u> to make quite enough to cover anything but it is better to pay & say nothing. So we asked (Mr Kennedy & I) to say goodbye to the superior monk in residence (the actual superior is gone to Cairo). We were shown into a miserable cell rather, but better than most of them in which the proud man was lying on a dirty bed on the floor resting. I suppose in place of the door at midnight given to service in the church. (118/869 15th Day Start for Akaba (sic)) He was just lying as he was, in his clothes: they all sleep, in fact live night and day in their clothes. We were taken through this small room into a sort of balcony outside, where were 2 chairs opposite a chair of state in which he sat down. We could not talk, & the few who talk in English (1) or French (1) or Italian were not present. So I could only express our thanks & goodbye in dumb show, with some mumbled words & lay a few dollars upon the table & then we bowed & said goodbye & hurried away. The Fentons joined us outside & we walked up the valley to W. Sheikh when I let them go on & sat down to make another sketch. I took 2 in the morning from the Convent roof. I overtook the younger Fenton round in W. Sheikh the greater & we walked for sometime & all 4 joined at the Tomb of Sheikh El Saleh where our road turned off. We went into the Tomb. It is a round building of rough stone domed. A coffin shaped box lies across the middle covered with green stuff- over this hang from a rough wooden frame white cloths with 2 Arabic inscriptions worked on them. I believe the usual "God is great etc." While in there our Sheikh came in. It is a sacred place among the Arabs. He kissed (119/870 15th Day Start for Akaba) each side of the door way & stood a second in prayer. Then he came in saluting us, with his usual kindly smile & Arab salutation of laying his right hand on his breast & brow & then went to I suppose the head of the coffin or sham coffin for the grave is below the earth- knelt- kissed the crescent at the head. Then the frame. Then said a prayer & kissed his 2 fore fingers, & finally stooped lower & reaching his hand under the coffin bought out a handful of dust or sand from the top of the grave & rising laid some of it on the top of his head with a few words of prayer & left the Tomb. Outside he gave some of the dust to another Arab & ran away after the camels. Several other Arabs came in & did much the same. We {soon} got into the W.{Sahl}. Curious for its basaltic rocks. We left the Fentons sitting down to luncheon- rose a very steep rocky piece of ground & descended on the other side to a small plain of sand. We led on from this into another part of the Wadi & our road wound along all the rest of the day amongst granite & basaltic rocks- not so remarkable for their height as for the colouring on them. Kennedy & I rode on ahead, reading & talking & making out the history of the Nabataeans who once owned Petra as their Metropolis (120/871 15th Day Start for Akaba) They were descended from 'Nebajoth the son of Ishmail'. They got possession of Edom about 300 BC & there was a line of kings. We hear of them as Christians in the 4th Century & Petra was the see of a Metropolitian Germanus Bishop was present at Council of Seleucia AD 354. Theodorus AD 536 at that of Jersulaem. We rode on till 4 when we encamped still in the Wadi {Sahl}. I walked across the valley which the tents were being pitched & tried to sketch. Then came in- rejoiced to get back into our tent again. Washed my hands & wrote journal etc. till dinner. Read & for some time puzzled over some of Abraham & the Israelites etc. Ishmail the son of Hajor Abraham's concubine has 2 sons, Nebajoth etc. These are said (in Gen XXV. 12 etc.) to have dwelt from Havilah "onto Shur, that is before Egypt, as thou goest into Assyria".

Probably all Arabia Petrea. Years after the sons of Esau had Seir or Idumea given them- the Horites being driven out (Duet 2. 12. 22). Esau was to be subject to his younger brother but finally to break off the yoke & oppress Israel. This he did several times (2 Kings VIII. 20. 2. Chron XXVII 17. They assisted the Chaldeans in the destruction of Jerusalem. Lev xlix. 7-22. Eg: xxv. 12-14. xxxv. 15. Divine vengeance persuaded them- Malachi 1. 274. but they were finally conquered & reduced by <u>Nabateans</u>. These are supposed to (121/872 15th 16th Day) be descended from Nebajoth, Ishmael's eldest son. They had possession of <u>Petra</u> as their capital about BC 300- when Antigionus successor of Alexander the Great sent 2 expeditions against them. They became subject to the Romans & their being [*** (?)] was reduced to a Roman Province, after several dynasties of their kings had reigned, by Trajan's lieutenant Palma. The last we hear of the Idumeans ie. the descendants of Esau, was the introducing of 20,000 of them by the Zealots into Jerusalem for its defence before the siege of Titus (Josephus). Petra became afterwards a Christian See & their descendants became Mohammadans after all Syria had been over run by Moslems about AD 630.

<u>Wednesday. March 15.</u>

I haven't much to say about today's journey. We were off be times & walked away by ourselves. It was a dull morning & pleasantly grey. It was an opportunity of seeing the hills under different circumstances. They were <u>not</u> the less beautiful. Turn after turn of the valley bought out new changes. I walked till about 11 & then rode for an hour reading- till a halt was called for luncheon. I rode again afterwards. We gradually changed the stile (sic) of country. Sandstone appeared in the distance. At last we came upon the sandstone, & (122/873 16th Day) going down a slight descent over some sandstone rocks, we came to a large flat valley bounded by high sandstone cliffs. At the first corner we came to we found Sinaitic & other inscriptions. I spent sometime there, examining & copying one or two. I found the name of {Cosmos} again. In the afternoon we came upon a curious sort of <u>island</u> of sandstone rising from a sea of flat sand inscribed all over. There was much heavy deep sand. Mr Kennedy & I got down & walked for more than an hour again. We passed up a hill the deep sand & down into a large basin more than a mile across with hills round- with appparently no exit but where we got up to to the great cliff in front we found a narrow way out to the right passed under some very high cliffs eaten out into a regular stone network, round another point & found our tents just pitched. I tried to make 2 sketches- made myself comfortable. We dined, read & went to bed. A few drops of rain just before dinner but they came to nothing.

<u>Thursday. March 16</u>

The American lady Mrs Rogers, it appears, gave their Dragoman to understand last night, that it was not right they sh(oul)d be always last- & gave strict orders that they sh(oul)d be off first. They weren't to be behind the 2 English parties. So this morning (123/874 17th Day to the shore of G. of Akaba) before we were out of bed I believe she was seen all ready to start outside their tent. John was all impatient to get us out of our tent & we could not imagine what it all meant- till as we went to breakfast we saw their tents down- & in a very short time their camels were loading. As usual however we with the Fentons quietly walked off the ground, followed however immediately by all 4 Americans, their camels beginning to move at the same time. We then saw how it was & were amused at the way they & their camels pushed on. We were among the sandstone for some time, plodding through deep sand & among curious sandstone rocks, covered in places with numbers of Sinaitic & Greek Inscriptions. By & bye we left the sandstone & got among basalt & granite. We passed from the Wadi {Sahl}, into Wadi {Searr} (I think) & thence into a long narrow Wadi, Rasali, between hills of granite. These rose higher & higher & became very fine. We were a good deal behind- our camels long gone on, while we walked. I walked from 6 to 9- when I was glad to get my camel. The sand was up to my ankles most of that way & very tiring. We stopped to luncheon, finding that John had made the usual preparation near the Americans. So that we quickly (124/875) eat our luncheon & past (sic) on before them. Our camels too we found had as usual left them far in the rear. From the W. Rasali, the last part of which was very grand, we came to the W. El Ain or Valley of Springs, where were some date trees & a delicious stream of water- at which our camels drank eagerly. The entrance of this Wadi was from a sort of small plain with magnificent granite & porphyry rocks round- into a narrow gorge down which we descended winding backwards & forwards round point after point of fine rocks from ½ past 12 till near 4- when we came out upon a plain beyond which was the Gulf of Akaba- with the Arabian Mountains beyond. We camped about ¼ past 4 after passing along at the foot of the hills on our left for about a mile. The Fentons' tents & ours were all pitched & half furnished before the Americans' camels came out upon the plain. So much for "going ahead"! It was too far across the plain to the sea. I intended to have had a bathe. So I went out & made 2 sketches. Home to wash my hands & feet an Eastern luxury after walking which I fully appreciate. Dinner, journal, reading & to bed. We must have made 30 miles today. Much laughing, before going to bed, outside. At last it was so exciting that I turned out to see. I found the younger Fenton, pipe in mouth, (125/876 18th Day along the G. of Akaba) drawn out of his tent too. One of the Americans' Arabs who is always playing pranks, had gone away & came back dressed as a woman & they did not found (sic) him out for some time. The discovery had caused the amusement.

Friday. March 17.

We were up early & ready for a start. The Americans again determining to go ahead & accordingly after walking 2 or 300 yards & turning about to see if Mr Kennedy was coming, I found them all following on foot. We 4 Englishmen waited till the camels were gone, & then followed sauntering along. The morning was very beautiful. The sea was smooth & the hills soft. It was enough enjoyment to walk quietly along & watch the scene. I have felt again & again that I had no idea our journey would be half so pleasant. I am enjoying it most thoroughly. We came to a grove of some few palms & thorns just where the track ran in rather, making a small bay. There we saw numerous tracks of gazelles & hares but none of the originals. It was here that a few years ago a horrid scene took place. Some people, I think Mr Fisk &L(or)d Castlereagh,[17] passing along this way stopped for the night. A party of Arabs came up & shot or killed somehow one of the sheikhs of the party. It seems it was 'blood for blood'. This sheikh (126/877 18th Day. 19th Day to Akaba) had killed the brother of one of the Arabs. I fell behind walking slowly & about ½ past 9 got my camel & mounted him. The large sandy plain or wide track narrowed & we passed under some fine sandstone cliffs of many colours that ran down to the beach. Our journey all day was along the sea. It was most enjoyable. The sun was hot but there was a fresh breeze. The sea was blue & the Arabian hills grey & hazy; while the mountains on our left rose high into the air- soft & beautiful from their many colours & materials. I rode till past 2 when I got down & walked, thinking the sun w(oul)d be less broiling while I was taking exercise. We passed a man fishing, with a sort of casting net & bought some fish. We came to a halt about 4 near a sandy beach. I got a towel & walked along the beach for

---

[17] Both in 1842

some distance, where I bathed & back to the tent. The evening as usual. Tomorrow we hope to reach Akaba.

Saturday. March 18.

We started soon after 6- the camels that is- Mr Kennedy & I were off by 6. But after about an hour he sat down & I went on alone. The track led us along the sea till about 9- when I saw the path was [***(?)] up by a steep headland & that it left the sea & ran up a Wadi between the hills. I toiled up among the stones & granite debris & in about ½ an hour took a wrong turn. Coming to a valley running down again to the sea, I thought that <u>must</u> be the (127/878 19th Day to Akaba) way & followed some camel tracks down. But still I felt uncertain & seeing nothing of the camels I at last went back a toilsome way in time to catch a glimpse of the whole train making their way up a steep pass across the head of the valley. Wadi Negabad- pronounced by the Arabs- Negab. I followed them. It was a nasty place for camels or for anything. Very steep & rough. It took the camels some ¾ of an hour to go up & down the otherside where we got on to a rough plain again &{down} upon the shore. Here we came in sight of Akaba but so far away that it was quite plain we should not reach it before the Evening. We had luncheon under a great rock {feeling} well the expressiveness of the "great rock in a weary land". A small island, Graie stood some ½ mile from the shore- with ruins on it of some fortifications made it is said by the Crusaders. I never saw much more striking views than 2 or 3 I had today. The nearer distance,the warm hot glowing colour of the sand & rock. Both sandstones & granites of colours. Then the sea a deep blue. The island in one case:- & beyond, across the Gulf a range of fine mountains, high &bold. I w(oul)d say from their appearance granite. Their colour I can't describe. It was perhaps a sort of French grey, with shadows upon them but in the soft haze that had been over everything (128/879 19th Day to Akaba) on the gulf since we struck it, they looked more like dreams of mountains. After luncheon we started at 12 (after I had tried to make 2 sketches which were most unsatisfactory) & the way seemed endless. The gulf ran up some way & we could see the Palm trees of Akaba on the further side of the head of the Gulf. Our way lay all round the Gulf & point after point & bay after bay opened up as we passed the last. At last we got to the head of the bay & then I got down to walk & Mr Kennedy followed my example. We thought we were at Akaba- but we trudged along in the broiling sun for a good hour before reaching the place where our tent was pitching. Then we fell in with Mr Ross & Mr Freeman- who gave us the unpleasant news that Sheikh Hussein was away & that he had not been sent for till yesterday. He is 24 hour's journey distance. So we shall be here a day or two. I went out & had a most delicious bathe- & home to dinner. After dinner while sitting writing (my) journal we had a visit from Mr Eames, the American of the 1st party- he came to express a wish that we might have (a) service tomorrow- to which of course I most gladly assented. Reading- a stroll along the shore- & to bed. We have reached Akaba from the Convent at Mt Sinai in less time than either of the 2 preceding caravans. (129/880 20th Day. At Akaba)

Sunday 3rd in Lent. March 19

We have all been glad to have a quiet day. I got up about 6 & went out to bathe. After breakfast I sat reading till Mr Eames made his appearance. He found there had been some mistake as by their watches it was then past 11. I was only ½ past 10. We assembled in the tent in which Mr Ross & the 2 Americans live. 16 in number. One of the Americans was a Baptist, & Dr Bryce

& Mr Wakefield are Dissenters[18] of some sort but they all asked to be allowed to join. I said prayers & young Fenton read the Lessons. It is not often that 16 people join in the service of the English church at Akaba. We sat talking tog(ethe)r after service & on breaking up Mr & Mrs Eames asked me into their tent where I sat some time. Before long it turned out Mr Eames is an American clergyman. I asked him why he had not himself said prayers, instead of asking me- & regretted that I had not known. I had a good deal of talk- about the American church- & the English deputation to their convention etc. I am glad to come across him. After luncheon I sat in our tent reading- & then went for a stroll along the beach with the Fentons, past the village (if it can be called such) & along the beach to the East. It was very pleasant. We got the breeze there, which curled the waves (130/881) upon the beach & made it cool, whereas where the tents were pitched it was very hot. The sea was blue & the hills very fine & we strolled along, picking up shells & bits of coral, till I be thought me that the omnipotent John had told me that dinner w(oul)d be earlier so we turned back. The palm trees were out in blossom & very sweet. We stood looking at some as we came back. Dinner- reading- a stroll & to bed. No news of Sheik Hussein yet.

Monday. March 20.

A quiet day- the 2nd- & a great pleasure. After all there is nothing like a quiet life. I have thought with much pleasure & satisfaction of being settled in a home. But that is no word for me. Still I may look forward to a settled life somewhere & perhaps Anna & Alice[19] may have a fancy for coming to live with me, or one of them. I wish such might be the case but they will hardly think I care for them enough, after running away from them for so long just at so sad a time.[20] Every day brings me nearer home & now we shall be moving on again soon. On coming in from my walk this evening I was glad to hear that Sheikh Hussein had arrived & that we should start after 2 days. Mr Kennedy was in when he came & he sat & had a pipe & coffee. I am sorry I missed seeing him but I shall have other oppor- (129 (a repeated page number in error)/882 21st Day. Akaba) -tunities. Four days journey will take us to Petra & 5 from Petra to Hebron. News has come too that there is no longer Quarantine at Hebron. I went out at 6 in the morning & bathed & had a walk. We had breakfast about 8:30 and after that I sat in my tent till 12- reading & drawing- finishing up some of my sketching attempts. I went out & tried to make another sketch or two- but I want instruction. I can't get on. I settled to walk with the Fentons after luncheon. I had some bread & cheese & dates& at 2 we started- up to the hills- where we dawdled about & then climbed down into the Wadi that runs up & made our way up a narrow gorge, which was very fine. {Slabs} of greenstone & red stone amongst the red granite. We climbed up one or two steep places & followed it up some way. I tried to make a sketch in one place where the rocks were very {fair} but I c(oul)d not manage it. We heard Arab voices over our heads & wondered who was coming, when suddenly there were some loud shouts & I saw that they came from 2 Arabs of my party, who came clambering down the rocks- & greeted us very warmly. They had been in search of some prickly stuff for their camels. Beans are very dear- 7 piastres for what you would only give 2 at Cairo. So they

---

[18] i.e. Protestants. High Anglicans like Bowles tended to use the term in an unflattering fashion. Both Badger whom Bowles met in Aden and William Palmer he encountered in Egypt, were very willing to express pungent views about Dissenters (Coakley 1992: ch. 1 *passim*).
[19] TB's two youngest sisters.
[20] The death of their sister, Frances Martha Woodyer in childbirth just a few months before TB left on his grand tour (above Ch. 1).

(130 (a repeated page number in error)/883 21st Day. "The Arabs") not only have driven out the camels during the day, but have been foraging about themselves for them. Tomorrow they will go- as Sheikh Hussein has come. I have been trying to find out about the Arabs- but it is difficult. As far as I can make out the Bedouins consist of a great many different families or tribes. The word Ioanne [= Giovanne?] says means "those who live on (or by) camels"- The Arabs with whom we came belong to the "Tawárah" tribe & are Bedouins. Their territory runs from Suez or perhaps nearer Cairo to very near Akaba. But in the midst of their territory, viz at W. Feiran, they have allowed a small tribe or part of a tribe who are called Fellahines (sic) to occupy the more fertile district there-who I suspect pay rent for it. The Tawarahs cultivate no soil, but buy their corn etc. at Cairo. The tribe we are amongst now of whom Sheikh Hussain is the principal person are Aloines. Bedouins also. North of them are another tribe running down into the W. Akaba & beyond this at Petra & northwards are the Fellahines- Bedouins again. The Fellahines I think extend to Jericho & Jerusalem. The Bedouins look down upon those who are not Bedouins. They will never sleep in a house. It is said that Sheikh Hussein's eldest son a most gallant young fellow died in con- (131/884 22nd Day. Akaba) -sequence of his living indoors a good deal at Cairo. I cannot vouch for this. After dinner I wrote journal- read- took a turn along the shore by starlight & to bed.

March 21 Tuesday

A rough windy night & morning but I got up at 6 & bathed & had a walk along the beach after. After breakfast I was sitting writing to F. Naylor when Young Fenton came, & then Mr. Ross & Mr Freeman & while we were sitting talking a request came that we would all go to attend a Divan with Sheikh Hussein. So we walked down to one of the other tents, where we all took our seats & presently in came Sheikh Hussein & his son. We all waited to hear what it was about & Paul[21] came in to act as spokesman & Interpreter. It seems to be the Sheikh's custom to make all agreements with the Dragomen in the presence of their masters. A wise plan. In the present instance there seemed to be some difficulty in the way & it turned out that the dragoman of the American party misled to them the burden of Baksheesh contrary to contract upon his masters. So it ended in a wrangle. But it was a striking scene. The dignity & calmness of the old man, as he sat smoking his pipe & the grace with which he spoke, when he did speak, waving his hand the while. It was (132/885 22nd Day. Akaba) determined at last the Sheikh sh(oul)d settle the matter with the dragomen, as all the parties had contracts. The whole trouble was made by the Americans' dragoman. However it was all settled amicably at last. & Sheikh Hussein & his son have been dining with John this evening. We managed to settle one thing- viz: that our camp should be pitched in Petra & not as was proposed near. This difficulty too, John has discovered was created by some of the other Dragomen. This business detained us till past 12. After luncheon I went for a walk with the Fentons. Down & back East of the Sea- to an old tower, up to which we climbed but could make nothing of it. When then the top of the nearer chain of mountains looked tempting & we thought we sh(oul)d get a view of the country beyond. We climbed up with considerable labour & were rewarded for our pains. The country beyond is a mass of hills & wadis. We went down into a sandy plain within & thought to find our way out &down some {easy} valley,but the one which looked likely led us to the mountain again & we had but last nothing for it but to climb down. It was a nasty steep place but we got down all safely- walked home quickly, to find dinner ready, being last as usual. (133/886 23rd Day. Akaba)

---

[21] The Fentons' dragoman

March 22 Wednesday.

Bathing as usual at 6. After breakfast I sat down at once & wrote 2½ sheets to Frederick Naylor. Just as I was finishing the Fentons came in & John brought in the bread & cheese & dates. They sat sometime & then went away to have luncheon, agreeing to walk afterwards. I went out after luncheon & made a sketch & then we 3 went out. Kennedy afraid of the heat. On the plain we saw several, 7 or 8, Arabs on Dromedaries, who as soon as they saw us came streaming down towards us. We saluted them & looked at their camels & suggested "W. Musa bokkara"- to w(hi)ch they answered "iwa" [= yes]- & asked "Sheikh Hussein hina" [= is Sheikh Hussein here] & then we went on. By & bye we saw a train of 20 or 30 more coming. It was too hot to walk very far & we turned & sauntered back along the plain. Near the Palms young Fenton & I sauntered away to bathe & the other one went home. We had the most delicious bathe I have had & then walked homeward- it being near 6. Near our tents we found our new camels all lying down. During dinner a few more came & so there is no doubt we shall start tomorrow. Journal- reading- a stroll along the beach among the Arabs & their fires & to bed. But before bed (134/887 24th Day. Start for Petra) the Fentons came in to say that we must go & <u>see</u> the talking round the Sheikh's fire. He had been complaining of the demands of Abu Djasir I think, or the old Chief whose territories dovetail in between his own & the Fellahines- who demands so much for every caravan of travellers. He began by saying that they were brothers now & he was very glad to give him a fair share; but let him <u>ask</u> for it: not <u>demand</u> it- and so he ran on- but by & bye he grew excited & rose upon his feet & said he & his sons w(oul)d not care if they left alone. "I & my sons defy the world". Which earned great applause- & as the newspapers would say "great sensation". I went out & strolled about watching them but I had no interpreter & c(oul)d not understand a word they said. One of Sheikh Hussein's black slaves sat in the ring round the fire near the old Sheikh, making coffee & bringing it to him & the principal persons. Before going to bed I packed up & made all ready for an early start.

Thursday. March 23.

Giovanni called us soon after 5. We got up & had breakfast & then the tents were struck but there was great confusion & delay about loading the camels. Before loading the old Sheikh came riding round on horseback with a long spear in his hand, looking at the burdens wh(ich) were all arranged. After he had left (135/888 24th Day. Start for Petra) us there was a great dispute between 2 Arabs about a camel saddling for me. The question in dispute (sic) was whether it sh(oul)d carry me or some baggage. Of course its' owner wished to have a rider- as there are peculiar little advantages in the way of remains of luncheon & the hopes of baksheesh perhaps at the end from the "hauaja", falling to the share of the owner. I must say my sympathies were all with the owner- for it was a fine camel & a nice clean saddle & rein. The dispute ran high. I never saw 2 men much more angry & wild looking. Such attitude & action. I saw at once what it was about. Presently, the attention of Mohammad, Sheikh Hussein's eldest son, the Sheikh in command of the Caravan was drawn to the scene. He intereferd in vain. Presently he stooped suddenly to the ground & gathering up a handful of gravel & sand he put some on the head of each disputant. What it meant I have yet to learn. John would not tell me. Presently he left them, for his words had no effect. They still stood one on either side the camel- vociferating & struggling the one to remove, the other to keep my things on the camel's back. Before long however he turned round again, & [***** (?)] them to his father, to settle the question before him. They went at once- & the end of it was (136/889 24th Day. Start for Petra) that my things were shifted to a Dromedary, wh(ich) I had been admiring on the beach yesterday & I lost the small saddle & bridle. By 8 the Fentons camels moved. Ours were all ready before but were waiting the Cook & some water. The Sheikh in command of the whole Caravan is Mohammed- Sheikh Hussein's eldest surviving son. Each party has also a Camel Sheikh, Sheikh Hussein's 2nd son goes as Camel Sheikh to the Fentons, to bring them back from Petra. Mohammed rides a bay horse- & carries in his hand a long spear. The head has a large fringe (like a mop) round the bottom of the metal. These Arabs, (Aloines) are very different from those we have had before. Taller- wilder looking men. Young Fenton & I walked on ahead. It was very hot. He stopped at last for his camel. I walked on alone & walked till 12- when I waited for the Caravan- who were all bent on luncheon when they came up to me. I found Mr Kennedy rather in trouble about his Camel. After luncheon I mounted my Dromedary & found him much easier than my former friend. I read & talked- till about ½ past 3, Mohammad brandished his spear & galloped forward till he came to a fit spot when he dug his spear into the ground & we halted. The wind had got (137/890 W. Araba) up & clouds of sand were flying. I left the crowd & walked away to the hills on the East. We have been in the broad valley between the 2 ranges of moutains all day & shall be for the next 2. On our right or the East- the mountains are Granite- on the West Granite &Sandstone. I can't help thinking more & more that the Jordan flowed down this valley once. I went up towards the Eastern mountains & sat down. Four sheep following me like dogs. They have been {running} with the Camels all day. Poor things, they don't know that they are for mutton. Back to the tent after 5. Washed & had dinner. Heat very great. Journal- reading & to bed.

Friday. March 24.

Everyone up early. Our tent was taken down round us as we had breakfast. Much complaining on the part of the 2 parties who preceeded us to Akaba. J(ohn)& I think all of our former caravan {enjoyed} it. It was pleasant walking ahead up the valley before the sun rose. We came to a large salt marsh & some water at last with palms & thick bushes. It was brackish & swarming with insects. After passing this, about 9:30 I mounted & rode all the rest of the day. The heat was very great. Yesterday the thermometer stood at 125 & 113 in some ones' tent in the even(in)g. I don't feel the heat as I used to. I have a fairish turban (138/891 W. Araba) on & perhaps this saves me, but I never feel it on my head as I used to do. The Caravan is too large & people are not of one mind & there is much confusion among the baggage camels. These Arabs too are in & out among you- never caring where they ride. It is certainly not as pleasant as it was before Akaba. Dr Bryce pointed out a curious fact near out last night's camping ground viz- that the black trap rocks had shot up among Granite rocks. A remarkable thing he said. A dispute rather about our halt this afternoon. Some people were for stopping before 3 but others of us urged the young Sheikh on & we went on another mile, amongst a good deal of grumbling. People don't consider the extra expense to those Dragomen who have made a contract. After halting I got away to the hills & after walking some way I found a large rock which cast a deep shaddow & was very glad to sit down out of the burning sun. I made a failure of a sketch & went back to the tent, where I found Mr Kennedy asleep on his bed & preparations for dinner going on. Giovanni in trouble. The Arabs very idle & awkward & his sheep lost through their carelessness. Dinner, journal writing, reading, & to bed.

Saturday. March 25. Lady Day.

We have had a hot day, with much dust, a hot wind, or what next month would be a regular hot (139/892 26th Day) wind, (kamseem (sic)), blowing up a cloud of dust all day till the atmosphere was a sort of [***** (?)] sand. We started off before 6. I walked on ahead, ploughing through the deep sand. I saw traces of some animal in the sand, who had evidently been round & about though not venturing very near the sand. I asked an Arab what it was but, I could not make out quite the name he gave. It turns out that it was what they all translate into a wolf or wild dog. They say it will attack a man at night. {Rifat} says that a sheep was carried off in the night. I can't make out whether truly. We came to a valley to the right, called the Wadi Gharandel. Here was a spring, to wh(ich) all the Camels were taken.[22] The Sheikh {begged} we w(oul)d not walk on. Last year some of the Karak tribe came down as low as this- & robbed some people, besides carrying off Camels. The Aloines are afraid of them. I was not very well- so rode all the rest of the day. The Desert air makes me quite unwell, unlike the Sea air. Most of the day was spent in a very uninteresting way. The air quite thick with sand- & no talking- one's throat got full of sand. We made ½ an hour's halt about 11. After that we passed over a very stony bit of ground, & then broke into the hills to the East. We began to get among the sandstone again- & passed some marvellous (140/893 26th Day) islands of sandstone- on one of which the upper part had been carved square- & a square doorway led into a small chamber. There had been a Cornice over the door, but it was worn away. No inscription. We camped about 3 by my watch- under the hills outside the skirts of Seir- with Mt Hor looking over the hills at us. I went out & made a sketch & lay in the shade of a rock, where 2 Arabs came & talked to me. I can't say I talked to them in return- not understanding a word they said- but I looked at their swords & matchlocks etc- & then came home to dinner. While sitting at the tent door waiting for dinner, the Sheikh of the Fellahines came by, dressed in a light blue cloak lined with white fur. He saluted me very courteously. Dinner, journal etc as usual. Always ending with a turn round the camp about 10, before getting into bed. The camp at night about 8, before people go to bed, is a very pretty sight. The lighted tents, cooking & Arab fires- together with the Arabs gathered round the different fires, with their camels. It is not describable. Just before going to bed, we heard some confusion outside & Paul came rushing in to say that one of the Arabs had been bitten by a <u>serpent.</u> It turned out to have been a scorpion. He was taken to Dr. Bryce at once- who lanced it, put on a tight bandage up the leg- & got one (141/894 Mt Hor and Petra) of the Arabs to suck the wound. He them rubbed in {sweet} oil- applied caustic- & gave the man a dose of Calomel. Strongish measures for a man who had never tasted anything stronger than coffee.[23]

<u>Sunday 4th in Lent. March 26. Petra</u>

Here I am, sitting waiting in Petra, having just said Evening prayer in one of the tents- with a congregation of 14. We made an early start this morning. By 6 or a very few minutes after all the Camels were starting. Having Mt Hor before me to climb I started on my Camel. We struck into the {pass}at once & began ascending- cool & pleasant after the suffocating light before the sun had got above the hills. I got out my Bible & Prayer book & read the Psalms &Lessons as well as the Epistles & Gospel at once. Before long the climbing was so severe that I got off & walked & did so all the rest of the day. How shall I make a memorandum of the {way} I can't describe it more than by just saying that it was a steep zigzag way among rocks up a steep pass. Sandstone rocks rose abruptly on each side. Especially on the right or East. But this gives no idea of the scene- yellow, rose red, {purple} (nearly black), purple, violet,

---

[22] If they had had time to explore, they might have noticed near the spring the mounded sand which covers the ruins of a small Roman fort (Kennedy 2004: 209-211).
[23] See Chapter 7 Bryce for the occasion and treatment.

green, grey etc.- all colours. It was the most wonderful piece of colouring all there said they had ever seen. We wound & zigzagged up & backwards & forwards to the top of the high range- & then Djebel Haroun/ Mt Hor stood before us. We descended some (142/895 Mt Hor and Petra) way & then stopped- as the baggage Camels & those who did not mean to ascend Mt Hor were to go one way, & we, who ascended, the other. Mt Hor is one of the only places whose tradition has never been disputed. It has never been doubted that it was on this very mount that Aaron died. We took some luncheon with us & an Arab each to carry it, & our Camels went down to wait for us on the otherside & we began to climb Mt Hor. It is steep & hard climbing in one place- but the rest not very difficult- still it was hard work. We got to the top as I thought but when there found a large flat & that the top was some way higher. We passed round some fine rocks, with some old arbor or Lignumvita trees about, from wh(ich) I gathered some seeds to take home. Steps had been cut in the rocks above- I suppose to help pilgrims up to the mosque built over Aaron's Tomb. We passed over some old arched work where had been a cistern for water & very steep steps led us up through a great crevice to the top.[24] We looked down all the while upon the rocks round about Petra & caught sight of tombs hewn out in the rocks. Below us lay valley & rock & gorge & mountain & the W. Araba- to the South- a hill to the North lay the continuation of the 'Wadi'- but we could not see the Dead Sea in consequence of the haze. (143/896 Mt Hor and Petra) A door led into the Tomb- but it was locked, or we could not find the key. Some of our Arabs officiously tried to force the door & one to climb in by a hole. It was well we stopped them- last year that door was forced, & the Fellahines came down & were very angry about it. I c(oul)d not help being struck with the apparent difficulty of an old man's getting up Mt Hor even on to the plain below the peak. It is not <u>said</u> that Aaron went to the top- though the Arabs have there built the tomb {still}, Stephen[s] says, there under the Mosque is a large & old cavern [Stephens 1837: 2, 95].[25] We came down on the opposite side- over & among masses of sandstone the colour of some of wh(ich) answers to Miss Martineau's description of the "Mahogany Sandstone" [Martineau 1848: 2, 317]. Numbers of the trees mentioned before relieved the eye, as well as a good deal of scanty vegetation & great numbers of flowers- anemonies, Geraniums, Rock lillies, {Mattry}, blue pimpernel, rock roses or {systus} pink & host of others. We found our Camels below- near some great caves & tombs. They were merely {square} chambers in the rocks. We mounted & rode down the winding valley for some distance passing temples & tombs in the rocks & one solitary built pillar & remains of ruins & suddenly dropped upon the heart of Petra & our tents pitched facing the Grand Temples [= Royal Tombs]. It was very beautiful. I think (144/897 Petra) the only place that ever quite came up to description & does not disappoint. After a cup of coffee, a (thimble full rather) & a rest, the 2 Fentons, Kennedy & I walked out with 4 guards past the Theatre & up to the El Khasnee, the Great Temple opposite the narrow gorge which faces the original entrance to Petra. With the exception of the bottom of the pillars injured by water, one whole pillar gone, & figures partly worn away above- the Temple is as fresh & sharp as the day that it was cut. We sat & enjoyed it. Even with its' awkward broken or interrupted pediment it is beautiful. I can't describe it so as to help my memory when I forget

---

[24] Neither TB nor most other 19th century visitors noted the ruins of the Christian church on the platform below the summit (Figure 9.4).

[25] John Lloyd Stephens, the first American to reach Petra, was there in March 1836. He subsequently published books on his travels in Poland and Russia and in Central America. In the latter he was active exploring the archaeological remains of the Yucatan Peninsula together with the British architect/ artist, Frederick Catherwood, who travelled extensively in Egypt and the Levant in 1834 and may have reached Petra that year.

it, if I ever do. We left it, glad to have seen it quietly this evening & to be able to come to see it again tomorrow & we walked slowly back among the carved cliffs to our tent, where dinner was waiting for us. During dinner I sent a note to Mr Eames asking if we sh(oul)d have evening prayer- & shortly after we most of us met & said Evening prayer. Journal & to bed, after a walk among the Tents. I have forgotten to note that I heard a Cuckoo this morning which took me back to England, where I daily long to be. We killed several scorpions on & descending Mt. Hor & numbers more were killed in the encampment. I gathered some nettles, thinking only of the time of England (145/898 Petra) & only afterwards thinking of the prophecy of briars & nettles. Thistles are abundant. The Arabs have patches of grain growing, green & fresh. Coming down Mt. Hor we came upon a patch of turf near a delicious spring. I was the first to lie down on it, but my example was followed by all the party.

Monday. March 27.

I have spent the whole day wandering & climbing among Tombs, Temples & rocks. I got up at 6 & we had breakfast soon after 7 but other people were later. It rained most of the night & the early morning & I feared we were to have a wet day- but after strolling about a short time by myself without caring for the rain which was only now in short showers, the Fentons finished their breakfasts & came to ask what we were meditating. About 9 we 3, started for the El Deir- or "the Convent". This is a Temple among the hills to the N(orth) part of which we had seen from Mt. Hor. We started across the centre of the valley Westward- along the course of the Wadi or dry brook. The whole of this had been confined by massive walls. We passed the remains of pillars & a Temple on our left [= Great Temple] & then came to some high ruins of a Triumphal Arch it is supposed. It must have consisted of a centre & 2 side arches. The remains are carved very richly with flower & leaf tracery- & some of the capitals (146/899 Petra) lying on the ground are beautifully carved. I have noticed all the day that the stone in some places has worn well, while in others it has all melted away. Beyond this are the remains of a large Temple [= Qasr el-Bint]. Towards the stream was a propylon with massive pillars- these have all fallen. A high arch leads into the interior. Along the top of the further wall is still an arcade, for what purpose I don't know. The building is made of solid stone work- but the whole surface has been plastered & all the ornamentation except the cornice (wh(ich) is very deep & fine) has been in plaster- small bits only of which now remain- under the cornice is a good deal of ornament like this. There was one thing which I noticed. To keep the plaster firm on the walls, the whole surface has been chiselled much in the way that people sometimes chisel the stone work in churches. They were going to the foolish expense of this at the Cathedral at Sydney- & the new St Phillips (sic).[26] I was always convinced this was wrong in Gothic work & now I begin to think that what is called <u>Rustication</u> is wrong generally. Certainly here it is only to form a hold for the plaster. I suspect this is the meaning everywhere. Behind this in the face of the rock is a specimen of a Tomb or Temple begun but never for some cause finished. As usual a smooth surface has been made by digging away(147/900 Petra) into the rock some feet, thereby leaving a great buttress as it were on each side. They have made an opening below & worked in so as to have excavated one great chamber & above the capitals of the pillars are roughly cut out- but there the work has stopped. Near it the rock has been cut in, & 2 open square chambers have been cut. The curious part about this is that the whole face of these chambers, as well as the face of the rock outside, has been chiselled out into small square compartments leaving the appearance of lattice work. These are cut out square at the bottom, but only bevelled down from above & many of these compartments have been filled in again with small stones & mortar.

---

[26] In his previous Journal TB had comment on the Cathedral at Sydney.

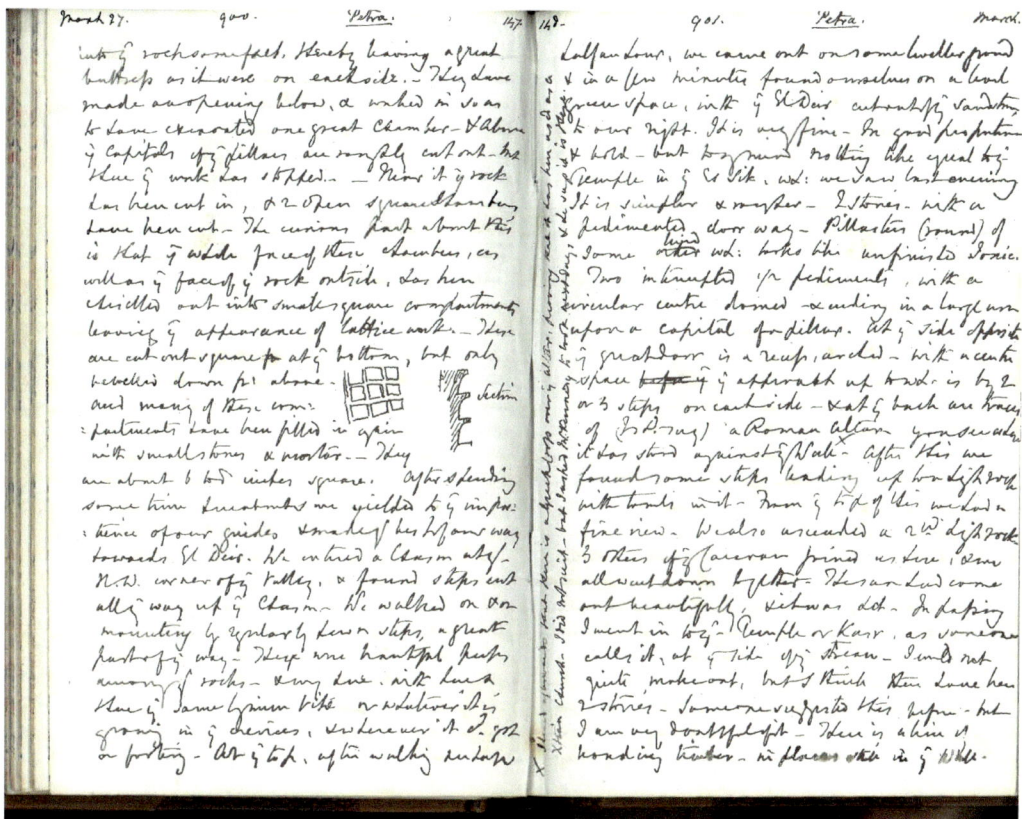

Figure 14.2: Sketch of features of a tomb at Petra.

They are about 6 to 8? inches square. After spending sometime {thereabouts} we yielded to the impatience of our guides & made the rest of our way towards El Deir. We entered a chasm at the N.W. corner of the Valley & found steps cut all the way up the Chasm. We walked on & on mounting by regularly hewn steps, a great part of the way. There were beautiful [*** (?)] among the rocks & my [*** (?)] with here & there the same [***-like (?)] or whatever it is growing in the crevices, & wherever it c(oul)d get a footing. At the top, after walking perhaps (148/901 Petra) half an hour, we came out on some leveller ground & in a few minutes found ourselves on a level green space, with the El Deir cut out of the sandstone to our right. It was very fine. In good proportion & bold but to the mind nothing like equal to the Temple in the Es Sik, wh(ich) we saw last evening. It is simpler & rougher- 2 stories- with a pedimented doorway. Pillasters (round) of some kind wh(ich) looks like unfinished Ionic. Two interrupted ½ pediments, with a circular centre domed- & ending in a large urn upon a capital of a pillar. At the side opposite the great door is a recess, arched- with a centre space the approach up to which is by 2 or 3 steps on each side & at the back are traces of (I w(oul)d say) a Roman altar- you see where it has stood against the Wall. [Insert from margin:] (I heard at convent that there is a Greek cross over the altar, proving that it has been used as a Christian Church. I did not see it- but I asked Mr Kennedy to look next day & he says it is there.) After this we found some steps leading up to a high rock with tombs in it. From the top of this we had a fine view.[27] We also ascended a 2nd high rock. 3 others of the Caravan joined us here & we all went down together. The sun had come out beautifully & it was hot. In passing I went in the Temple or Kasr [= Qasr el-Bint], as someone calls it, at the side of the stream. I could not quite make out, but I think there have been 2 stories. Someone suggested this before- but I am very doubtful of it. There is a line of bonding timber- in places still in the wall. (149/902 Petra) I then walked up the dry bed of the water course to the tents. The masonry on each side is very fine. I had a biscuit & some cheese. Then we all met & determined not to go tomorrow. We had to arrange this with Sheikh Mohammad- who agreed- though he was anxious to get away. The Fellahhines have come down in great numbers &there were violent debates. The Dragomen said there w(oul)d be a fight. Sheikh Suleiman, one of the Fellahhine Sheikhs said that were it not for his respect for Sheikh Hussein he w(oul)d kill every European here this year. I then went out alone to the south & while there saw the Fentons starting off & with a guard northward. So I ran down the valley & finally overtook them. We passed the Large Temples on the East Cliffs [= Royal Tombs] in order to get to the Temple with a Latin Inscription [= Tomb of Sextius Florentinus] & so begin then & work back towards the Es Sik. But after examining this we were led on northwards & so worked gradually up the valley to the head- at the N.E. Corner-nearly to where it goes out upon the hills beyond & so home again in time just for dinner. Journal & reading.

Tuesday. March 28.

A day of hard work, but very pleasant & equally wonderful. I had a turn before breakfast, between 7 & 8. I took my Bible with me & read the Psalms (150/903 Petra) & lessons etc. out among the ruins. After breakfast wh(en) (the) Fentons were ready to start & we 4 went first to see the Khasnee again & so up the Es Sik- the Chasm. Two more of the 4 Fellahine Sheiks were about the Camp & as we walked up by the Theatre they overtook us on their horses. Gipsy looking fellows. They stopped & demanded rather than asked for some powder. Their Servant was carrying Fenton's gun & he gave them a small quantity of powder. We looked at their horses & seeing us notice them they began to show off, dashing up & down with their horses, brandishing their long spears etc. Mr Ross had come so far to show us a Tomb opposite the Theatre in wh(ich) he had found inscriptions [= Tomb of Unayshu]. It was a square room with a square recess at the opposite side ending in an arch. In the further recess were 2 small & one large grave, or hole(s) for the coffins. In the outer room were 15. On the left hand side over one of the Tombs nearest the outer wall were 2 pyramids (outlines) chiselled on the Wall- below was a long inscription in one line- not Sinaitic. I copied it. I half think the letters are Nabataean.[28] Further along the wall, over the centre tomb was another pyramid, ending square, with on the lower part of it an inscription in 2 lines which I also copied- Somebody, I think Miss Martineau calls these Tombs baths [no]. Certainly there are steps into one- but into one only. An aqueduct runs all along the cliff but you see the same everywhere. The Romans (151/904 Petra) (I suppose) seem to have saved every drop of rain that fell & to have conducted it down & along the face of the cliffs in aqueducts- you trace them everywhere. On to the Khasnee. It grows on you more & more. If I staid (sic) long at Petra I should be in danger of thinking Italian architecture very fine. I don't dare say finer than Gothic. But I have seldom

---

[27] The location today of the 'End of the World Café'.
[28] The inscription reads: "Uneishu, brother of Shuqailat, son of ..." (Zayadine 1974). Although seen and probably copied by Ross in 1854, it was not published until at least four decades later - cf. Wenning 1990.

seen anything come up to the Khasnee. I have thought of place after place, but I have not thought of any building that I think comes up to it. Of course I have no means of comparing the best- having never been in Italy. This morning the sun was partly on the Temple- & the rich rose colouring was much more beautiful than it was the Evening before. I tried to make a rough sketch of it, swiftly in order to <u>learn</u> it. I then followed them up the Sik, turning to look back at the Khasnee from the ravine. It is more & more beautiful from every new place you see it from- perhaps most, when you catch bits of it between the dark rocks of the Es Sik. We walked along the ravine. It is perhaps 10 or 12 feet wide in the narrowest parts. The height of the rocks I can't at all tell. I should think about from 300 to 700 feet. Others have thought them much higher. In places we traced what I suppose was the Roman paving & in places the aqueduct at the sides. To avoid (152/905 Petra) having the water running down the Ravine as it w(oul)d do naturally in rainy weather interfering with the passage of the ravine, they seem to have made an aqueduct all the way. Here & there were small niches in the rocks. In one place, on the r(igh)t hand side, were a succession of niches & inscriptions. I <u>think</u>, in Greek. I c(oul)d not make them out. A person used to old inscriptions w(oul)d perhaps read them easily. Festoons of ivy hung in places down the cliffs. The rocks grew lower & presently we came in sight of the arch which spans the ravine. Young Fenton & I climbed up the rocks near & succeeded in getting to the top. When, of course, we found we might have gone by a much easier way. I did not get down onto the bridge- for it was difficult- & a slip would have killed one- & the end of the bridge on that side was worn & broken very narrow. We passed away over the rocks above. Even up there were tombs & the rocks all bore the marks of the chisel in places, where the causeway had been widened. We found a curious Tomb. A highish rock had been cut down into a sort of flat pyramid or pedestal. On one side we found a low door & crept in. There was a step on each side & a deep place below & on one side the rock was excavated below, rising from the floor about 2 feet or rather more & extending 12 or perhaps 15 feet under the rock. Not (153/906 Petra) far off were graves everywhere. At the top of every small rock graves had been dug in. We also came to a well. A large rock chiselled out round standing over apparently a hole in the rock,wh(ich) made a reservoir. In the stone were the marks of the rope, with wh(ich) the bucket had been drawn up. We went down into the way leading fr(om) the ravine again- & on the other side we found a large Tomb- or perhaps 2 Tombs one above the other. The lower had pillars & a pediment & from the Terrace above this rose again other doors & windows- ending above in 4 pyramidal stones [= Obelisk Tomb]. Near this was a deep excavation- with a large chamber below. Other Tombs were all about. It is impossible to particularize. We passed back again down the ravine. I think the arch is simply for ornament or to mark the beginning of Petra- or an arch of Triumph. It was neither for a road, nor for an aqueduct. Another delay at the Khasnee & so home for a mouthful of luncheon. After wh(ich) Kennedy started for the El Deir- & we 3 to examine the great Temple or Tomb(s) whichever they are to the East. First the one with 3 tiers of 18 columns each [= Palace Tomb]. Inside are 4 chambers- what used for I don't know. Next a Tomb with a broken pediment [= Corinthian Tomb] & then on to the high Temple with columns [= Urn Tomb]. This has been used it is said for a Church. There were originally four (154/907 Petra) (4) recesses at the further or East end. Two of these have been broken into one- it is said & probably, to make an apse for the Church. In the left hand further corner is an inscription. It is now illegible. I could not make out even whether it was Greek or Latin or what. Wilson gives a note saying that it gives the date of the Consecration of the Temple to be a Church [Wilson 1847: 316 n.1]. On coming down from Terrace in front of the theatre our eye fell upon the pyramid up on the top of the mountain & we were s(e)ized with a desire to visit it. We went off to the tents to get the guide directed (sic)

& Paul, the Fentons' Dragoman proposed to go with us. We passed at the rising ground south-round the rocks etc behind the Theatre & up a valley passing many Temples & Tombs [= Wadi Farasa]. It was a beautiful little valley- with a Tomb at the further end. The rocks most picturesque. They were for going back, there w(oul)d be no way there. However I was not so sure & went to the end. I was just turning too- when my gaze fell on some worn stairs in the corner. I went up & found others & on & on. I called them up. There was one difficulty- a few half steps led up a steep piece of the cliff &ended at the line of the aqueduct. Whether this was the way was the question. However I was in front & climbed round & over a piece of Roman Wall & so we got on up the Mountain. At the top we found 2 pyramids- cut down out of the solid rock [= High Place Obelisks on Jabal al-Madhbah] & on looking about, discovered (155/908 Petra) the ruins of a large building further West. I climbed up & found the whole top of the rock there cut [= High Place of Sacrifice]. First I came to a square cistern cut in the rock- about 8 feet square. Beyond this the rock had been hollowed down about 18 inches in an oblong form & in the centre stood a raised slab- 6 feet x 3. A small drain or duct led out over the side of the rock near one end. Beyond again the rock had been cut, apparently making 2 more narrow cisterns or depressed places. The length of the breadth of the larger one. On the left opposite the slab in the centre 2 rocks had been carved. The first had a long trough in front, with a drain from it. 3 steps led upon (sic) one side, with a small excavated trough by the steps &on the top a circular basin had been hollowed out- also with a drain. Next & exactly opposite- the ruined slab 3 or 4 steps led up to another rock in the top of which a shallow oblong hole had been cut & a passage cut so that people would pass round this. From the further end of the hill [= the north and looking north] we had a fine view of the Valley of W. Musa- or Petra & to the W. Djebel Haroun, Mt Hor. We had to hurry back to get down before the sunset. But we had time in the valley below to go into the several tombs. We found 2 large Tombs opposite each other. Both very striking in their way. The one on the south side of the little valley had (156/909 Petra) four pilasters & inside 2 large chambers [= Tomb of the Soldier]. The Tomb on the northside had no exterior- but in passing in through the broken wall, we were astonished to see that the whole was surrounded with fluted pilasters [= Triclinium (235) Tomb]. The flutes (if that is the word) being alternately large & small. Between each pilaster was an upright recess- 4 or 5 feet x 2 ½ & cut about 8 inches back into the rock. At the middle of the side opposite the door, was a small recess. This would not have been a Tomb I think. Indeed it is impossible to make out the caves of Petra. What was all this {for} Enormous facades carved in the Cliffs to an enormous height & only one chamber cut out below. Some square & plain. Some with recesses. Some with Divans round 3 sides. Some with windows. Others having nothing but the door and all these without a vestige of anything in them as a clue to their use. It is very mysterious. We stumbled home down the hill & I was right glad of some dinner. After which Mr Ross came in & then the 2 Fentons & we sat talking till 10- when I went to bed without writing journal. Cold again.

<u>Wednesday. March 29.</u>

I awoke at ½ past 4 by my watch- & hearing voices got up. We had some breakfast & found the whole party doing the same thing. We four mounted away behind the camp (157/910 30th Day. Leave Petra) passing most of the people sitting out in the grey dawn round their tables at breakfast, the Tents being nearly all down & packing. We went to the West side- & down a ravine round a high rock behind the Kasr or ruined Temple near the stream bed & then clambered down & followed a stream course [=Wadi Thughra] all round behind the great &

highest rock of Petra [= Umm el-Biyara]. Here we found some inscriptions, one of wh(ich) I copied. We passed on through the Oleanders growing in the course of the stream. But finding it getting more precipitous & not knowing whether there was any way out, we turned. I found regular English brambles growing, with blackberries on them & tore my hand. When we got back into Petra we found by the silences & the new marks along the sand that the Caravan was gone. We hurried on & overtook them at last. I c(oul)d hardly get along. Yesterday & today I have been wearing the Sydney {button} shoes. All the rest being quite done for & they were in pieces. Fenton & I determined to go over Mt Hor again & started across to the right with a Fellahine for guide. Sheikh Mohammad taking & keeping his gun as security. We had a severe climb. The way we had come down & got to the top in about an hour. We hoped to have seen the (158/911 30th Day. Start for Hebron) Dead Sea. But, though clearer than when there before, we could not see it. The view was much more extended than the other day & we knew where we were & where to look. We descended on the North side & found our camels & the 2 Sheikhs waiting for us. Mounted and rode on. About ½ way down we parted. The Fentons to the left down the way we had ascended on their way back to Akaba &we northwards. I was very sorry to say goodbye. They have been very pleasant companions. The usual story in travelling, you make friends to see them again perhaps no more. I went on down the mountain & overtook the rest- who all greeted me as most energetic, to have climbed Mt. Hor again. We halted for luncheon about 11 & at 2 arrived at our camping ground- on the plain again- under a great black hill- in W. Araba. I was glad to change my buttoned shoes & put them away as useless & got out the thin boots made in Galle- wash- dress and sit down to journal. Then dinner soon after 5, more journal, reading and bed.

Friday (sic) [should be Thursday] March 30.

We have been journeying most of the day across the W. Arabah & into the hills on the West side. All day we have been in the Karak's Territory & accordingly very much on the alert. The Caravan has had (159/912 31st Day) scouts out on each flank & after getting into the hills about 2:30- an alarm was given by one of them of (sic) Arabs. Sheikh Mohammad heard the man's report & then was on the alert instantly. He tore off his abba & threw it to Giovanni & then was off like a shot- his red garments streaming in the wind & his long spear waving in his hand. It was very fine. Other Arabs at his order were climbing a high hill on our right, guns in hand. The Camels were driven tog(ethe)r & other men seized their guns & ran forward. I thought it was a false alarm- but thinking I should walk to see the fun I drove on my camel & on coming round a corner, found Sheikh Mohammad quietly waiting for us & in answer to my looks of inquiry said 'Mafish'- nothing. We started before 6 this morning in order to clear the plain in good time & get hidden among the hills. We have in addition to our own caravan between 80 & 100 other camels, young & old. They belong to the Aloines & have come to meet us here in order to go under the protection of the Caravan to Gaza, where there is plenty of food. There is none hardly in the Aloines' Territory, in consequence of want of rain. It has been a lovely day & it was a pretty sight early in the morning, to walk about ½ a mile on one side of the Caravan & see the 160 or 170 camels streaming along. (160/913 31st Day. 32nd Day.) I had a walk in the morning with Freeman & then with 2 of the Americans- Yateman & Fish. They are certainly very unlike Englishmen. I have ridden all the day since about 9- to save my shoes. After Camping I went out & wandered among the Valleys near, where as usual I knelt down & said Evening Prayer. Dinner, journal, reading & to bed.

Friday. March 31.

All the camp were a stir very early this morning. We were some of the last I think & Giovanni accordingly, very black. I started fr(om) camp for the first time mounted- for economy & knowing that we were to have a steep hill- Mar {Saffed}. I rode in front of the caravan most of the day- reading & enjoying my own thoughts. It is mostly a great nuisance when people will talk. The camels never keep tog(ethe)r. We saw 5 gazelles tog(ethe)r & soon after 2 others were seen near. After stopping for luncheon about ½ past 2 we got near the steep hills. I jogged on- to get ahead of the baggage Camels. Dismounted & walked up the hill- it is very steep- but not so bad as everyone proclaimed it to be. The formation is curious. A rift seems to have been made where the hills (sandstone) was driven up by the igneous rock below. The dip on one side is S.W. & on the other N.E. The day had grown cloudy & now threatened to rain. A few drops fell. (161/914 32nd Day) The camels reached the top at last & I walked on- round the head of the ravine & down along the otherside of the range. The descent was not great. We came out upon a sandy plain- with a great many flowers & more vegetation. I fancy the desert, at all events the waste wilderness is nearly past. Tomorrow we cross a long plain gradually increasing in life & vegetation & on Sunday by noon we hope to reach Hebron. I shall be glad & sorry too when I part with the camels. It has been a very pleasant time. Now that I have got so far I may honestly confess that I think perhaps it is a mistake trying to do so much at once. I believe people who have just come fr(om) England w(oul)d enjoy more keenly the free roving life, than a man who has been travelling about for 18 months. All the time my thoughts are principally running on getting back to England. But it was the only chance as far as I see of ever crossing the Desert & therefore it was wise to si(e)ze it & I do not regret doing so for one minute. I must get to work as soon as I can after getting back. I sometimes wonder how I am to live. When we got to (the) camping ground, I walked away & strolled about in the neighbourhood of the camp till ½ past 5. I have seen numbers of English plants & flowers today. The common (162/915 33rd Day. Leave the Desert) red poppy,[29] wild mignonette etc. Evening as usual.

Saturday. April 1st.

An early start. I walked all the morning from ¼ before 6 to luncheon time, past 11. Cloudy & a coldish wind- people shivering on their camels. Our camping ground last night was a change from the desert & after getting across the plain in which we had camped & passing Jebel {Garib}, by a break in the hills, where I made a mistake climbing the cliff at the side instead of going up the valley, we came into W. Kurnub. Here the grass was growing & the scenery was more like down scenery, though of course the vegetation (green as it was) was but scarce. On the left were some ruins of a place called by Wilson Kurnub [= Nabataean to Byzantine town of Mampsis]. Here were some Arabs encamped away to the left & a very green patch of some corn or others. The change from the desert is most charming. The ground is covered with flowers & some grass & the larks sing as they do in England in May. We were passing through the territory of the Tiyahah Arabs & some of them came to us. A report was got up that they disputed the right of the Aloines to take us through their territory but- were to meet us later. Large flocks of storks were passing all day long, migrating to the North from Egypt. We passed patches of arable land & in one place a man ploughing with a camel. (163/916 33rd and 34th Day. To Hebron) I ran up & stopped him & got the plough out of the ground. It was a most

---

[29] Perhaps sumach rather than poppy.

ancient affair. Flocks of sheep & goats. We passed 2 wells together on our left- deep wells- with 10 stone troughs round one & several round the other. Wilson calls it Beersheba but this is I was sure further West. Sheikh Mohammad called it Beer el Minah. Soon after this we encamped upon a large plain near the foot of El Tellal?- from the top of wh(ich) most likely Abraham saw the smoke of Sodom & Gomorrah. Doubtless Abraham's flocks fed where we are now & Esau & Jacob & David fed his flocks. I went out alone, to get away from the tents & people. It was very solemn. Bitterly cold wind came on in the evening & blew all night with rain.

Sunday. April 2. 5th Sunday in Lent.

I was awake all through the night at times with the wind. Though I had covered myself so warmly that I was not cold. Miserable work for the Arabs sleeping on the ground, with no shelter. John called us late- it was getting light. He said during the day that it was too wet & cold to get up so early. He also kept the cooking tent standing, for us to have breakfast in- a comfort wh(ich) we have given up a long time to make the process of packing & loading easier to (sic) him. All this we felt as an additional (164/917 34th Day. To Hebron) attention, as our Cook had started at 3 in the morning for Hebron, to order horses. It was fine at first & I walked away across the green plain, enjoying the fresh morning just as the sun was getting over a hill. But it soon came on to rain & rained almost continually the rest of the day. I got on my dromedary as soon as the rain began & made myself up for wet- but my thick coat was put away & I had nothing but a thin paletot & a thin waterproof above it. It was so thick that we could see nothing. We passed a good deal of ploughed land with wheat & other grain growing. We passed a ruin. Dr. Durbin calls it the ruins of a Church. It was so completely a mere mass of stones, except round one end where were remnant (sic) of a wall, wh(ich) might have been the West & bits of the N. & S.walls of a Church, that I don't think any one c(oul)d say it was a Church or anything else. After this we mounted a long steep hill- covered with large white lichened stones & rocks among- which I could trace signs of terraces, probably the work of the old inhabitants whose labours the Israelites enjoyed the benefit of. The wind blew violently & every now & then the rain wh(ich) was beginning beat against one, as though it w(ould) be impossible to keep dry. About 10:30 Sheikh (165/918 34th Day. To Hebron) Mohammad made for what I took to be a ruined tower or fort, round wh(ich) were considerable signs of old buildings. Under the lea of this we dismounted & waited for the rest of the party & here we rested a while & had luncheon or breakfast as most people call it. The sun came out just while we were there. I could not learn what this place was- but the name of Carmin[30] was given me. Here lives a Turkish guard or two- who, when the quarantine was in force, conducted travellers to Hebron & deposited them in the Lazaretto. A guard came out & announced that the quarantine was withdrawn & that there were no disturbances in Syria. I can't describe the rest of our journey- but only add that it was cold & wet & very slippy for the the Camels, who weren't fitted for anything but a dry sandy soil. In mud their great flat feet have no hold. As we went on the country grew more fertile. Much of the land was terraced & green wheat etc. growing. Bushes too appeared & flocks of goats & sheep. We came in sight of Hebron. It stands in the valley & on the sides of the hills & has all the land almost round cultivated- vines & olives & corn. We went to the Town, arrived in pouring rain. I got off my camel & kissed his nose & said (166/919 34th Day. Hebron. Jerusalem) a very hearty goodbye. Excellent in the desert, they are compartively useless in wet & slippy weather. They had taken advantage of

---

[30] Presumably Carmel, 14km south of Hebron.

the offer of the Lazarett(o). John had got us a room & before long we had got out of our wet things & our beds, tables sheets etc. were b(r)ought up. I afterwards strolled out through the mud into the Town. It was muddy & dirty. I found a hilly way along outside the walls & stopped to see some men at work preparing cotton in a sort of wretched house. Then I poked in through a gate but there was nothing except mud & filth. We found ourselves just outside the mosk (sic)- but there is no getting inside. Home to dinner- passing on the way a large square pool surrounded with masonry. This is called David's Pool. Dinner journal & bed.

Monday. April 3. Hebron to Jerusalem.

We were called early as John thought the Camels would be at the Lazaretto for our baggage. We were loth to get up- for it was raining & looked likely to rain & it seemed that we sh(ould) not go if the Camels did not come for luggage & their arrival depended on the weather. However John came in again & we got up & had breakfast & all our things were packed & we said our goodbyes to Sheikh Mohammad & there while other people were hesitating about going at all our (167/920 35th Day. Hebron to Jerusalem) horses came to the door & we started in the rain. I was sorry to leave Hebron without going about the place & seeing more of it but- it was impossible for me without shoes, almost literally, to walk about & therefore the thing was to get to Jerusalem as soon as ever I could. We went out of the way rather in order to see Bethlehem on the way. The first part of the way was a sort of paved lane- but the pavement was much broken & soon came to an end. The country all the way consisted principally of round swelling hills, covered with stones, in many places built into Terraces & small shrubs, principally of a sort of Ilex- with very small leaves. There is no timber, no trees of any sort but olives & these only near the Villages. It was very wonderful to look upon the old wall wh(ich) was done by the old inhabitants- fitting the land for the children of Israel. It looks v(er)y barren now- but you see what the land has been & what it might be again. The soil is rich & fine wheat is growing amongst the stones. We passed along some miles over a miserable road- such as few horses but those well used to it could get over. Here & there we saw ruins- on this hill & in that valley. Everywhere there are ruins. We c(oul)d not turn aside. It rained a great deal & was very thick. Even the rain (168/921 35th Day. Hebron to Jerusalem) was not disagreeable- for one felt how much the country owes to it. All was green- at last we got into a long valley. When we mounted a hill at the other end we looked down upon the pools of Solomon. These are large pools, made by Solomon & connected with Jerusalem by an Aqueduct which is broken & in ruins in part (they talk of repairing it). Near are the remains of a building said to have been built by the Saracens. Lower down in the Valley below the pools is the garden of Solomon. I have been since asked at dinner whether I had seen the old masonry near the bottom of part of the wall of the lower pool. The person who asked speaks of it as very old indeed. We passed on to Bethlehem where we went to the Convent. The Church has been fine. The nave has some fine pillars but it is unused. The skreen (sic) is handsome. The fittings are poor- stalls miserable. There was a handsome book of the Gospels on the altar. We walked round the convent. Below the Church we went down to the Holy Place, so much contended about. A star of metal is fastened over the floor- close by is the manger where our LORD was laid- & opposite, the stall where the cattle fed. All these places are covered with marble, & hung with bad hangings, except one bit of rock below the cradle. The north side of the Convent belongs (169/922 35th Day. Hebron to Jerusalem) to the Latins, who have a Parish Church. I just looked in. We rode on through Bethlehem. A curious place built upon the sloping hill. All round the country is covered with olives & you look away over hills & hills to the country wh(ich) is pointed out as the spot where the Shepherds were keeping watch over their flocks, when the Star appeared. We passed near another Greek Convent & on the next hill Jerusalem lay before us. I was strangely disappointed. It was merely a wall & (a) few houses & minarets in the bare undulating stony country- hills appearing above & behind it, but all bare, without a tree. It rained a good deal- but the sun broke out clear upon Jerusalem as we came up to it. We passed along by the edge of the Valley of Hinnom, passing bits of the old aqueduct & in by the Gate of David, the City of David being on our right handside. The walls of the old building, wh(ich) is inside the wall, are very old- with large blocks of stones. A filthy, dirty, narrow street, led us down to the Hotel. It was quite full. Lord Falkland & his family being due on their way fr(om) Bombay. However the landlord put us into a room in a house near wh(ich) he promised to furnish. The other hotel was quite full also. I went out at once with Giovanni to get a pair of shoes- wh(ich) (170/923 "Jerusalem") I was successful in. I went to the Salon in the Hotel & sat by a stove & had a long talk with an old gentleman, who turned out to be L(or)d Falkland. At dinner there were 11 men & 2 ladies beside ourselves & at a 2nd table by themselves L(or)d Falkland & his party consisting of 2 ladies & another man, Major Folay [= Foley]. Much talking & then I sat down to write journal & now to bed.

<u>Tuesday. April 4.</u>

Our luggage arrived just before 9 as I was going in to the Hotel to breakfast. The Camels had started late & could get not further than ½ way between Bethlehem & Jerusalem. After breakfast I went back to my room & unpacked- & then we started with a guide, stupid as all guides are, to walk round the walls. We went out though the Gaza Gate [= Jaffa Gate?],[31] the way we had entered from Bethlehem & turned to the left, skirting the walls. We were here passing along with the Walls on our left hand & the Valley of Hinnom on our right. The walls were evidently not very old. Over on the other side the Terraces were very perfect & the sides of the Valley cultivated. We came first to the Gate of David & opposite to it a Mosque over the Tomb of David. We then turned more to the south & went down to the pool of Siloam. This I did not measure, but it must be about 45 foot long by 20 wide I think- on one side are 5 pillars built in to the wall & (171/924 "Jerusalem") part of another stands in the water. A small arch at the further end with a few steps leads to an aqueduct underground bringing water to the pool, which issues by the same way at the other end. Near the pool we saw a very old tree, supported on brick & stone work. It is said that here Isaiah was put to death. We went down by some gardens on terraces, in good order, to another pool, called the pool of the B(lessed) Virgin. Dr Robinson maintains that this is Siloam. He as usual think it necessary to disbelieve what others hold. We passed along the Valley after this under the Village of Siloam & then up to the Wall again. Where at a corner were some enormous stones. We walked along under the Wall looking at these large stones. On our right across the Valley of Jehoshaphat was the Mount of Olives & a path leading down to the Valley by the Garden of Gethsemane & so up to a gate now stopped up, called the Golden Gate. We saw also in the wall some broken marble Pillars- said to have been in Solomon's House. Seeing some people in the Garden of Gethsemane we went down to catch the monk who keeps the key. We found L(or)d & Lady Falkland just leaving the garden & so got the monk to turn back. In the garden are (I think) 8 very old olive trees. I do not (172/925 "Jerusalem") suppose this small walled- in space was

---

[31] The Jaffa Gate on the west side of the Old City was the principal entrance. The subsequent description by TB of passing alongside the walls with the Hinnom Valley on his right would accord with exiting the Jaffa Gate and turning left. TB then makes his way anti-clockwise round the entire circuit of walls.

originally the whole of the Garden of Gethsemane- but this piece was walled-in to keep the memory of the place (& perhaps to make money). We went from thence down a long flight of steps to a Cave called the Burial place of the Blessed Virgin belongs to the Greek Church & not far off the cave in which the agony of our LORD took place. We went back to the Wall & traced it round Northwest, past the Gate of Damascus & so on to the Gate we had left by. It is a very interesting walk. We went home & I went out to buy a pair of slippers & on returning home fr(om) the Bazaar I found myself by the Church of the Holy Sepulchre & went in. Before you as you enter is a low marble tomb- said to have been the Cave where our LORD was buried. To the left you enter a circular or octagonal building in wh(ich) is the whole (sic) where the Cross stood & Eastward of this the Greek Church- a handsome building, with a carved stone in the centre 2 foot high said to contain the head of Adam. I was led up some stairs to the South & found myself where the rock was cracked in the Earthquake which took place on our LORD'S death. I then left the Church & came home. Dressed for dinner & went in to the Hotel. Dinner, journal writing, reading etc & to bed. (173/926 "Jerusalem")

<u>Wednesday. April 5.</u>

Before we were up this morning Giovanni came in with the tent curtain etc. & divided our joint {store} bedroom into two & b(r)ought in our carpets etc. & the tent table & very soon I had a private room to myself where I c(oul)d wash to my content. After breakfast we went out intending to poke about in the western part of the city & along the Via Dolorosa to St Stephen's gate. We found ourselves in the Via Dolorosa & poked our way along in the dirt & mud. At one place an arched way led along to the right & seemed to open into a square. I naturally pushed along it, though with some suspicion that I might be wrong in venturing. I got to the door at the other end & found it opened on a large green space leading up to the Mosque of Omar. A boy came running by me & making exclamations & when he had passed me he ran into the square & called to some boys there. I all the while stood innocently looking in. At the same time a little suspicious that I was the cause of the excitement. The boys ran up & in a minute a crowd had collected, shouting & picking up stones. I thought it was time to retreat, but did not like to be in a hurry, lest I sh(oul)d excite the very thing I wished to avoid. I quietly turned & was going to walk slowly away- when a shower (174/927 "Jerusalem") of stones came at me. I fronted round again directly, thoroughly indignant & stopped a boy from picking up a stone- those further in kept throwing & one hit me on the hand. Just then 2 men ran forward & warning me urgently away kept the small crowd back. So I turned to walking quietly back to Kennedy who was standing at the other end. I mentioned this at dinner & was congratulated on having escaped- people said I might easily have been murdered. We went on to S. Stephen's Gate & there fell in with our yesterday's guide. We took him with us to show us the way to the Governor's House meaning to have gone to the top to see the view of the City but we found that we ought to have bought an order from the Consul. So we turned back & went up on the Walls instead. We then came down & went on by S. Stephen's Gate- down across the Valley of Jehoshaphat & so slowly up the Mt of Olives- to what was the Church of the Ascension. It is now a sort of ruined Mosque. Here we were shown the print of our LORD's footstep, made when he ascended into Heaven. We went to the top of the little minaret from which we had a beautiful view of Jerusalem on the one side & the Dead Sea with the mountains beyond on the other. (175/928 "Jerusalem") We came down & went round the hill & so down to Bethany. The first thing we saw were the ruins of a building above the present Village. This we learnt is said to be the house of Mary & Martha & Lazarus. We sat down & enjoyed the quiet. The sun was out warm. The ground covered with flowers- the bees humming & the fruit trees coming into blossom. Below us & on the side of rising ground opposite was the Village of Bethany. We sat & enjoyed it & watched a large party, L(or)d Falkland first & then some Germans, visiting Lazarus' Tomb & Mary & Martha's house & when they were gone we went down the hill & into the Village. We went first to the Tomb of Lazarus- where he was raised from the dead. An Arab came forward & produced some candles & we went down a dozen steps or so & then by a small hole into the actual tomb. I would rather believe that it is than not. It seems in a natural position, some 80 or 100 yards from the house above. I staid (sic) behind in the Tomb & was left in the dark when the others had gone out- the thought coming across me that I must be alone in the grave for certain some day. We paid the usual backsheesh (sic) & went on to the house above. Halls of a house of 2 (176/929 "Jerusalem") rooms probably are still standing & when I came to see the thickness of the Walls, I would think they might easily have been standing since the flood. They were 8 feet thick in places. On one side is a door, arched, but filled up- at one corner signs of a vault having covered it in above. All the houses here are built in this way. The sitting room in the little Hotel, ("Mediterranean") has 2 pillars running down the room, supporting six vaults & my bedroom has the same vaulted roof. I can quite fancy that the ruins are those of a very old house. One great stone just above the ground was eleven feet long. We then came back towards Jerusalem by the way along which it is thought our LORD rode- when the multitudes cut down palm branches & laid them in the road. Below us on our left my [*** (?)]of the way by a valley with fine olive trees in it. Beyond this a hill was crowned with the village of "Abu Dies" [= Abu Deis, Abu Dis]. When we got to the Valley of Jehoshaphat we turned to the left & went up to the Walls & so along, passing the Gate of the Prophetess Hulda & in by the Gate of David & so home along the streets through the Jews quarter & into the Bazaar, past the Church of the Holy Sepulchre to home. I have been most struck with the fair hair & (177/930 "Jerusalem") freshness of the Jews here. For the most part they are quite unlike the Jews we have seen in England. Very fair- bright red & white- with light hair. I never saw anything so fair. They are quite like Francis's pictures. At dinner I sat next to Mr Graham, who was dining with one of the 2 officers (I think) tog(ethe)r with Mr Ross. He talked both long & loud & made himself very conspicuous. He talked to me about the work among the Jews & ended by asking me to go & see Miss {Cooper}& her school of Jewesses tomorrow- wh(ich) I agreed to do. He also offered to show me anything he c(oul)d & again mentioned his connection with the Jews Miss(ionary)Soc(iety). I don't like him, but I am prejudiced-I allowed myself to be very angry today when the boys & men threw stones at me- but I had no right to be. Though I would not know it, I was where I had no right to be. And I felt afterwards angry with myself instead at having allowed myself to feel any annoyance at an insult offered to me as a follower of him who was spit upon & buffeted & reviled for me. Strange that even in Jerusalem one is not influenced to good. I have just had 2 or 3 interesting rides sketched out for me round Jerusalem. (178/931 "Jerusalem")

Thursday. April 6.

I went out for a walk before breakfast- both because it was fine & because I was determined no longer to lose so much time we are obliged to do by the late breakfast. I went by the Bethlehem gate, turned to the left & so went round by Zion & so that way all round the Walls. We had hardly swallowed breakfast when Mr. Graham came with Mr Roudewald (sic) to take us to Miss Cooper's Establishment. We went first to the Convent of the Knights of St. John. The building is at present used by some Tanners I think. The remains consists of a cloister below- plainish vaulted cloisters with plain pointed arches. A stairs outside in front leads you to the upper cloisters & from them on the further side into the old chapel which is whole & entire. It might be fitted for use again in a few days- & it would almost be a good speculation to buy it- for it is so full of guano, from having been a haunt of pigeons for years, that the manure would almost repay the purchase money. In the front wall of the upper part of the outer stairs is a window of 2 lights, with horse shoe arches. Below is a vaulted room in which it is said St. Peter was confined when released by the Angel. In it is a pillar with a very fine capital. (179/932 "Jerusalem") plainly much older than the rest- where it comes from I can't think- unless it be part of some old Jewish building. The arch over the door inside is round & worked: also apparently bought from some older building. At the end of the open space outside the cloisters is the apse of a Church, the only part left. A small window in the apse is Romanesque. From this we went on to a house fitted up in workshops where the Soc(iety) for the Conv(ersion) of the Jews have catechumens at work & carpenters etc. This was a place where men were working as carpenters- making boxes etc. of olive & oakwood. The upper room was fitted up as a show room & we were plainly meant to buy. Thence to Miss Cooper's. Miss Cooper is a lady who employs Jewesses sewing & mantua making & the clothes made are sold at a sort of bazaar & the percent applied to the support of the institution- {as} they receive payment for their work. I heard some things about the Jews which I never knew before. All Jews are divided into 2 classes- the <u>Sephardim</u> & the <u>Ashkenaz</u>. In any country where Jews are found, you find these 2 classes & that their Synagogues are distinct. The Sephardim are considered of the Tribe of Judah & were called 'Royal'. They (180/933 "Jerusalem") were principally resident in Spain & Portugal from- whence they were driven by Ferdinard & Isabella when they came to the Holy Land. But so distinct are they from all other nations, that in Jerusalem at this day they still talk Spanish. The Ashkenaz are dispersed through Germany, Hungary, Russia & England etc. These also came to the Holy Land. But on arrival they were obliged to pledge themselves to live & die on the land & on giving this pledge they are on an equality with the Sephardim, & keep but the one Festival, each in its own season. If they refuse to give this pledge they are obliged to keep the Festivals double. We saw the Consul's Wife, Mrs Hardy,[32] then- I can't make out that any other advantage arises from the insitution besides the relief this afforded to the Jews. Besides this relief the Chief Rabbi receives from Europe, principally from England, between £3000 pounds & £4000 a year to distribute amongst them. Kennedy & I went to call on the Consul & unfortunately found him just going out. We asked him to send us a kuwass to take us up to the Turkish Government's House to see the view- wh(ich) he will do. Thence to the French Steamboat Office in vain- as it was (181/934 "Jerusalem") closed. We then started with a guide to visit the Tombs of the Kings- Judges & Prophets - as they are called. The first are on the Damascus Road about ½ or a mile from the Damascus Gate. You descend into a square space cut down into the ground like a quarry, about 50 y(ar)ds square. On the West side is a portico or oblong grotto cut into the rock, at the South side is a low door wh(ich), you creep through on your hands & knees & within are 3 chambers- containing altog(ethe)r 37 spaces for sarcophagi. The portico has been handsomely ornamented with grapes & palm branches etc.- much worn away. We walked another mile to the Tomb of the Judges- a pedimented doorway admits to 3 chambers containing 23, 21 & 12 recesses for Sarcophagi. Near this I saw a curious cistern & near it at the side of the same rock an arched recess with a tank for water 4 foot by 2 perhaps, with an aqueduct down to it, from the rock above & running along

---

[32] TB only mentions Mrs Hardy (and her husband) once more, when they boarded the ship at Jaffa as he sailed from Beirut to Alexandria (below, under 9 May).

this upper rock- as though to catch all the rain water that fell. Thence we went to Jeremiah's cave, near the Wall of Jerusalem, where he is said to have been confined & to have written the "Lamentations". This is now in the hands of the Dervishes. We went up the Mt. of Olives halfway & were (182/935 "Jerusalem") taken down by a hole into a large cavern which wound & twisted about in a wonderful way under ground- giving room for a great number of graves. What all these Tombs were used for it is hard to say. We came in by St. Stephen's Gate &having a better guide than usual, we got good information about the places on our way. We passed the pool of Bethesda & up the Via Dolorosa- passed Pilate's House, where the old dooway was pointed out & the remains of "Scala Santa". Dressed & went to dinner- where I had much talk with Mr Heneage- who has been out in Tanjiere [= Tangier] & now away to bed.

Friday. April 7.

I was up & out soon after 7. I went out at the Jaffa Gate & round to near the pool of Siloam- where I tried to make a sketch & then down to the pool of Nehemiah. I sauntered about in the Valley enjoying the lovely morning. The gardens there, formerly the "King's Gardens", were full of people at work & the whole Valley was a charming Spring scene. I was to have gone with Mr Graham after breakfast to the large Cavern outside the Damascus Gate- but while I was out of the room he went & on going to my bedroom I found Mr Kennedy at work with Giovanni settling accounts & we set (183/936 "Jerusalem") to for some time. Much discussion about the way we sh(oul)d go to the Jordan. John insisting on the neccessity of going to Jericho & the Jordan first & then on to Mar Saba & the Dead Sea. Whereas we want to go to Bethlehem on our way. At <u>least</u> I want to go from Jericho back to Jerusalem. Dragoman-like, he thinks it necessary to go by the stereotyped way & no other. He seems to have no idea of doing differently to other people. The next thing was a notice that in consequence of the high prices here he can't keep his contract & wants 5/- [= 5 Shillings] a head per day more. But the most uncomfortable thing of all is that a steamer is expected at Beyrout on the 9th or 10th of May that would secure my being in time at Alexandria for the P. & O. Steamer home leaving on the 22nd & it is impossible I fear to get there in time. I bargained to be in time when I arranged to come with Mr Kennedy & he wishes now to wait for the post- which will delay us longer in starting. I fear I must give up going to Damascus. I think if I can get to Beyrout direct & persuade Mr Kennedy to go this way, I can then manage to make an expedition to see Mt Lebanon & back in time for the steamer. We took John out to see if we c(oul)d find the Agent of the French steamer but of course he was (184/937 "Jerusalem") out & we could learn nothing. So we gave it up. We went thence to the Armenian Convent & saw the church- a great deal out and we c(oul)d learn nothing. So we gave it up. We started with a guide to the "Wall of Wailing". This is a part of the wall which by general consent is thought to be part of the South Wall of the Old Temple Platform. For certain days the Jews come & sit over against this wall & wail & lament & pray. Today this happened. It was an interesting sight. We found at the time we were there from 10 to 20 Jews & Jewesses sitting there. They were reading from books- some very earnestly- some less so. Others stood with their mouths against the Wall praying- others passed along touching each stone & kissing it. The stones are large & have the same bevelled edges which all the old stones about Jerusalem have. We went thence to see a pretty fountain in the vestibule off the Cadi's Court & then on to a part of the wall further East where is the beginning of an arch. It is a great question what this has been. It is thought to have been part of the bridge which connected Mt Moriah & Mt Zion. Others again think it part of the vaults under the Temple & some have fancied it part of the substructure on which

Justinian built a Church in honour of the Blessed Virgin. The stones in what is left of the arch (only just the uprising) & in the wall below, are wonderful. 25 feet long two of them must be. We (185/938 "Jerusalem") went out by the Gate of "Hulda" & down to the Mosque outside- where David & most of the Kings of Judah were buried. I knew we sh(oul)d see nothing & so it turned out- we are only annoyed as usual by the insolence of the Turks. I have said before I w(oul)d go no more to see Mosques. We went next from this to the Armenian Convent - said to be built on the spot where our LORD was scourged on his way from the House of Caiaphas. Here also the altar is made out of or upon the stone which covered the mouth of our LORD's Sepulchre. It was high and thin. Most of it is covered over with plaister (sic)- one small piece only being exposed- at the N. End. The old monk however uncovered the front to show me other parts of it. It is of the common limestone. The whole convent was in beautiful order. The nicest I have seen. We walked round above the Birket Es Sultan or lower pool of Gihon & then passing down into & crossing the Valley of Hinnom we went up the other side & along the ledge under the "Hill of Evil Council" to "Aceldama". Here is an enormous vault into which the bodies of pilgrims were put. The walk down above the valley was very pretty. We came back- went down to the Birket Es Sultan, (186/939 "Jerusalem") which is now empty & dry & walked through it. It is an enormous pool- rebuilt by one of the Capliphs A.D. 600 odd. Thence back to the Jaffa Gate. The Consul [= James Finn] had kindly invited me to a meeting this evening of the "Jerusalem Literary Society" a sort of Conversatione at his house. But I feel that I can't go in this wild man of the woods state- with beard & black neck cloth. The Bishop will be there & a number of other people. I have been worrying myself all the Evening notwithstanding at not having shaved & gone. I would have heard about a great deal. Wrote journal & went to bed.

<u>Saturday April 8th</u>

I went out early, across the Valley outside the Jaffa Gate, where I tried to make a sketch of the Gate of Hippicus. After breakfast most of the party started on horseback for Bethlehem & Solomon's pools. We were just starting too in the opposite direction, when Mr Freeman came up & proposed to join our party as far as Beyrout. We then started on horseback ourselves. We went out by the Jaffa Gate & so into the Damascus road. We passed from the road a few miles from Jerusalem & struck across the country & passed through the Village of Hanina. I cannot yet learn the ancient name of it & don't know what it can have been. There is nothing to see there. It stands (187/940 "Jerusalem") as most of the Villages I have seen stand, on a hill. We passed through it & went down a valley & rose another & so up to Nebi Samwil. Here is a mosque, wh(ich) was evidently once a Xtian Church. The Mosque seem now as much neglected as the Xtian Church. In the S. Transept which is built up, stands the Tomb of the Prophet Samuel. There is nothing to be seen of it- but in its' place stands a wooden case covered with a cloth. {Ragged} & filthy. The whole place is most miserable. The Church must have been a good Village Church. It has single light pointed windows, very high up above the ground. We went to the top of the Church & then of the Minaret. The view round was very striking. The first thing that caught our attention was the Mediterranean, apparently not far from us. To the East, (I think) on the next hill was a Village called {Chane} & on a hill near to us the remains of a Castle- said to have belonged to the Macabees. On the otherside was Er Jib. This too stands on a hill. On a hill north of us was what looked like a <u>small</u> house or two. This turned out to be Er Rama. We stood sometime trying to learn the different places, & then we remounted & rode on. We passed "{Nauina}" again & struck (188/941 "Jerusalem") across the country. We came upon "Er Ram", said to have been "Ramah". We passed down into a valley, which was very wild & dreary. This is to me the appearance of the whole country. If there were only some timber it would be different. Round hills, with bleached stones & bits of cultivation on the terraces. Our road led us at last up the right hand bank, so that we left Anata, the old Anathoth at the end of the Valley on a hill we had had before us. We came over the hills several miles & then the view of Jerusalem broke upon us & we found ourselves on "Scopus", the hill I believe from which the Roman Legions first looked upon Jerusalem. We meant to have gone round by Bethany & to home- but finding ourselves here we rode on into Jerusalem. After getting home I went down to the Church of the Holy Sepulchre & spent sometime there. In the Latin Church service was going on- miserable singing, all out of time to a bad organ. Outside the Church I fell in with the guide we had had 2 days & walked through the Town part of the way with him. He tells me his family have been Xtians for one or two generations. Much pleasant talk at & after dinner. Mr Heneage is the man who used to have the "Sparrow Hawk". It (189/942 "Jerusalem") was pleasant having a talk about Cowes. My left hand neighbour at dinner too, Mr {Leichardt?} is nephew to Lord Yarborough. Journal & then away to bed.

<u>Sunday April 9. Palm Sunday.</u>

I was up early, but did not go out as I usually do. At 10 I went to service at the Church built here. It is adjoining to the Consul's house- in fact built as his chapel. It seems odd that the house built onto the Church should be the Consul's. I expected to find that it was the Bishop's. The Church is built of stone & is solid & good. I can't say much else for it- a bad ground plan. 2 large transepts & a small sacrarium, no chancel. Enormous windows in the transept & none in the Church. A reading desk facing westward. The Bishop & another man sat at each end of the altar & during the Sermon, the only time when they should have been within the altar rails, the altar was deserted & the Bishop came down to a seat outside facing the pulpit. Banns of marriage were published after the 'Jubilate' & the good man who preached kept us 45 minutes by the watch listening to a sermon preached with so strong a German accent that I for one could not understand half. They chanted the "Venite" very strongly to (190/943 "Jerusalem") a very high chant & sang (or didn't sing) Jackson's Te Deum & somebody else's Jubilate- <u>without a choir</u>. The font stands close in front of the altar rails. At the same time everything about the Church is good & nice look as far it goes- intently a good deal of pains have been taken. Pity that they didn't employ a grand architect. I sat as I always try to do close at the West end & as soon as service ended went out & walked away past the Armenian Convent & out by the Gate of David & so across the Valley of Hinnom & up the Hill of "Evil Counsel" & then down again- along Aceldama, down into the Valley of Hinnom & to the Pool of Nehemiah & so back up by the Pool of Siloam, in by the Gate of David & so home. In consequence of this I missed a scene at the Consuls- going home from Church Col. & Mrs West met a file of soldiers. One took the trouble to step aside from the ranks & jostle Col. West - while another followed his example & did the same to Mrs West. Col. West recovered himself & struck at the soldier with a small stick he had in his hand- but missed him & hit the pouch on his back. The soldier dived into the line or ran in & out among the soldiers. Col. West following (191/944 "Jerusalem") him. He complained to the Officer & the man said he had only called the Inglesi a "Giaour" (Infidel). The officer pooh poohed the whole affair. Col. West at once went to Mr Phinn (sic) [= Finn]- our Consul. Mr Phinn heard his story & sent to the Colonel of the Turkish Regiment - who came at last & made light of the whole affair. However it was insisted that an apology should be made & that in public & in presence of all the English that would be gathered together. The Col. refused- but at last yielded with a bad grace & the whole party (several English men had followed) adjourned to the barracks where the company or whatever it is was called out & the Col. made then a speech telling the men that they were sent here to keep order & <u>prevent</u> people being insulted. That they might only call Russians Giaours not English nor French & that the next time such a thing happen he would have a pair of scissors & slit the offender's tongue. Mr Heneage who was present said this was no figure of speech: he had been in Morrocco (sic) & known this done. I went to my room & said Evening Prayer & then went down to the Church of the Holy Sepulchre expecting to find service going on- but the (192/945 "Jerusalem") Church was shut. Mr {Leichardt?} followed me & we went back tog(ethe)r & then onto the Church at the great Armenian Convent- where we found service going on consisting of bad chanting. An old priest came to the Chancel rails at the end & preached a sermon (we thought), of 5 minutes- during wh(ich) he seemed to be facetious, as several of his audience smiled & then all pressed forward & kissed the altar or something on it & then proceeded to a small chapel on the N. side of the Church- wh(ich) they entered on their knees & kissed the floor. We then went to the Latin Convent & went into the Chapel & {over} the Convent. After that we went for a walk out at S. Stephen's Gate & down the Valley of Jehoshaphat & so up by the Valley of Hinnom where we found several Tombs with nearly illegible inscriptions, but made out one.

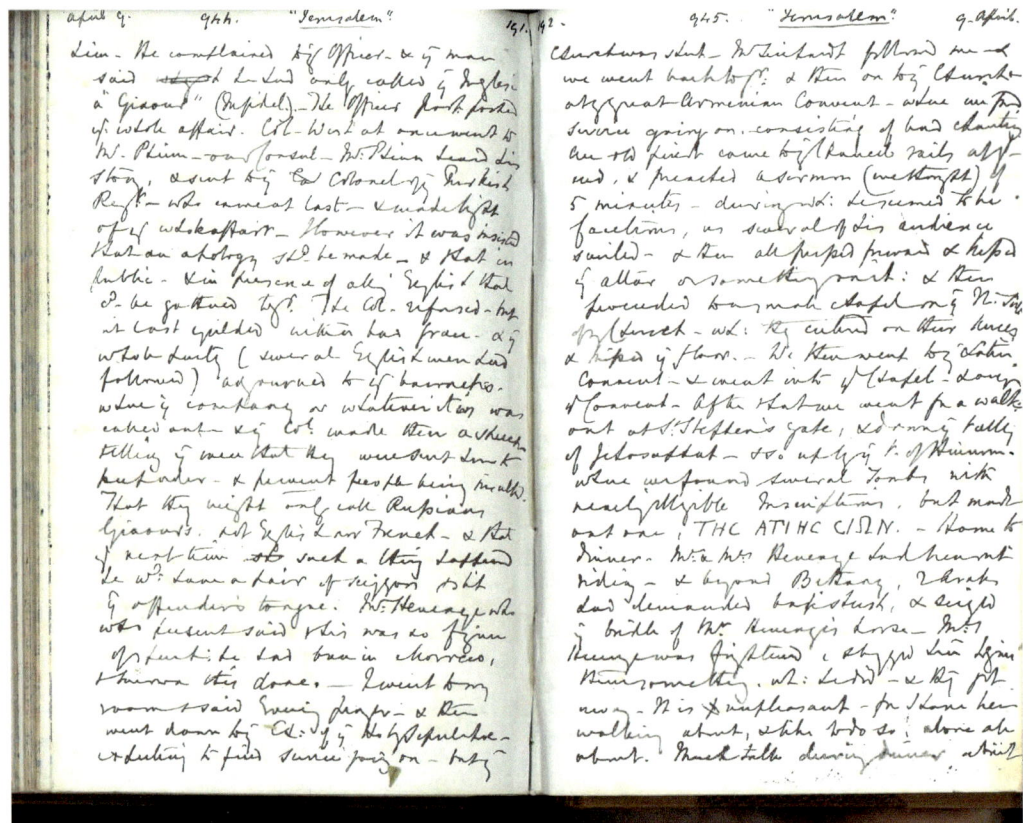

Figure 14.3: Greek Text.

Home to dinner. Mr & Mrs Heneage had been out riding & beyond Bethany, 2 Arabs had demanded baksheesh & seized the bridle of Mrs Heneage's horse. Mrs Heneage was frightened, & begged him to give them something wh(ich) he did & they got away. It is unpleasant- for I have been walking about & like to do so, alone all about. Much talk during dinner about (193/946 "Jerusalem" "Mar Sabba") Cowes & the Falcon etc etc.

Monday. April 10th. M(onday) before Easter.

While at breakfast the Consul, Mr Phinn (sic), came in & discussed the affair with Col. West. The offending soldier had been sentenced to a fortnight's imprisonment & 2 men <u>who were thought to be near the man who had pushed Mrs West(!)</u> to a week each. I think I was the first to exclaim openly at this piece of injustice & then we all expressed our feelings & it ended by our all feeling that Col. West w(oul)d himself go & get all the sentences remitted. I also told the Consul what Giovanni had said- viz that it had been all accidental altog(ethe)r the soldier having slipped. This however was doubted. The Consul said the man had <u>not</u> made this excuse himself, as he would have done, had it been the case. I suspect the Consul came to feel rather whether there was much wish for severity on the part of the English visitors & finding that the feeling was all the other way, he said he would go to Col. West & suggest his interference on behalf of the offenders. A doubt was expressed about Col. West & I quietly said that probably the Consul's interference would be more influential than Col. West's & he looked across the table & said at once, he would speak to Col. West & if he was disinclined to move, he sh(oul)d (194/947 "Jerusalem to Mar Sabba") feel that he had done his duty so far & would take the rest upon himself. I went out to the French Packet Office & succeeded in catching the agent there at last. Thence I went to see if the Church of Holy Sepulchre was open- expecting to see some Latin Ceremony going on- but it was all closed. I returned to my room & found John ready to start. I am sorry to come out on an expedition this week. This is the 2nd Passion week that has been very unlike what it should. Last year I landed in Sydney on this very Monday & all the week was excitement & bustle. And now this week I am making an expedition to the Jordan. Last week we could not come (sic) as I wished. Our tents were all wet & beds & everything & the sheets etc gone to the wash. And next Monday (Easter M(onday)) we start on our way towards Beyrout. I am very doubtful of the good efforts of pilgrimmage. Try as I will to feel the sacredness of every spot about Jerusalem & to stir up holy thoughts. I find one goes to everything as a sight. I cannot feel the strong emotions others have described. But I have not been with a person whom I can sympathize with, or who sympathizes with me & that puts a restraint upon me. Certainly I felt the sacredness of our service yesterday (195/948 "Mar Sabba") as a great contrast to what I had seen since being in Jerusalem. It made me wish the Holy Land was in the hands of England. And yet it was very dull & dry, with a sermon of 45 minutes. We started on our horses, I myself with a miserably uncomfortable saddle, out by the Jaffa Gate, down into the Valley of Hinnom in wh(ich) we overtook Mr & Mrs Heneage going out for a ride with {Leichardt}, met the Valley of Jehoshaphat at En Rogel & so down the Valley of Kidron. We mounted (sic) out of this & got into another. Here we came to a well with a number of Arab women drawing water for the flocks of goats & sheep & a herd of Camels which were grazing on the hill sides under the care of their brothers & sons (I suppose). A Sheikh also was watering his horse at the same place, his long spear stuck upright in the ground. I rode quietly along behind the others most of the way, trying to be quiet & to use the time as it ought to be used- but presently we came to the encampment of the Arabs, with their black tents & then turned sharp round to the right, following the edge of a deep rocky Valley, which deepened as we ascended up the hill side above it & at another turn came in sight of the Convent. This Valley was once filled with hermits. (196/949 "Mar Sabba") Fourteen hundred years ago about Mar (or Saint) Sabba lived in a cave in the Valley, & numbers of men followed his example. They led the lives of Hermits, considering this a fitting place to retire to away from the world. Why it was chosen I don't know- for it is not the scene of our LORD's Fast & Temptation. The Story runs that they were attacked & numbers killed by the Persians & by the Arabs & consquently a Convent was built to protect them & they moved within the shelter of its' walls. It must have been originally more a mere wall in closing part of the rocks than anything else- but many of the cells even now being scooped out of the rock within the Walls. The Wall runs where it can down the face of the precipitious side of the Valley & you go down from ledge to ledge, sometimes by stairs cut in the rock, sometimes by built stairs or ladders- ½ way down there is a broader ledge, on which stands the Church or Chapel & lower down again the Refectory or Kitchen. They lead a strict life- eating but twice in the 24 hours & then no meat ever. There are 60 altogether, Priests, Monks & lay brothers. The Chapel is a fairish building- with a great many middling pictures & a skreen (sic) handsome from (197/950 "Mar Sabba") the amount of carved & gilded work. We walked about the Convent, saw the Chapel & the small Chapel where Mar Sabba was buried in the level space opposite the W. Door of the Chapel & the Refectory & Kitchen & an old cave in the rock, the original Church, still used as Chapel, with a great Font in it & then went out at a window & down by a ladder into the Valley below to see the spring, to which Mar Sabba is said to have been guided by an antelope. All along the Valley the rocks are full of small caves, once venerated by men who retired here for the LORD & I hear that for a mile or more the Rocky Valley is full of these caves & that there were several chapels. When we came up again, as we passed the Chapel we found service going on. I went it at once & took possession of a stall. The service seemed to consist of a long reading from some book & then 2 men sang, (if singing such a noise would be called), what I thought were verses of a Psalm antiphonally. Then a priest passed into the Sanctuary behind the skreen (sic) & drew aside the curtain behind the choir doors & having put on a stole which I had seen hanging on the altar he came out with a thurible & incensed the altar, the choir doors, steps & the whole chapel generally & then the (198/950 (repeated in error)) seat of the [Hegamenos (?)], & then passed down the North side incensing each person & me amongst them seperately & up the South & then outside by the N. door & in at the West, incensing those without. He then put away his Thurible & Stole & after a short service again, all prostated themselves 3 times & the same priest or perhaps the hegamenos himself put on another stole hanging at the entrance of the Choir (the only vestment they seemed to use) & gave his blessing from in front of the choir doors. He then went & stood in front of the Throne on the S. side of the Church, when each in turn came up, prostrated, kissed his hand & received his blessing & the service was over. I sauntered up & down the small level space, enjoying the perfect quiet, a great contrast to the noise & confusion of the Convent at Mt. Sinai & then was called in to dinner. One of the monks bought us a salad. I eat (sic) some of it, but it had so much oil that I can't say I liked it. Journal writing & walk on the roof & to bed.

Tuesday. April 11. T(uesday) bef(ore) Easter

We were called & got up about ½ past 4. A lovely cloudless morning. About 6 we got well off. We followed down the gorge again & where we had turned off yesterday crossed it & went on over the hills beyond. The ground everywhere covered with flowers. We mounted a highish (199/952 "San Sabba to Jericho") hill & there saw the Dead Sea. Soon after this our Arabs began one of their usual diversions- a cry of "Arab" was raised & they begain to make great preparations- scouts being sent out & the Sheikh dashing off waving his spear etc. We only laughed & cried "Arab Mafish". In a few minutes we discovered several Arabs sitting on a hill & our Arabs dashed up, shouting & they disappeared together over the brow of the hill. Of course they were our own Arabs. They made exhibition of their horsemanship, riding after each other as hard as they could tear, brandishing their spears etc. About eleven we came to a steep valley leading down to the sea- along the ridge of one side of which we got down- passing at some distance on our left on a hill a Musulman Mosque & convent, which they maintain is the Tomb of <u>Moses</u>. Unfortunately Moses died & was buried across the Jordan without being allowed to enter the Land of Promise. We came down upon a plain- with a few bushes upon it, all covered with salt & then down on the shore. The sea was high & before us lay a small island, which generally is connected with the main land. It was so shallow that we rode across to it & dismounting there 3 of us bathed. Kennedy, Yateman & I. (200/953 "San Sabba to Jericho") All that has been said of the saltiness & acridness of the water as well as of its' density is perfectly true. On getting into the water I put my head under & drew it out again in real pain. Eyes & nose smarted sharply & it was sometime before I could see. I got into deeper water & found myself carried off my legs & there I was literally floating like a log. Yateman who was not far off at the same time called to me saying that he could not keep his legs. I never knew anything more curious. I got out of my depth & swam but found it very unpleasant- my legs by the buoyancy of the water were thrown out of the water, the effect of course being that my head was lowered. I dressed in a burning sun- but the water on me didn't seem inclined to evaporate. I dried at last covered with a coating of white salt. I walked about upon the beach- c(oul)d see no bitumen. Some gulls or some white sea birds flew up the lake towards the head- where I hear they feed on the fish brought down by the stream & which die as soon as they get into the lake. It was a cloudless day, but a thick haze obscured everything on the sea- so that you could not see far. I cannot say that there was any particular appearance of desolation. It was desert about- but not (201/954 "Mar Sabba - Dead Sea - Jericho") so desert or desolate as we have seen. The water was perfectly smooth- except now & then when there was just a motion against the beach. I got away by myself & the silence was extreme but there was no particular sadness about it. There were one or two places when my horse refused to pass. I expect that below must have been a bed of bitumen perhaps. The sand & gravel were all caked & cemented with salt & the Tamarisk & other shrubs were both salt to the taste & covered with crystals of salt. I believe the analysis of the water gives 3 principal ingredients, {nitrates} of lime, of magnesium & of soda. We mounted & rode on. I was annoyed to find that we were not going to the spot where the Jordan empties itself into the sea & here I had to give Giovanni a blowing up. I asked why we were not going there & he answered without much civility that we could not. So I called him aside & blew him up- during wh(ich) process he said he w(oul)d leave at Jerusalem- of which I took no notice & it ended in him being very pertinent & the rest of the day he took every opportunity of being exceedingly civil. We came across to the Jordan- according to him & the Greeks the spot where our LORD was baptised (202/955 Jordan - Jericho) & the Israelites crossed. It might have been- though it must have been lower & quieter if so- for now the turbid stream was flowing. I should say at 6 or 7 knotts (sic) good & no man would keep his legs. I mean this supposing it to have been the general crossing place- because of course their passage would not be affected by the stream- the water standing on an {heap} & the people passing through dry shod. The stream was so strong that we were prevented bathing as we fully meant to have done. It would have been unsafe for any but a very powerful swimmer. I washed my face & hands & drank some of the water, which thick as it was, was very sweet. Three or four of the Mar Sabba monks were there spending several days in a sort of pilgrimmage- a custom they repeat every year at the end of Lent. Tomorrow the pilgrims are expected from Jerusalem. The French Pilgrims, I heard, are to have mass celebrated on the banks. Most of our own party seemed to have no thoughts of any sort- but a wish to cut a stick to take away. Between the actual bed of the stream (wh(ich) is <u>much</u> narrower than I expected, perhaps 50 yards at the most) & the outer bank, is a thick mass of trees & underwood. Once the covert of some wild beasts or other- in the Bible (203/958 (sic) Jordan - Jericho) mention is made of the "lion coming up from the swelling of the Jordan"/ "coming up like a lion from the swelling of Jordan" ([Jeremiah 49.19])[33]. The stream overflows & fills the wider bed & drives out everything sheltering in the underwood. I gathered bits from some of the trees & was surprised to find them eucalyptic, like the Australian though there was no aromatic smell like the gum trees.[34] I got away quietly by myself down the river & knelt down & meditated on our LORD's Baptism & the Baptism of the Children of Israel in this stream & their entrance through it into the Promised Land & renewed my own Baptisimal Vows- God help me, & give me grace to serve him better than I have done. I have been dwelling much lately on the hope of getting home & going to work again. I look forward daily to feeling myself afloat on my way home. I shall give up Damascus & the Lebanon & any other place if I find no steamer but the one on the 8th of May. We left the Jordan & struck across the plain, past some ruins, which Giovanni said was once a house or building to which the Jews used to come in order to visit the place where their forefathers entered Canaan. Vegetation increased & in 2 hours we came to a square ½ ruined Tower in which a Turkish Officer lives with a few troops & near wh(ich) we found our tents (204/957 Jericho) pitched on some mounds & ruins, where have evidently been extensive buildings once. It is doubtful whether these or some others a little further towards Jerusalem are the actual remains of Jericho. It is the site for a "City of the Plain". I have looked in vain for signs of Palm Trees- from which Jericho also derived a name, the "City of Palms". Near the mound on which our tent stands a fresh brook runs, I believe from Elisha's Fountain. Soon after getting to camp, I went down & washed myself well free from all the salt & impurities of the Dead Sea. My hair was stiff & sticky & my whole body cased in a coating of salt. While sitting writing the Jackalls (sic) have been crying out far from our Tent & the crickets in the bushes along the stream remind me of the Cape or Sincapore (sic). A party of Arab girls have been dancing & singing at out Tent door. We shall get back to Jerusalem tomorrow early. I shall make a point of riding by way of Bethany: the route our LORD & the 12 went as on last Friday & tomorrow evening I must go out by St. Stepehen's gate & up to the place on the Mt. of Olives where after finally leaving the Temple on Wednesday afternoon He sat down & wept over Jerusalem on His way back to Bethany for the last time. (205/958 "Jericho to Jerusalem")

---

[33] TB never got round to filling in the reference himself

[34] TB would have been very familiar with eucalypts from his lengthy stay in Australia. Officially there is no evidence, however, that any eucalypts were imported into Palestine until the 1860s - for timber and also to help soak up water in marshy ground. Of course, not all eucalypts have a scent. However, the unpublished diary of the Consul James Finn (1846-1863) notes that he took a keen interest in trees and seeds: "Eucalyptus seeds were brought in by him and planted; and it would be interesting to know whether this represents the first introduction of a tree which has since become naturalized in the country" (Abrahams 1978-80: 47).

Wednesday. April 12. W(ednesday) before Easter

We took a slightly round about way after leaving the Camping ground this morning to visit the Spring of Elisha. We passed over masses of ruins & mounds where ruins had been- among quantities of the Nakh, or Thorn. The spring comes out from under the rock, into a basin which was once built of masonry. It is a beautiful spring & has been & might be again, the means of fructifying the whole plain. Jericho must have been a large place, much larger than one has fancied- remains spread over a great space of ground- but a good many of these were connected with irrigation. The hills above are pierced with caves in which in early days Christians lived, in order to be on the spot of our LORD's Temptations. We rose [up] the hill, from which we had a beautiful view back over the plain. There is not much to say about the way back. The country was rough, but not so wild as I expected from what other travellers have said. We met the Greek Pilgrims going down to Jordan. A regular holliday (sic) making- no solemnity about it. We passed through Bethany & over the Mt. of Olives & crossing the Valley of Jehoshaphat entered by St. Stephen's Gate. I sat in my room till about four when I went out & passing by the Church of the Holy Sepulchre & seeing (206/959 "Jericho to Jerusalem") it open I went in & found a Latin Service going on. I don't know what it is called- one of the Passion Week services- consisting of Psalms & Antiphons etc. etc. It was celebrated by a party of Franciscans who sat in a double row from the gate of the Choir to the building in the middle of the Church. I looked over one of the monks' book & joined in the Gregorian Psalms. The chanting was decidely bad- but yet the slow solemn Gregorian chant was very fine. The finale was the extinguishing of all the candles one by one the meaning of which I don't know. After dinner much laughing beginning with Mr Graham at the idea of the Bishop of London rebuking the Queen for attending Presbyterian sevices when in Scotland. The set are unanimous I daresay here. I heard at least that the Bishop gave a generous party on Monday. A mail has come in.

Thursday. April 13. Th(ursday) before Easter

News came to the Hotel by breakfast time that an English Man of War had been sunk at Varna & that the Russians had crossed the Danube.[35] I had my arm pricked with a Jerusalem cross after breakfast & went to the Tailor. When I went back to meet Mr. (207/960 "Jerusalem") Briggs I found he had got tired of waiting & was gone out without me. I went to the Ch(urch) of the Holy Sepulchre, but found no service going on. I went to look for the Latin Convent & went into another convent, wh(ich) I don't know. It had a solid stone skreen (sic), with 3 peculiar arches in it- within the centre of wh(ich) the altar stood. I thought it must have been a Greek Church- but there were several altars. Nevertheless they spoke of themselves as <u>Catholic</u> in distinction to the <u>Latin</u>. I afterwards found the Latin Convent & learnt that the services at the Ch(urch) of (the) H(oly) Sepulchre began at 2. I went for a walk up Mt (of) Olives & sat down on the spot marked as the place where our LORD sat on Wednesday Ev(e)n(in)g & wept over Jerusalem. Back to the Hotel & then went to the Church: but after waiting about for sometime I found the doors were not opened & there was some mistake. So I went back to my bedroom & sat down till 4 when I went out for a walk with Kennedy. We went out to the upper pool of Gid(e)on & across the Country in that direction. Much talk during & after dinner, the whole party dining together.

---

[35] That had happened several months earlier.

Good Friday. April 14.

I got up early & sat quietly in my room till (208/961 "Jerusalem") nearly 8, when I went to see what was going on in the Church of the Holy Sepulchre. I found the doors shut & nothing doing. So I went to St. Stephen's Gate & out up the Mt. of Olives, where I sat down for ½ an hour & then round among the tombs in the Valley of Jehosaphat & in by the Lion Gate. I had a short time in my own room & then went to Church. A Jew was baptized. I can't say I sympathised with the Baptism as much as I might at another time for this reason- viz: that one knows the common tone of {Puling} amongst the Missionaries on the subject of Baptism & that one of the sponsors was a man who the other Evening in this very room was boasting of being a Presbyterian & though he went to the English Service, he didn't like it. In the afternoon I went to the Holy Sepulchre, where I found the Latin Service going on. Four or 5 monks sang along piece very well- I think the Lamentations & the "Deus Misereatur" was most beautifully sung; to a Gregorian very slow. I believe for short anthems, or for the Canticles, nothing would be better than a few verses of the one or the other, sung to a Gregorian in this way. Moreover a good effect was produced by the alternate verse being (209/962 "Jerusalem") merely recited not chanted. The chanted verse was in harmony. After the service everyone as on Thursday sprang a rattle, or made a noise in some way or other. I fell in with an Irish priest who had fallen in with Mr Ross last year. He seemed to doubt about the <u>sacred</u> spots. He <u>also</u> seemed not to know quite what the extraordinary noise made at the end of the service meant. I walked up to the Latin Convent with him. After dinner I went down to the Church again. The Latins had great services. A procession went round & round the Church to different spots & sermons were presented in different languages- Italian, English, German, French, Arabic & Spanish- I think one other as well. The English I unfortunately missed as I was with L(or)d Falkland's party & we were taken up at once to Calvary, in order to be in good time there. The scene there was very extraordinary. The crucifix was brought up, with a wonderfully well executed figure of our LORD on it, with a painfully agonized countenance & a large crown of thorns, made I think from Nabk thorn- said to have been the thorn used in the crucifixion of our LORD. After the German service, the Crucifix was borne to the altar over the actual hole in which our (210/963 "Jerusalem") LORD's cross is said to have stood & here it was planted & after a sermon in French, the nails were withdrawn & the figure lowered from the Cross & placed on the altar, whence it was borne in a linen cloth by 2 priests & taken downstairs to the <u>Stone of Unction.</u> Here a sermon in Arabic was preached & then it was taken to the round Church again & after a sermon in Spanish, loud, with some very beautiful chanting, in the Sepulchre & then the doors were closed & after more very effective chanting the door was locked & the procession slowly & apparently sorrowfully left & went I think to the Latin Chapel, but it was then near eleven & I went away. People in the Hotel spoke of it next morning as disgusting & blasphemous. I cannot say I thought it so. I looked upon it as a piece of very curious sacred acting & probably much too solemn & sacred a subject to be treated in this way- but it certainly was a <u>Sermon on the Crucifixion</u> to the poor & ignorant: it was so to me. I felt it powerfully.

Saturday. Easter Eve. April 15.

I was up not so early & did not go out (211/964 "Jerusalem") before breakfast. John came while I was at breakfast to ask if I was ready to go to the Greek Convent & we went at once. He has got Horses etc & we are to start on Monday for Damascus & Beyrout- leaving out Carmel- I am glad this is settled. I am to be in time at Beyrout for 8th May Steamer. We went to the

Convent. It is a large rambling modern building- upstairs & down. Some of the priests took us to the Library. They have some valuable books & showed us some {manuscripts} which I should think very valuable. Among others they showed us a S. Chrysostom & a very curious {manuscript} Book of Job with commentary. Kennedy soon got tired & went. Then I got them to produce some of their musick books & got some account of it & they sang for me. Next they took me to the Chapel- a small building- but containing some very curious things. Amongst others a picture sent from Russia of which the whole dress etc. of the 2 figures of the Blessed Virgin & our LORD are all of silver gilt except the faces & hands. On the robe of the Virgin are 70 very large diamonds- which must be of great value. A star of large Emeralds & another of Amethysts. I had some talk through John about their services & the celebration of mass. They pour (212/965 "Jerusalem") warm & cold water into the Wine before Consecration. They found out that I was a priest. After one o'clock Hayward [= Rev. J. W. Hayward], whom I met to my astonishment in the street yesterday, came to walk with me. We went to the Holy Sepulchre & found the Greeks & the Armenians just filing in, in procession. In the Church were also the Latins & the Copts. The 3 priests I had seen in the morning at the Greek Convent came forward at once & took us in by a side door in the Choir & placed us against the skreen (sic) & explained everything. It was a very fine spectacle. The gorgeous church filled with lamps- some 50 or 60 Bishops, Priests & Deacons as well as Vice Patriarch & the 2 choirs in galleries on each site. The chanting & singing was execrable. I never heard some (sic) thorough discord in my life. After all the priests etc had come forward & prostrated themselves & kissed the Vice Patriarch's hand, {while} they gained permission to officate- they went behind the skreen (sic) & vested & came out attired very grandly in cape & stole. There was a procession & sweet incense. The Gospel was read- Psalms & Hymns chanted a chapter from Zechariahread (213/966 "Jerusalem") & then came a curious ceremony. A small table was brought forward & placed in front of the skreen (sic). On it were 5 loaves & above this a candlestick with 7 small candles. I could not learn the meaning of this- but that it had some connection with the miracle of loaves & fishes & that the Candles were the 7 spirits before the Throne of God. The Bread was blessed & taken within the skreen (sic) again. Some was brought for us & also lighted candles. Before the end of the long service the priests took us out at the side doors to see the place where the vent in the rock is shown before the place of the Cross. It is I believe an undoubted vent in the rock. We took leave of them- much struck with their kindness to us & willingness to fraternize- a strong contrast to the Roman Church. But, their service was utterly unspiritual to me. I never saw such a mass of symbolism & ceremony. The meaning is I am sure quite hidden to the uninitiated & they are the priests only. We went for a walk out by the St. Stephen's gate & round the walls to the Zion Gate. It began to rain. Hayward is living with a young clergyman [= Rev. W.J. Beamont][36] who is come out to take the management (214/967 "Jerusalem") of a College [= the English College], wh(ich) is now in {Imbitus}. It is suggested to me to go home by Marseilles & I think I shall. I have asked the French Steamer Agent to write to Beyrout & secure me a passage to Malta & on to Marseilles if I like. I have forgotten to speak of the sad Batel [= Battle (?)] in the H(oly) Sepulchre today. The Greeks in their chapel, chanting aloud in their whining nasal way- close by them part of the time in Calvary the Latins. In the round Church the Armenians, chanting loud & striking bells. It was discord- of sound & of mind.

---

[36] See above, Chapter 10.

Easter Day. April 16.

I got up at 6 & after dressing leisurely & reading etc. I went down to the Church. There I found the Latin Service going on: but the part I wished to see was over. The Greek Service was about to begin &{vigorous} striking their wooden bell, (beam) & making (I believe) as much noise as they could, to interfere with the Latin Service. The Armenians also came in & the Copts began their service & the Abyssinians I believe. The Latin Service ended & the Greek began. It was a gorgeous ceremony- multitudes of priests & 3 Bishops- one representing (215/968 "Jerusalem") the Patriarch who is at Constantinople. It was a pretty sight- for it being their Palm Sunday they had Palms & Flowers. Many of their vestments very handsome. The Vice Patriarch carried 3 candles joined in one & then blessed the people with 2 like this, crossing them & again with a small cross most beautifully set with jewels. There was a great crowd & much noise & confusion. It was a painful scene altogether & I was glad to go away to our own quiet service. The Bishop preached- a long sermon & the H. Communion was administered in English, German, Arabic & I think Hebrew- according to the nation of each communicant. After lucheon I packed up & I went down to the Church- where I fell in with Hayward & Rycroft. We found nothing going on. So we perambulated the various parts of the Church & saw the sword of Godfrey de Bouillon- etc. Then a Latin procession began- going round to all the parts of the Church marking the various part of our LORD's passion- by going with a Service in the Latin Chapel. Hayward & I went with them. The services during the procession & before each altar consisted of the old Gregorian Hymns & mostly a prayer (216/969 "Jerusalem" Start for Damascus) at the end of each wh(ich) I was familiar with. I went afterwards to the Convent of S. Salvador to get 2 of the blessed Rosaries for Emily & Fred, but failed. Then Hayward & I walked up & down in front of the Armenian Convent- having pleasant talk. Dinner- Journal- finished packing & to bed. Tomorrow en route for Damascus.

Easter Monday. April 17.

I awoke at 5 & got up- as our baggage was to be off very early. It ended in not getting away till ½ past 9. I sent out for a man to come & cut my hair- he came & a young fellow of the house with him. I explained that I only wanted just the ends cutting & made it plain as I thought- but in a minute I found the fellow literally shaving me & all the right side of my head was cropped so short that I could hardly take up a bit of hair anywhere. The man had brought his basin & razors & I have no doubt that unless stopped, in about a minute he would have whipped off all my hair & lathered & shaved me into Musulman. I went out to try & get a couple of Rosaries for Emily & Fred at the Latin Convent & was told the monks were chanting mass & I must come in an hour. So I went home to breakfast. All the party is broken up. (217/970 "Jerusalem to Ain es Yebrud") The Heneages were gone this morning- for Jaffa & {Pagde}[37] too. We go- The Rogerses- Eames- Rudeswaldt (sic) Ward & Yateman etc. Hayward came at breakfast time to say goodbye again. After breakfast I went back to the Convent & managed to make my wants known- but the monk who keeps "Magasin" was out. The superior directed a boy to take me to the bazaar & he led the way to a man who makes them- or had some to sell. I brought 2 very bad ones- but in passing home along the street I came to a shop where were some very good ones & brought 2 or 3 more of olive wood & of Mecca seeds etc. I had some little discussion with the landlord- but ended in paying his charges & then we went to get our horses, which were waiting near the Tower of Hippicus. The Rogerses started with us, but we left them behind.

---

[37] According to Andrian (2011: 2, 696) a Mr Page rented him a Nile boat in December 1853.

It was a most lovely day after yesterday's rain- a soft fresh air, with a warm sun- a regular "growing" day. I was sorry to leave Jerusalem - but glad to feel myself once more moving towards Home (moving or standing still, let me remember, I am ever moving on towards the end of my journey. God grant it may end in HOME, not in the place of the outcasts). We passed out by the Jaffa Gate, round the NE Corner & turned off into the Damascus road & up the hill of Scopus. Here we turned & had our last (218/971 "Jerusalem to Ain es Yebrud") look at the Desolate City. I have been nowhere more saddened with Church matters than I have been at Jerusalem. It is a concentrated picture of the world. The Church distracted & divided, one part of our LORD's Body (if they be such) fighting against & hating another. Fancy the scene in the Church yesterday morning- Latin, Greek, Armenian, Coptic & Abysinnian (sic) Franciscans, all at once & order only kept by the presence of a body of Muselmen (sic) & a strong guard of Turkish soliders, who hate & despise the whole & how should they do otherside with such scenes before {them} How can they bear to think of the words, "Love your enemies, do good to them that hate you, & pray for them that despitefully use & entreat you" or the command to be {"One"} But with all its' present degradation, it is still the place where our Saviour trod & taught & suffered & died. "…..[blank]……" "dear is every stone of hers'; for thou wast surely here. There is a spot within this sacred dale. That felt thee kneeling- touched Thy prostrate brow" and I was sorry to be going away. We rode quietly along the wretched apology for a road. We passed Shafat, a small Village to the left. Then a hill with ruins on the right. Behind this at some distance lies [****, ***] [smudged ink] (219/972 "Jerusalem to Ain es Yebrud") which we rode near to the other day. On our left again lay Haninah & over this Nebi Samwil- the place to wh(ich) we rode. Some say "Ramah", where Samuel was buried- others "Mizpah"- below this again was El Jib , "Gibeon". After this we passed close under a hill with Er Ram on it to our right. A half ruined place- wh(ich) I also rode to the other day. "Ramah"- Mukhmas, "Michmas" & Geba , "Gibeah", lay to the East of this & then we came to a hill on the top of which are the ruins of Atārah, "Anathoth" & then El "Birah", "Buroth". All these places are mentioned in the book of Joshua as cities of Benjamin. Josh ix.17- xviii.25. 2 Sam iv. 2. Egrn. ii.25. Neh. viii. 27. The last, Buroth, is believed to be the place a day's journey from Jerusalem, where our LORD was missed by his parents returning from Jerusalem. Beitin, or "Bethel" was not far off. Here Jacob has the vision for the Ladder from Earth to Heaven & Jacob built the pillar. There are some ruins. Jufna or "Gufna" or Ophni, [*** (?)]. Josh xviii. 24. I did not remark several of these places- for Mr. {Hayes} came up & insisted on discussing the Greek & Roman Churches & their faults. We descended a steep rocky hill & came upon a very pretty fertile scene- a valley with vines & figs & olives & almonds with cornfields & the hills terraced to the top & a small village on a hill to the left & just beyond (220/973 "Ain es Yebrud to Nablous or Shechem") this to a large well & our Tents pitched near & <u>2 girls or women drawing water for our horses</u>. Kennedy & I went for a walk over the hill to the East- flowers & freshness & all spring- most delicious. I shall enjoy all this in England soon. Dinner & reading & journal & our tent pegs turn up by a rampaging beast of a horse that broke his tether & nearly let our whole Tent down & then to bed & so ends Easter Monday.

<u>Easter Tuesday. April 18 [blank] to Nabulus (sic). Neapolis.</u>

I had barely got into bed last night when a brute of a horse again came stumbling across the Tent lines, kicking at another horse & let down all that side of the Tent. In starting up in his bed Mr. Kennedy tumbled out, bed & all, upon the ground. It was sometime before our misfortunes were remedied- but I stayed quietly in bed. I did not sleep very comfortably, perhaps in consequence of this disturbance & at ½ past 4 or rather later we were called. We had breakfast & walked up the Valley & back to mount our horses. After passing up to the top of the little Valley we were in, with a village on each side, containing inhabitants of very bad character & who last year were fighting against each other, John says, when he passed along here, we turned to the left & began to (221/974 to Nablous/ Shechem) descend into a Valley, the sides of which were covered with olives, vines, figs & corn. It was well cultivated. The morning was cold, but very fine. At the bottom at the further end, we passed the ruins of a church & the rocks opposite had caves in them & a short distance beyond a spring called 'Ain el Haramiyan' or the fount of the 'robbers'. We then passed "Sinjil". The whole country is now very different to what we saw about Jerusalem- ie. much more cultivated- the character of the ground the same- terraced & stoney hills etc. We went down a steep hill & came at the bottom to a ruined Khan, the Khan Lubban. Lubban, the ancient "Lebonah" was near. Here were a fine well & tank of water. We rode on through a rich valley, which was busy with oxen & ploughs. All along today we have seen great signs of industry. Everywhere ploughing & weeding were going on. Near this 'Shiloh' must have stood. In Judges XXI.19 it is said to be "on the north side of Bethel, on the East side of the highway that goeth up from Bethel to Shechem & on the South of Lebonah". I did not know of this till getting here & reading Wilson in the evening, or I would have gone aside to see the place where the Ark so long rested. Nothing remains but ruins. "But go ye now to my place which was in (222/975 to Nablous) Shiloh, where I set my Name at the first & see what I did to it for the wickedness of my people Israel". Jer. vii. 12. We passed on through a pretty valley, of rich soil in which the Fellahs were busily at plough- ascended a hill at the other end & descended to a well at the bottom nearly on the otherside: where we had luncheon. Mt. Hermon was before us far away completely covered with snow. It looked well- at our feet & running for some miles before us lay the Valley Makdnah- bright & green with crops just bursting into ear & fields now ploughing for Indian Corn. To the left was a range of hills running northwards & ending in Mt. Gerizim- on their side were "Hawurah" & other villages. On our right was "Awartah" where Eleazar is buried. Josh xxiv 33. A little further on, north, "Rejib". All this plain was very rich. On coming under the end hill, which is Gerizim, the Valley opened up to the East, on our right hand. This is most likely the plain of "Moreh", to wh(ich) Abraham came when he left Haran "And Abram passed through the land unto the place of Sichem unto the plain of Moreh" (Gen: xii. 9). On a hill to the N.E. lies "Salim". This is the "Shalem a city of Sychem [= error for Canaan?]" to which Jacob came on his return from Padan Aram, "& pitched his tent before the city" (Gen. xxxiii. 18). The South (223/976 "Nablous") was Azmut, "Azmerath" & further in the east part of the Valley Ed. Deir & Bet dejan. We turned aside from our road to visit Jacob's Well where on our LORD sat & talked with the woman of Samaria. It was covered up with stones. A man came & rolled away a great stone from the top & pointed to me to go down, wh(ich) I accordingly did, through a very narrow hole- landing on my feet below. I lowered myself down & found myself in a small arched chamber & at the east end from the well- the top of the well has evidently been removed in some way & a great stone has fallen in, resting across the well. This has perfectly prevented the earth from falling in. I tried to catch sight of the water but could not. I dropped a stone in & from the time it took before it splashed in the water, it is probably as deep as they say- viz. 75 feet. It is this one circumstance, agreeing with the account in S. John iv. 11- "the well is deep" together with the universally believed tradition, that makes one believe it is Jacob's Well. About are ruins. A church is said to have been built over it- 2 or 3 broken pillars still stand. A few hundred yards further N. is Joseph's Tomb. Here he was buried after the Israelites reached Canaan (Josh xxiv. 37). An open mosque now stands over it. We rode on up the valley which branches to the West, between Mt. Ebal (north) & Gerizim (south) (224/977 "Nablous") with a delicious stream running in an aqueduct from a spring which we came to. Further on the valley was full of Olive trees, many of a great age. We came to the Gate of Nablous & passed through the town. We rode through the Bazaar, much better than that of Jerusalem. Water running along the middle of the narrow street- making the place quite clean comparatively. On our left we passed a very fine doorway of pointed architecture & a large building, once a Christian Church, now a mosk (sic) & out at a gate the other end of the Town- onto a green space covered with Olive Trees- where we dismounted to wait for our camels. I was very tired- but I determined to go up Gerizim. So I got a boy & started. We turned up to the South & began to ascend a valley, rich with gardens & fruit trees- as we rose we had a fine view of Nablous. It is a pretty place. As we were going along we met 2 men- the younger of whom addressed me in English & said he was a Christian & native of the place. He had been at school at Jerusalem & was now schoolmaster with the other man here. He offered to go with me & I was glad to have someone who could speak some English. On the top of the hill we came to the place where the Samaritans of whom there were still about (225/978 "Nablous") 40 in Nablous, yearly sacrifice at their Passover. The place of sacrifice was a sort of well, perhaps 6 ft. deep, of rough stones black with fire. After eating the lambs, the bones are burnt in a narrow trough of 10 (or 12 originally) stones built in the ground near. Last Monday week was their Passover. We went on to a higher top some way further on. Here were great remains of some prior buildings- with large stones & walls in places 9 f(ee)t at least though. This was a fortress Robinson says- in it are the remains of the Temple. At one place I was shown the place where are buried it is said the 12 stones brought by the Israelites from the Jordan, where they crossed. I went up to the top of the small mosque & had a beautiful view all round. The Mediterranean to the West- N. of me lay Mt. Ebal- East- Salim & the plains of Moreh etc. I came down & went in by a gate at the S.W. into the Town & so to the Samaritan Synagogue. It was the last day of their Passover & they were assembled reading the Law. They all wear <u>red</u> turbans- except their high priest, who wears a white one. They received me very civilly & I took off my shoes & was admitted to see the valuable Samaritan {manuscript} of the Pentateuch. It was produced from behind a curtain. It is kept in (226/979 "Nablous to Jenin") a curious brass case, embossed. This was opened & the roll partly unrolled. It is put on a velvet cover embrodied with gold. It is genuine & above 3000 years old. I c(oul)d not talk to them. I hurried to the School room to please the young fellow who went with me & then back to find the Tents pitched- after a hard walk of 3 hours. Dinner- reading & journal- to bed late. Wind got up & some rain.

<u>Wednesday April 19</u>

After eating a mouthful of breakfast during wh(ich) some heavy drops of rain & dark cloudy sky made us anticipate a wet day, I went out & showed Mr Rudeswald (sic) & the spot from wh(ich) I was so much struck with the Town yesterday even(in)g. When I got back I found Giovanni waiting to start. We up (sic) down the Valley, westward of Nablous, formed by the continuation in that direction of the hills on each side, of w(hic)h Gerizim faces the [*** (?)] on the S. side & Ebal on the north. Streams of water ran down the Valley- which was full of gardens & trees. The green grass & trees were very pleasant. We have evidently got into a richer country since passing Khan Lubban. There were villages on each side. We struck up the hill to the right & then cut off a corner. From the top we descended to another Valley & got sight of <u>Samaria</u> (227/980 "Nablous to Jenin") standing on an isolated hill opposite the end of the Valley. We passed down among thick & {luxuriant} trees, turned a little to the right & passing what I thought were the ruins of the gate, wound out into "Seba(s)tiyah"- the ancient Sebaste, built by Herod the Great in place of Samaria. We came first to the remains of a very solid old church- that of S. John Baptist. The inhabitants use it as a quarry. The floor of the Church is covered & filled up with rubbish etc. except towards the West end, where the Musulmen have cleared part of it as a Mosque & here stands the Tomb of S. John the Baptist- (as they say) & other modern Tombs. I got on to the vaulting over the N. aisle & stepped the church roughly- I make it about 150 x 70 feet. There are 4 small pointed windows very high in the wall, on the S. side & 4 corresponding on the North. The wall was about 2 foot wider than I c(oul)d stretch. The Church had consisted of a nave & side aisles extending apparently to the East End. The centre aisle had been apsidal- but all the East end had been ruined within the last year or two. We passed on higher up the hill on w(hic)h Sebastiyah stands & came to a great number of monolithic pillars- standing about 8 foot apart, in 2 long rows. The pillars have no capitals remaining. We (228/981 "Nablous to Jenin") c(oul)d not count them- but their number is variously stated at from 70 to above a hundred. It seem to terminate in a gate- what looks like a gate is probably the ruin. The situation is magnificent- a detached hill, separated from the other hills about by a deep valley. We rode on down the hill & crossed a valley & mounted a steep hill on the S. side. Multitudes of Olive trees & much cultivation. We passed several villages. At the top of one of the hills or passes- we saw the Mediterranean & then descending again amongst a great many olives, we sat down to luncheon. We found Mrs Eames & Mr Rudeswaldt (sic), who had come a shorter way & while resting after luncheon, Mr Eames, Fish, & Adam came up. They had stayed behind to ascend Gerizim. They too had come the same shorter way & so had missed Samaria. We came after this into a fine broad Valley, a perfectly flat plain. The soil was dark in place of the red soil we have had all the way. The native Arabs were very busy ploughing. On the left was a Village crowning a steep hill. It's name is Sanur. It was destroyed by the Turkish forces who were obliged to bring some Albanians against it. It ended by being completely destroyed. It is now built up again very strongly. Not far from it (229/982 "Nablous to Jenin") I saw a large shot[38] lying on the ground. We tried to pass up the middle of the plain, but were stopped by a bog & after Giovanni who was leading the way had extricated himself & horse covered with mud, we went back & made the circuit of the plain. Passed over a very nasty ridge, through a large village, whose name I can't recall & along some very rich country on the other side. We came to a narrow valley- green with grass a foot high & one mass of flowers. It was like spring in a field in dear old England waiting to be cut for hay. I have not seen such green grass since I left England. About 5 o'clock we came out of this upon "Jenin" & found our Tents pitched & dinner almost ready. Most people call Jenin "Engennim". It stands near the edge of the Plain of Esdraelon. I walked out & looked over the rich fertile plain- across which I could see Carmel running down into the sea. I took a stroll into the Town & found a wretched Bazaar & ill looking people. They bear a bad character. Giovanni maintains that this is Jezreel. I believe we are to pass through that tomorrow. Zerain is thought to be it. Tomorrow we reach Nazareth. We have learnt today that the Holy Land is indeed a (230/983 "Jenin to Nazareth") rich country. Excepting timber & woodland it seems to want nothing. Contrary to first appearances it has turned out a lovely, cloudless day. Warm but fresh. I have enjoyed every minute of it. And now to bed.

---

[38] Presumably a cannonball.

Thursday. April 20.

We had breakfast & while the things were being packed I sauntered out a little way to see the Plain of Esdraelon by the morning's light. I looked over a wide expanse of cultivated land & rich vegetation to Carmel in the far distance where it ran down to the sea. We started off, uncertain who of the party was coming the same way. We passed outside the Town, under an aqueduct & soon upon the plain, which stretched away before us for miles. I c(oul)d never have realised the richness & bounty of the Promised Land if I had not come up through the country from Jerusalem. As far as we could see on all sides all was rich- a great contrast to the country round & south of Jerusalem. A still more charming change to men who had been 33 days in the Desert. In something under an hour we came to a Village standing on a hill in the plain. We determined to go up to the Village & so see the rest of the plain. Passing round outside the Village we stumbled upon an old sarcophagus. Open. The 2 ends & outside were carved. The other side was (231/984 "Jenin to Nazareth") plain or worn so. An Arab came up to us from the Village & pointed out the next Village as "Zerain", the ancient Jezreel. He also said there were more sarcophagi. And as we rode away he called out to us to beware of the Arabs. We rode on again- the soil dark- & the remains of last years' cotton plants in the ground. On our right on a rising ground we passed a Village which an Arab at plough told us was "Zéllami". Further on on the right again was "El Mazraah"- w(hic)h possibly may have been Meroz. About ½ past 8 we came to "Zerain" also standing on a hill in the plain. This every one seems agreed is Jezreel. We fancied the arrival of Jebu & the "portion of Nabeth". But Jebu must have come from the East side- probably up the Valley, or continuation of the plain, w(hic)h we looked down upon from Jezreel. Down this Valley the waters flow eastward to Jordan. While from near us the "Kishon" flows Westward to the "Great Sea" or Mediterranean. The plain before us was a mass of Vegetation, flowers rather than grass, up nearly to the horses knees & dotted all over with Bedouin Tents & flocks. It was a beautiful scene & on seeing it, one culd realize how peaceful & happy a life God had in store for the descendants of Abraham when HE promised the land to his sons & bought up (232/985 "Jenin to Nazareth") the Israelites from Egypt. The air was loaded with the scent of the flowers. Thousands of wild bees were at work (there are no Tame ones). Large flocks of sheep & goats & herds of cattle & donkeys, as well as Camels & horses, were scattered about. And while the Bedouins had their flocks & wandered about pitching their tents where the pasturage was best, the Arabs of the Villages were busy ploughing & sowing. It was delicious. We found several broken sarcophagi & what we thought looked like the remains of a road that been cut or worn in the rocks. Chariots were used all over this plain country. Jehu, Ahaziah, Joram, rode in their chariots. As we went down the hill to the plain again we passed a well at which the women were drawing their water. This is apparently an old well. It may even the very one mentioned in the Bible. Before us across the plain lay a hill called Djebel El Dahi- this is most probably the "Little Hill of Hermon" as distinct from Hermon in the Lebanon. Away, on over East was another range of hills thought, from the fact of a Village called "Gelbon" being situate on there, to be the Mountains of Gilboa. On the slope of Djebel El Dahi lay a village, "Sulam"- the ancient Shunem, where the Shunamitesh women used to entertain Elisha. And when (233/986 "Jenin to Nazareth") the son given her for her kindness to the prophet fell sick as he was with his father & the reapers in the field, (probably this very plain), she passed over this plain on her way to Carmel. We rode along over the plain- passing the flocks &Tents of the Bedouin- whose dogs were most noisy- but I hear their bark is worse than their bite. Giovanni took a right hand for a left hand path & we did not find out our mistake till we were near upon "Sulam": so we rode on & passed through Shunem. It is but a miserable place now- a Village of mud huts. There were olive trees & gardens, hedged in by the large caches. We kept away to the left then, having "Hermon" on our r(igh)t covered with cattle & asses- "Cattle upon a thousand hills". Such is the general characteristic of the country we have passed through- hills & valleys. The Valleys for Corn- the hills for cattle. And during the last day or two we have learnt that the Valley may in truth "stand thick with corn". To our left in the plain on small hills stood 2 Villages El Afulah & El Fulah (or Furah?). The latter is the "Village of the bean", which Napoleon visited. We passed over the end of the hill Hermon down into another large branch of the plain. On our right under the Hill of Hermon lay Nain & Endor & opposite the hills among & on which Nazareth stands, but (234/987 "Jenin to Nazareth") we could not see it yet. At the foot of the Nazareth range a little to our right lay "Debūrieh", "Daberath" & "Iksal", "Chisloth Tabor", the latter under Mt. Tabor & among the hills more to the left stood "Yafa", "Japhia", all 3 places mentioned in Josh xix. 12- as on the boundary between Zebulun- & Issachar. Mt Tabor is one of the range a high round hill, rising 1000 feet above the plain, 1700 above the Sea. This is thought to be the Mount of the "Transfiguration". An hour's riding took us across to the foot of the Nazareth hills. Here we rested our horses for an hour & had luncheon & then walked on up the hill by a very steep rocky way. On the top Nazareth burst upon us. It lies in a sort of basin, or Valley without an outlet at the further end. The land is cultivated in patches & Olives & figs grow about the Town. We rode in past the Latin Convent- through the narrow dirty streets, if such they can be called & out upon on a delicious bit of flowery grass on the other side, the N.W. where we camped among some trees. In a few minutes time, after lying on the grass & referring to Wilson's book, we took an Arab (a Christian) & went up the hill at the back of our Tents & the Town. From here we had a glorious view- overlooking the Plains of Esdraelon on the S.E & S. & the country towards the Mediterranean as well as to the N. over to the Lebanon & the sea of Gallilee. (235/988 "Jenin to Nazareth")

Many a time must our LORD have been on this hill. Here in Nazareth HE condescended to pass childhood & youth & manhood till HE was 30 years of age- waiting to do his work- away from Heaven. Nazareth struck me more than any place I have been (236/989 "Jenin to Nazareth") at. We went down to the Tent, as John was very anxious to take us to the Greek Church. It has a handsome skreen (sic). It has a well in it called the Well of the B. Virgin- in a grotto, when she was supposed to live. Thence we went to the Latin Convent, where we also saw 2 grottos under the Church. One of which was the Virgin's Bedroom & in the other she was praying when the Angel announced to her that she should be the Mother of our LORD. Why all these things should always be in grottoes I can't imagine. There is nothing to see in the Church. We were next taken to a small chapel belonging to the Maronites, who were at service. This was the school in which our LORD was taught; or the synagogue from which the Jews thrust HIM out & were about to cast HIM head long down a hill- when He passed out of sight. Next to the house & Carpenter's shop of Joseph. But it was locked & we did not go inside. I saw that it had been turned into a chapel, with a Tawdry altar, through the window. Home to dinner. Wrote journal- read & to bed.

Friday. April 21.

I had a very uncomfortable night- continually disturbed- with diarrhoea. The horses too were very troublesome- getting loose & running foul of the Tent lines. In the middle of the night one of our horses got loose & carried away 3 or 4 of (237/990 "Nazareth to Tiberias") the Tent lines. I shouted & jumped out of bed & held up the Tent post, fearing all would come

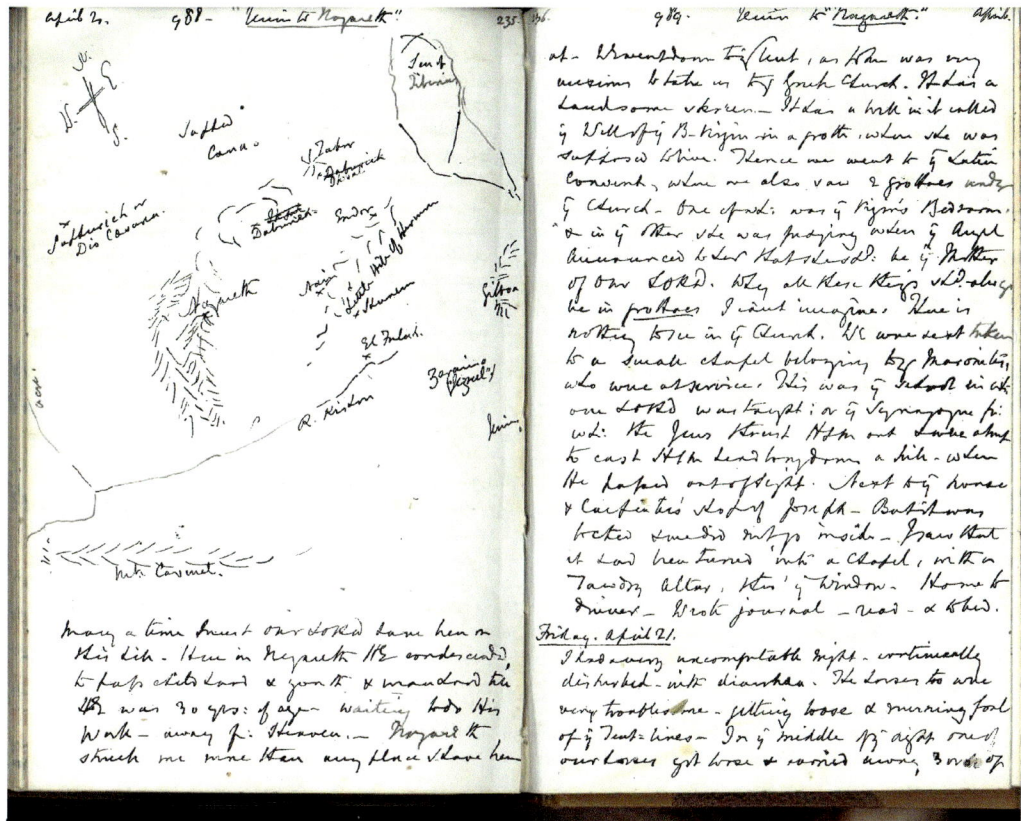

Figure 14.4: Sketch of the Convent at Nazareth locality.

down together. The whole camp was roused & we were saved from an overthrow. I found that our Tent was pitched near a burial ground, belonging I think to the Christians of the Greek Church. At all counts it was a place of burial & just after our breakfast, while the packing was going on, several women came out to the graves & sat down & began to wail a lament. They came up, be it rembered (sic), talking & laughing, but on getting to the grave, each women begain swaying her head to & fro & waving her arms & wailing & then prostrating herself & touching the grave with her forehead, each sat down & continued, waving her hands & singing in a sort of wailing, doleful way, breaking out into a loud wail & ending in sobs. It was a strange scene. I went into the Greek Church, till summoned back by a shout. The priest was sitting near the door reading. The lamps were being lighted & preparations making (sic) for the usual morning service. We rode up the opposite or east side of the Valley, mounting the hill on that side. At the top I stopped to take a last look at Nazareth. It lay in the morning sun, looking very pretty & peaceful. It has been {nicer} to me than any place I have seen. I rode on after the others, loosing (sic) sight of Nazareth as soon as I had turned, dropping down a steep hill. We were (238/991 "Nazareth to Tiberias") on the road to Cana- where our LORD turned the water into wine. He must have passed along this very path then at all events. Most likely all this country & hills about here had been trodden by Him. We had a pleasant ride in a fresh light morning- coming first to a small Village called Erana. We found a large sarcophagus outside the Village, near a fountain- used probably as a drinking trough. Leaving this Village we passed on & arrived at 9.30 at Cana- 1 hour & 10 minutes after leaving Nazareth. We stopped at an old well outside- which bore marks of great antiquity. The women of Cana were washing here & filling their large pitchers. From this well perhaps the water had been drawn which our LORD changed at the Wedding Feast. We did not into the Village- for they (are) all alike filthy & squalid. We turned to the r(igh)t & passed over a long sweeping hill. We came to a Village among the hills called "Er Meshed" an Arab Village. It was attacked & plundered & ruined by Bedouin a few years ago. It is celebrated for figs & fruit gen(eral)ly. MtTabor now stood before us [*** (?)]. We began to descend. The grass & flowers were high up our horses' legs & [*** (?)] & other [*** (?)] abounded. We got to the bottom of the Valley & began to ascend. The path so very steep, with large slippery pieces of rock. The top is a Table land, about ¼ mile in (239/992 "Nazareth to Tiberias") circumference. There are a great quantity of ruins & amongst others our guide led us to 2 small ruined chapels. I <u>believe</u> said to be built in consequence of the expression of the suggestion- "Let us build here 3 Tabernacles" etc. The ruins were very extensive & massive. The stones were large & like the Temple at Jerusalem the edges are bevelled. Whether this denotes work of the time of the Jews I don't know. We sat down & learnt the names of all that we could see. To the N.E. lay <u>Kafere</u>- {Kennaer}- E. "Aulan" or "Ulama". Kapher {Numssah} Kawkab al Hawah- etc: South. Eulen- Nain- Hermon etc. & the great plain of Esdraelon- S & west. The sun shone on the Mediterranean on one side & the Lake of Galilee on the other. We could trace the Valley of the Jordan & on the hills beyond saw a fine plain, apparently almost level but having one or two elevated spots on it. A heavy thunderstorm was laying on the hills to the North & presently the rain & hail came down. We put on our coats etc. I started to ride back down the hill- but we had to get off & lead them. The hill is covered with small but old timber. We came down into a Valley, more like a park than anything- so much so that one continually felt ½ inclined to ask where the house was. (240/993 "Nazareth to Tiberias") I can't discuss the arguments for & con, as to Tabor being the Mt. Of Transfiguration. I think that it probably is not. After riding sometime in this Valley- we came out on the plain again still part of the Plain of Esdraelon. It was less cultivated & consequently did not give one so much pleasure. We passed a ruined Khan called Es Suk- with a ruined fortress opposite to it, with round towers. This was a great market or fair & indeed a fair is held there still by the Arabs. We came to a Village- passed it & still ploughed along the great plain. After a long ride we got sight of the Sea of Tiberias & about 5.20 in the evening rode down to the Town of Tiberias on the waters' edge. Our former companions we found all pitched. Before dinner I went & had a bathe- just where I undressed I found 2 bushes of Rhododendron in bloom. I think they are Oleanders. Keble mentions their growing here in Xtian Year in a [*** (?)]. I am not much struck with the scenery round the lake. The hills are too tame & their (sic) is not a bit of Timber- nothing to beat the monstrous green of spring & parched up land in Winter. The less [**** (?)] are more than the perfect. Dinner. Discussion with John about route. Journal, reading & bed. (241/994 "Tiberias to W. Matabeh")

<u>Saturday April 22.</u>

We were not called quite so soon as usual- it was rather past 5. However, I jumped up & ran down the lake a short distance & had a bathe- dressed- had breakfast & we walked down the lake to see some old Baths both Roman & later, with a hot spring. We had to walk more than a mile & saw after all only the modern bath, built by Mehemet Ali. The old Roman were just beyond. I saw the ruin, but trusted to Mr Kennedy. When we got back we found Tents & baggage all gone & the last party just starting. They are gone back to Nazareth & so to

Carmel- Tyre etc. We walked through the Town- a wretched place- the only decent thing or person we saw were some decent looking & clean women- German Jewesses: dressed almost in European clothes. Such a contrast to the dirty women we generally see. There is nothing to see in Town- it took but some 5 minutes (or 10 at the most) to walk through. Outside at the north end we passed a great many remains of ruins of which we could make nothing. Dr Wilson says the ruins of S. Peter's Church are here. We could not distinguish anything like a Church. We had had great discussion about leaving the beaten track in order to visit- the 2 places which have been thought by different people to have been the site of Capernaum. We had however insisted on going. Our Dragoman (242/995 "Tiberias to W. Matabeh") was not pleased & had thrown all sorts of difficulties in the way. We wished him to take a guide- but he said he would get one on the way. Kennedy & I rode on wishing to make the best of the day, but were detained by his sauntering behind with the muleteers. The first place we came to was "Mijdal". This has been thought to be Magdala & perhaps Migdal of the Old Testaments (Josh. [19.38]). It is now only a small collection of most miserable mud huts. It stands at the southern extremity of a small plain which is called by Wilson & Robinson etc. the plain of Gennesaret. The Arabs have another name for it- which I asked, but could not catch. If it is Magdala it is the place where our LORD & the Apostles landed after feeding the 5000 on the N.E. side of the Lake. Not far from here a Valley runs up from the plain in a S.W. direction, called the Wadi El Hamam, in which or at the head of which is a very remarkable fort excavated & cut out of the rock called the "Kalat Ibn Maan". This we could not go to see. At the Western side of the plain is a village "Shushek". We kept along the water's edge. It was very pleasant. The wind which was disturbing the water yesterday had died away in the evening & the water was almost perfectly smooth. The whole margin of the Lake was bordered with Rhododendrons, or should I say <u>Oleanders</u>. These were all in flower. Some birds I have not seen before were all along the lake. The sky was (243/996 "Tiberias to W. Matabeh") cloudless & above all we were on the shore of that Lake w(hic)h is associated so much with the earthly life of our LORD. We now left the Damascus Track & our object was to see a stream, the "Ain et Tin", marking the remains of an old Town, all travellers have agreed & some ruins called "Tell Husn". Where the mountains again opposite the Lake at the North end of the plain is a ruined Khan, called Khan Minyeh & the first {round} spring & stream are near this. South of the stream is a mound evidently a ruin- traces of masonry being visible. This has been reputed to be the site of Capernaum. The great argument being that it is on the W. side of the Lake & also that Josephus mentions a spring & stream which watered the plain & which was called in his time "Capernaum". This seemed a reasonable ground for calling the ruin Capernaum. However this stream does <u>not</u> water the plain. It is at one corner & runs into the sea ten minutes after leaving its basin. We saw the mound & mill & the old Khan & passed on. We climbed the hill partly, which here runs down to the Lake & found a way cut in the rock, about 4 feet wide & 3 deep & paved partly, which led round the headland. On the shore beyond this we came to a mill turned by a stream called the Ain Tabiyhah. The water is brackish- not far from the mill is a fountain enclosed (244/997 "Tiberias to W. Matabeh") in a circular wall. About ½ an hour beyond this we came to some ruins. Wilson says they were an hour beyond "Tabiyhah" & so they were for pressing forward. However I persuaded Kennedy to stop & we found first a bit of old masonry. Here I left him & pushed my way among the high walls, in many places over my head & over stones & remains down towards the Lake making for a wall [*** (?)]- then I found a large piece of building & near it, parts of marble pillars & carved stone etc. I have no doubt that remains might be found over a large space from the appearance of the ground. But the weeds & vegetation generally were so rank that I (could) not search further. Kennedy pushed his way down too & then got sick of it & said 'let us get back to the horses'. So as it was getting late & John was very impatient about coming here at all, we pushed our way back some ½ mile or more & got our horses. Dr Wilson gets over the difficulty about the name in the following way "Capernaum is a compound and, "Kaphar Nahum" meaning the Village of Consolation" according to Origen or "of Nahum". On the supposition "that the word "Kaphar", the original form of Caper, has been exchanged for "Tell", a mound, on the place becoming a ruin, we were disposed to agree with those who think that Husn is (245/998 "Tiberias to W. Matabeh") a contraction for "Naham". There is a strong circumstance in its form. Josephus relates there he hurt his wrist by an accident close by the entrance of the Jordan into the Lake & he was carried to a Village called [Cepharnome].[39] They would naturally take him to the nearest place & this is not an hour's distance from the Jordan & thus nearer than the Khan Minyeh. Again, before the miracle of feeding the 5000- it is said that our LORD crossed from Capernaum to go into a desert place & the "multitudes seeing it, ran apart thither & out went them".[40] This w(oul)d be easier for "Tell Husn", than for "Khan Minyeh". At all events having seen the 2 I feel pretty sure that I have seen the site of Capernaum. We retraced our steps to Khan Minyeh. John being afraid to strike over the hills in the direction of this place. We took a turn for Khan Minyeh which after some nasty climbing for the horses brought us up into the Damascus track. We had some luncheon & rode on. We are now on high ground again. Below us on our right we still saw the Lake. The ground was covered with grass & flowers & black basaltic stone amongst them. No signs of inhabitants. But Bedouin Tents in the distance. Away to the right we had a fine view of the country across Jordan. A high Tableland, which appeared to be one vast green plain. (246/999 "Tiberias to W. Matabeh") Be it remembered however that this vegetation only lasts for a month or two & then all is burnt up into sand for the rest of the year. It was hard to keep our track- it was so obscure- we passed the ruined "Khan Jubb Yusuf", where Joseph is wrongly supposed to have been placed in the pit. Though it is just the Country for the patriarch to have sent his flocks to. Here or a little beyond this we lost sight of the Lake. Before and below us lay another large plain covered with vegetation & to the right many Arab tents with cattle. In it lay the "Waters of Merom". Beyond, above the Eastern hills, stood "Hermon", 10000 feet high- a considerable part of it covered with snow. We wandered in this plain for sometime looking for our Tents. Some Arabs told us at last where the Ain es Mellarah was & we reached our Tents exactly at 6. They were pitched by the side of the Ain es Mellarah- a clear, rapid stream & I at once got a towel & clean shirt & went up the stream till I could find a place deep enough to bathe, or at least wash in- a great luxury, after the heat & fatigue of scrambling among the ruins of Capernaum. Dinner- journal & bed.

Sunday. April 23. 1st after Easter.

We left our encampment among the long grass by the stream with our baggage mules this (247/1000 "W. Mullalah to Banias") morning as we were in no hurry & meant to saunter quietly along with them. It must have been ¼ to 7. The Arabs about on this part of the plain are not Bedouins, but Turkman- Persians originally they came westward & settled on the Jordan & have quite forgotten their own language & have adopted Arab tongue as well as dress. A number of "cherks" John called them, Turkish Gens d'Armes [= gendarmes], passed

---

[39] Josephus *Life* 72.399.
[40] Mark 6:33 *And the people saw them departing, and many knew him, and ran afoot thither out of all cities, and outwent them, and came together unto him.*

us. They told Giovanni that we might quite safely have taken the route eastward of the Lake of Tiberias & Jordan. I discussed this with Giovanni & he had thought it unsafe. I am not sorry to have come the way we have. We have now passed through the land, from the South to the North. Half an hour after starting we came to a nice stream, rising just above where we forded it & turning a mill at once. This is the great Ain Malláhad. It becomes quite a river at once & flows down to Lake Merom. The plain or Valley had taken a slight turn to the left or W. on our rounding Djebel Malláhad. We kept along on the left hand side under the hills. After crossing Ain Malláhad the middle of the plain was all one great swamp & here they use buffalos instead of oxen. The Arabs too make cottages of the reeds & bulrushes, instead of using Tents- though often stretching a Tent or goats hair covering, above the roof. There (248/1001 "W. Mullalah to Banias") were regular Villages of these huts all down the Valley. I stopped at one & got some sour milk to taste. I had not tasted it before. You may get at most tents plain milk just from the cow- or sour milk alone which must be eaten with a spoon, or the sour milk beaten up in water- the usual drink all among the Arabs. The milk is boiled & left to cook- when it becomes a thick sort of curd. In every way it is good. We got to the end of the Valley, without seeing any stone village or old place. According to Dr. Wilson we must have passed "Kedes"- or it must have been near us. He thinks it is "Kadesh Naphtah". We tried to follow a path across towards the East side of the plain leading to Banias. But the ground was so wet so we could not go on. So we had to keep the foot of the hills all round. As we sat down on a stream to luncheon my eye fell on some Towers & old walls up on the hills which we were just leaving. They are the remains of "Kalat el Hûnin". This may have been "Hazor"- one of the cities belonging to Naphtah. Abil el Kamah he also says was somewhere near- perhaps Abel Beth Maachah. We had a very nasty road or no road almost all the rest of the way. Stony & rough & also very boggy. We were stopped for a long time by a chase after a horse belonging to a Damascene who has (249/1002 "W. Mullalah to Banias")

joined himself to us- it is supposed he is carrying money with him & is glad of protection. Then we were pulled up by coming to a wide river, very rapid. We had to follow it up till we came to an old bridge. I can't learn anything about it- the arches are pointed- it is paved- or has been & has apparently never had any parapets. The last part of our way was very pretty. We got on higher ground, rising up towards Banias among Oaks & Quinces & Olives etc. & got to our Tents, pitched among some ruins about 5. The great & principal source of the Jordan is here. There is a large cavern in the limestone rock. This is full of water, still & deep & apparently there is no exit for the water. However it finds its' way out through the rock in a beautiful clear stream & is a Torrent at once. On the right hand side of the cavern as you face it are 3 or 4 niches cut into the rock, with 2 inscriptions- one in Greek & the other in Latin. Herod the Great built a Temple here, after escorting Caesar to the Sea. Titus spent some time here. I have forgotten to say that Banias or Paneas is the ancient Caesarea Philippi & not far from "Dan", originally "Laish". There are numbers of bits of pillars & ruins about & part of the gateways. Above the Town on a high hill are the remains of a Castle (250/1003 "Banias") which Burkhardt gives some account of- but the history of which I don't know. I walked up to the cavern of the Source- then came back & sat reading in my Tent & before dinner went out & bathed in one of the freshest coldest stream I ever had the good fortune to meet with. I believe the stream, here though it is, comes through the rocks from the soils of Hermon. Who (?) has been before us all the day. Tomorrow we pass under his roots on our way towards Damascus.

Monday. April 24.

I rushed out of my Tent this morning to the stream close by our Tent & had a delicious bath- cold as ice. Before I was in a state to be seen an old lady came down to fill her pitcher & seemed amazed at my proceedings. But my presence did not prevent her quietly filling her jar & talking to me- not one word of wh(ich) of course could I understand- but I have no doubt she did herself good. While the mules were packing we walked through the remains of the Old Town. We found four or 5 ruined Towers- one with an Arabic inscription over the door. An artificial mote (moat) was carried all round the back, joining the 2 streams. I can't learn by the bye what the old travellers I see say, viz, that the 2 streams are called the Jor & the Dan & that the 2 uniting make the Jordan. But the 2 streams do unite. (251/1004 "Banias to Bejan") Above the Town on a high peak are the remains of a large fortification or Castle. Wilson mention it but gives no account of it nor of its' history. I suppose like ourselves he didn't fancy a 2 hour's climb. There is no Town- scarce a Village, left in Banias- mud & stone huts & a good many sheep. An old soldier of Mehement Ali came with us- he has married a wife & settled down here, tho' an Egyptian & has land. Our horses overtook us & we began to ascend the mountains & were climbing all day till 12 o'clock. Our muleteers are very uneasy. It seems that some of their people, (they are "Matawallis" of Saphed) killed one of the great Arab Tribe in this district, the "Anezehs or Anezas", sometime ago & revenge has never yet been taken- the law of blood for blood is never forgotten & retribution may follow even after generations. It was with difficulty they were persuaded to come on when they heard we were to come to Banias. The chief population about here are "Druzes". We met a woman on the hill with the high bun on her head- but I could not see how it was made as her veil covered it. The married women alone wear it. It is becoming disused. Giovanni told me that the Greek Bishop at Beyrout preached against it & the people v(er)y much left it off. I suppose he thought it conduced to idleness. At all events is would be impossible for (252/1005 "Banias to Bejan") a woman with one on to carry the usual great burden on her head. We had a fine view back- seeing "Kalat el Hûnin" & "Abil el Kamh" which Wilson suggest may be "Hazor" & "Abel Beth Maachah" very distinctly & another village which maybe "Kedes", which he suggests may be "Kedesh Naphtali". About noon we came upon a flat plain ¾ of a mile or rather more across, covered with mares & their foals & an encampment of Druzes, keeping them. I thought at first the horses were standing in water in one place- but when I came down I found it was a mass of a very small blue flower. We stopped here to have luncheon, close by 2 small streams of pure delicious water, which flow into a circular basin & disappear in a hole at the bottom- not far off, 150 yards or rather more, is a circular pool- the water of which is bad. It never increases or decreases but is always up to the edge of the pool. It is thought to be an old crater & filled & emptied by some subterranean means. Today besides water it contained an immense number of frogs. They were floating with just their heads above water & croaking loud- when as I was watching, an eagle came down & evidently meant to have some frogs for dinner. They disappeared instantly to a "frog" & the eagle after hovering over the pool some time in vain, gave it up. The snow was low on the hills- but we did not pass (253/1006 "Banias to Bejan") near any. The ground is barren again comparatively, great masses of black ({servia}) I suppose. We had some difficulty in finding our way. The muleteers had gone on. We got down into a narrow valley, in which I found 2 Tortoises, about 7 inches long each, taking a walk, (making 3 in all today) & at last got down to a Village called "Bejam". There were some goats & a grand spring, which poured out a torrent down the Valley. I felt as if I should wash myself as soon as we came to camp. Some trees very like willows grew all along the stream. We followed the stream for a mile & a half & at last came to our Tents on some rising ground, at the "{Mayarard} Bejan" that is a small hamlet on the cultivated ground belonging to the Village above- this is the explaination I got. I went down & got a delicious bathe in the cold stream. The people are all Druzes. A man had just come home with his flock. It was pleasant to see his wife come out to welcome him with 2 small children & they all sat down together on the top of the house. The elder child, just able to run, climbing about its' father & pulling his turban off. I went up & we did a good deal of conversation in the "Saláma" & "Taib" line.[41] From our Tent we have a fine view over a large plain- at one side of which Damascus, Es Sham lies. Dinner, journal & bed. (254/1007 "Bejan to Jidahda")

Tuesday. April 25.

The hot wind blew all night, as if it would tear our Tent down. However nothing happened. At day break we got up & started again with the mules about 6.30. The air was full of haze, so that we c(oul)d not see the distance at all & very hot. I have nothing to say. The road was dull & uninteresting- more like desert again. Here were plains & level places among the hills through which we passed to avoid the swampy ground below in the plain & here & there a little agriculture- but the soil seemed poor compared with what we have seen lately. Everything was covered with stones- large pebbles for the most part. We came upon 3 Druze shepherds with their flocks. One had a pipe with him. It was a reed- with 5 holes- a piece of leather was sown over it. The owner played for us. He made far more musick than I expected from the appearance of the instrument- varying it by harder or softer breathing & so getting beyond the one octave. A 2nd man, after a little persuasion danced. It was a curious performance. The country was full of streams- but when we wanted one to have luncheon by of course we could not find one & we had to travel up & down & up & down for (255/1008 "Bejan to Jidahda") nearly an hour- when we met some Druzes who pointed behind them & said "Mosa Kateur" at which we picked up & went on to a fresh stream, where we had luncheon & rested an hour. This is the 2nd time we have had the "Kumsein". It puts one in a sort of fever & you seem as if you could never drink enough. I expect water is good for one under these circumstances- but I have been afraid to drink too much & have found equal relief in rinsing out my mouth. We got at last among cornfields, with Villages standing in sight in the midst of trees: we came to one about 2.30 or 3 & halted for the mules- as soon as they came up we camped. We could not get into Damascus till late & might then have difficulty in getting into a hotel. So it is better to go on early tomorrow. The Villages seemed not to have seen Europeans often- we had a large {levee} round the Tent- but they behaved very well. I went to try & make a sketch & of course had a great concourse- amongst them a soldier, who was very anxious to hear news of the war. As the sun went down the Kumsein also died away & it grew cooler. I hope we shall have fine clear day tomorrow. (256/1009 "Jidahda to Damascus")

Wednesday. April 26. [Arrive in Damascus]

After repeated disturbances during the night from the horses running amongst our Tent lines & nearly letting our Tent down, we got up early & started about 6.20 for Damascus. A bright hot morning with every appearance of the hot wind springing up again- which accordingly happened soon after we reached Damascus. Our road lay over the plain which, if I had not just passed through the Holy Land & especially the plain of Esdraelon, I should call fertile &

---

[41] 'Hello' and 'good'.

so it is. It is well irrigated. Streams run in all directions & these they carry over their fields, wherever water is wanted. At the same time I can't say the reality come up to the flowery & glowing descriptions which travellers have always given of Damascus & its gardens & roses etc. A <u>garden</u> I have not yet seen. Plots of ground surrounded by mud walls & containing fruit trees there are but nothing equal to an Englishman's idea of a <u>garden</u>. Damascus lies among trees. Olives predominating. We came to it at last & I could not believe that it was the beautiful Damascus I have always heard of. I thought we had got to some dirty suburb or Village. The walls of the Town & the houses themselves are all built of unburnt mud hides, plastered with mud. I never saw anything so disreputable looking. The streets are dirty of course- I need (257/1010 "Damascus") not have taken the trouble to write that. The Town actually looked "<u>squalid</u>".We passed one ruin of a handsome mosque- built of coloured stones & marbles. We soon got into the bazaars & then the whole scene changed. The Bazaars are certainly very fine. We found room at the Hotel, when we arrived soon after 9 & found Mr Ward sitting waiting for the Falklands to come to breakfast. We got a room- I washed my hands & face & went down & joined the party at breakfast, so far as a cup of Tea was concerned- having had breakfast before starting. Before breakfast a plate of ice was bought (sic) me. At intervals too all through the morning the same delicate attention was repeated. I thought it was regular part of the entertainment- but afterwards found that each plate of ice cost 1 piastre- that is the 5th of a shilling. The snow is bought from Lebanon- Hermon principally. The consul had sent Lord Falkland Galignanis up to the 4 or 5th April- so I sat & read news.[42] After sending things to the wash etc. I sat & read till past 4- when we went out for a Turn. The Bazaars close at 4, generally, but many were still open. They are full of shops where sherbet & delicious edibles & drinkables were standing in most tempting array. I had a glass of iced (258/1011 "Damascus") sherbet as we went along- for the heat was intense. We passed a coffee shop, where a man was reading stories to the people, who were lounging smoking their pipes. We went into a very fine 'Khan'. A Khan is a place where merchants & bankers sometimes have their Offices & goods are exchanged etc- in fact it is half chambers, half an exchange. It consisted of a great square or round building, divided by pillars, supporting domes- a staircase led up to a gallery all round, out of which the different chambers opened. It was built throughout of alternate layers of black & white stone- very handsome. We passed the Great Mosque, made out of the old church of S. John's. No Xtian is allowed to enter & into one part no Turks hardly. We looked in through the door as we passed. I had no desire to go in. It is thoroughly decorated. Kennedy soon got tired. So I went for another turn with Giovanni. He took me to the Greek Church- where service was going on. I can't say I like their service at all. The nasal drawl is most hideous & the people seem to have nothing to do but stand & stare about. They are building a new church close to the present one. We went out by S. Paul's gate & to the stone where he is said to have been converted. Much talk all the evening & to bed lateish. (259/1012 "Damascus")

<u>Thursday. April 27.</u>

Before breakfast I wrote journal & read etc. & about 9.20 went down to breakfast- expecting to have perhaps an hour to wait. Lord Falkland however was there making arrangements with his Dragoman. So breakfast soon came. Just before we sat down who should arrive but 2 of our former travelling companions- Dr Bryce & Wakefield. They crossed the Jordan before Lake of

---

[42] Galignani's *Messenger* was a daily newspaper published in London from 1804-1884 under that name, then as the *Daily Messenger* until 1904.

Gennesaret & came up the Eastern route. Safely it has so happened but another party, mostly natives & Jews were attacked & robbed in the night while camping not above ½ a mile from where they were camped. The next day they fell in with one of the party stripped nearly to the skin. After breakfast & sitting reading a short time Kennedy & I went out to see the Bazaars. We wandered through some miles of Bazaars I should think- went into a Turkish Bath- bought some pipe sticks in one Bazaar- some "mish mish" or preserved apricots, in another- went to a Café & had a cup of Coffee, passed by the Castle, an old building apparently, with the stones bevelled in the usual way & finally went home to luncheon at 1.30. About 3 we went out again. First to see the interior of a house. The Jews have the handsomest houses & the one we visited is said to be one of the best in Damascus. It belongs to a Banker. Like all the houses it is built round a square. This square is paved with coloured marbles & has orange trees etc. with vases of flowers. In the centre stands a fountain (260/1013 "Damascus") octagonal or polygonal, or quaintly built & inlaid with coloured marbles. On one side is a high pointed arch built of coloured marbles (with by the by a skreen (sic) in front forming a sort of verandah, of 3 marble arches). Within the inner great arch is a room, a sort of large reception room- with Divans & marble floor & another marble fountain & the walls inlaid with marbles & mother of pearl. The ceiling, very lofty, is panelled & worked in wood & highly coloured & gild. Two doors lead from this, right & left with 2 rooms- these also are surrounded with Divans, for sleeping upon. On the other side of the courtyard a marble doorway with carved door, leads you into the state room- it consists of 3 divisions, like the 3 upper limbs of a Cross. This was the handsomest room in the house. The 3 divisions were surrounded with Divans. A very pretty fountain stood in the middle inlaid in very pretty patterns. The floors are marble in patterns. The walls covered with marbles, mother of pearl & gilding- 3 marble arches joined the centre compartment with the 3 raised divisions of the room & these arches (pointed) were incrusted with marbles. The ceilings of the 3 were alike & the ceiling of the centre over the fountain different to the others. It was as handsome as anything I ever saw. We were taken into another courtyard & up into another room upstairs wh(ich) had once (261/1014 "Damascus") been handsome, but was worn out. This I fancy is the worst of Eastern Houses. They never repair- a thing is done well once & then it must take it's chance. We saw 2 ladies in the house. One was I know the mistress of the house- who the other was I don't know. We couldn't talk, except through the interpreter- which is no means of conversation. We then went along the "Street called Straight" in which Ananias found St. Paul at the house of Judas & so out by the gate- the East gate. We then followed the wall south till we came to a Saracenic horse shoe arch, stopped at (sic) gateway, wh(ich) our guide pointed out as the place where St. Paul was let down from a window in a basket. From here we walked out to the stone which is shown as the place where "the bright light shone upon him & he fell to the ground" & so home by another gate & through the Bazaars to the Hotel. Here I left Kennedy & went out again alone. I went to look at the great Khan. I stood sometime there- examining it. The great doors especially struck me: they are plated outside with iron- {clenched} by ornamentally (sic). Dinner- Talking & bed- about 10. (262/1015 "Damascus")

<u>Friday. April 28th.</u>

This is a bad day. I have very little to say. My left foot was in great pain & I could not walk upon it. I sat in my own room reading till Kennedy came & asked if I would go into the Bazaar- I said I would try & went downstairs. Then he was not ready & we did not go out till nearly one- when we went into the Bazaar- my foot was so painful that I came back again. After luncheon

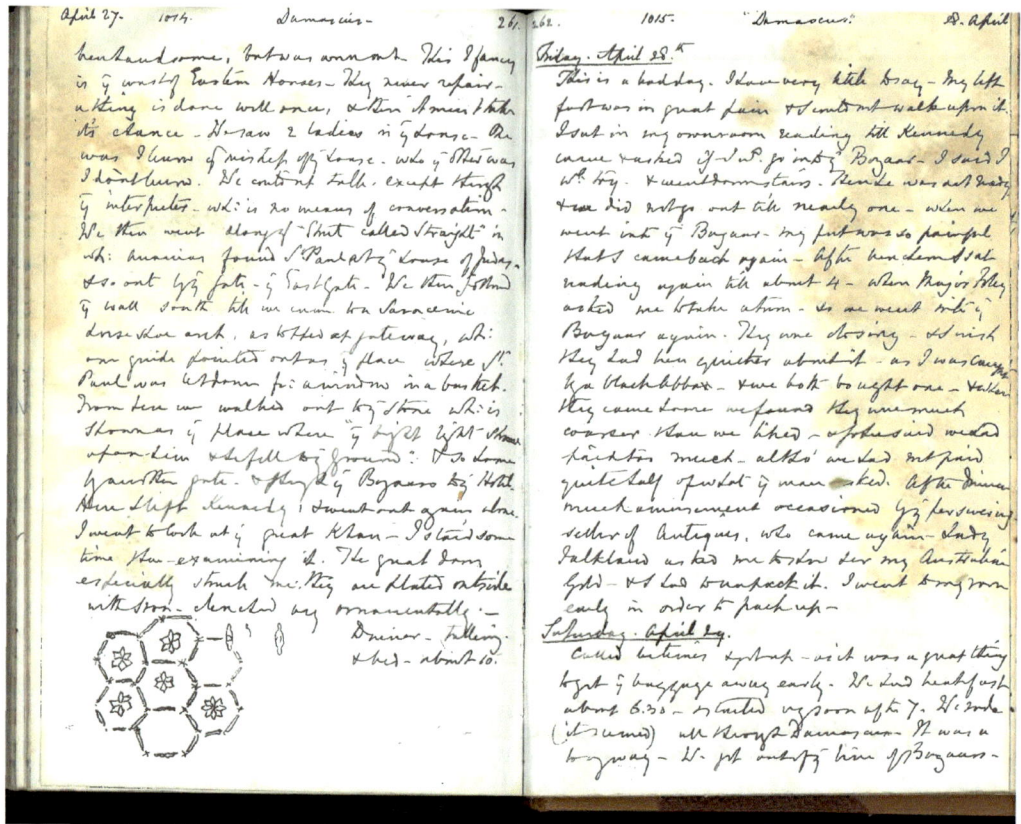

Figure 14.5: Pattern on the plates on the door of the Great Mosque in Damascus.

I sat reading again till almost 4- when Major Foley asked me to take a turn- so we went into the Bazaar again. They were closing & I wish they had been quicker about it- as I was caught by a black abba & we both bought one & when they came home we found they were much coarser than we liked & John said we had paid too much- although we had not paid quite half of what the man asked. After dinner much amusement occasioned by the persevering seller of Antiques, who came again- Lady Falkland asked me to show her my Australian Gold & I had to unpack it. I went to my room early in order to pack up.

Saturday. April 29.

Called betimes & got up- as it was a great thing to get the baggage away early. We had breakfast about 6.30- I started very soon after 7. We rode (it seemed) all through Damascus. It was a long way. We got out of the line of the Bazaars (263/1016 "Damascus to Ez Zebedani") & into the inhabited streets. Strange places- narrow & tortuous & dark:- mud walls & few windows. We passed through the horse bazaar- & so by the Governor or Pacha's house- one after the stile (sic) of the new palace Abbas Pacha has built outside Cairo. It seems to me he built upon the Wall of the Town. Outside we passed the beginning of some Barracks, begun by Ibrahim Pacha- but never finished. We passed through the luxuriant gardens & orchards outside Damascus

& came to the foot of the hill Sálheiyah. As we got near the top & looked back the view was very fine. I had no idea Damascus (the muddy) c(oul)d look so well. But standing as it does in the midst of green trees, wh(ich) spread for miles & crowned with numbers of minarets & the vast plain stretching away many a mile, it looks very well. We overtook a party of Kurds on their return home after selling the flocks they had bought down. They had quite a character of their own. They would speak nothing but their own language, which Wilson says is pure Persian. I didn't understand so from Giovanni. Our course lay the whole day along the Valley of the Baradá: supposed to be "Pharphar". Wilson says that its 'rapidity' (it is a perfect torrent) may favour its identification, (264/1017 "Damascus to Ez Zebedani") the Arabic equivalent of its name being Farfar, to hasten. We passed several villages- which all lay in the Valley- where the river's course was marked by vegetation & light green poplars & abundance of fruit trees of all sorts, the rest of the country being massed with stones. About midday we came to a Village where were some ruins, an aqueduct & tombs cut in the rocks. It is called Souk Wadi Baradá & is thought to be 'Abila' a Roman Town. Here we met a solitary English man- or American travelling almost alone. We crossed an old Saracenic Bridge over the roaring river & had luncheon. After about 4 hours journey (not including an hour's rest at luncheon) we came to Zeebedani, a Village in the Wadi Baradá. Here we pitched tent in an enclosure near a house. I sauntered up the Village, seeing by their long hair that most of the people were Christians. I was much struck by the beauty of the children & some of the women. I got a boy to take me to the Church & found a miserable building fitted up as a Greek Church. But though rough & rude it was pleasant to see that it sh(ould) have been build & afterwards I went to a house where a priest was saying service. I think he was "Maronite". I could not make out. We had dinner very late & I have just had time to write this. (265/1018 "Ez Zebedani to Baalbec") The Village is surrounded by a rich country well cultivated & enclosed, containing vines, figs, all sorts of fruit & plenty of grain. Top of the hill opposite is a Village in wh(ich) stands conspicuous a house, where the Consul comes to from Damascus in the summer. Now to bed.

Sunday. April 30. 2nd after Easter.

Called out at ½ past 4- started at 6. I had not gone 50 yards- before I discovered that my pocket handkerchief was gone. I rode back at once feeling sure that my pocket had been picked of it, as I had had it in the cottage a few minutes before. I insisted on its' being made forthcoming & a little girl produced it from her bosom. Father, mother & all laughed approvingly, as though it were a good joke. We followed up the river Barada to its head- came to the watershed in the valley separating the Eastern & Western waters & almost directly after came to a fine spring called Ain [note: space left where full name should be] wh(ich) is the source of the Litany [= Litani] river I suspect. We followed up the Valley due north- then turned to the west & mounted the side of a high & steep hill at the corner as it where (sic), from which we had a fine view of Lebanon beyond the Baalbec plain on our West & the Ante Lebanon, the last part of it, on the East. We wound round & over this hill, still & silent & lonely, with the eager young river roaring & fuming in the Valley below- reminding me when I shut my eyes of the (266/1019 "Ez Zebedani to Baalbec") noise of the weirs at Streatley, though louder. About 12 we sat down in the boiling sun & had some luncheon & I read Keble & then we went on again keeping over the ridges on the East side of the great Valley. It was very hot & my horse {seems} used up & I had a bad headache. We were overtaken by a man on a horse- he was a Mutawalli- he came on kindly to remind us to go round by the spring before going into Baalbec. He was a fine fellow- 6 foot good- & rode his horse well. We got down the hills & came to the Spring. It has been built over at some time- delicious water. Doubtless the deciding cause of the old Town being built on this spot. We passed back through the town & dismounted on some turf near the ruins of the Temple of the Sun. 3 tents were pitched not far off. It seems they are a French family. After resting a short time we started to see the ruins. It is useless trying to describe. "John of Antioch" says A(e)lius Antoninus Pius[43] built a great Temple of Jupiter at Heliopolis, near Libanus of Phoneicia & a nearly obliterated inscription confirms this. There was probably a Temple on this spot in the days of Solomon- called "Baal hamon" in Canticles viii. 11. The sun set magnificently while I was in the ruins. Home to dinner.

Monday. May 1. "Feast of SS Philip & James" [44]

We were called early, had breakfast & left the (267/1020 "Baalbek to Sahleh") baggage to be packed & proceed, while we staid behind an hour or two examining the ruins. We went first round the outside of the Temple to get a good general idea of the whole. We went along the West end, behind what has been the Sanctuaries of the 2 Temples. The outside walls have had additions to them & have been loopholed by the Turks, who seem to have made the whole building into a Fortress- {eddy}, Towers etc. It is very easy to distinguish Turkish from Roman. At the West End are some of the largest stones I ever saw in any building. The stones in the Temple at Jerusalem are nothing to them. Along the W. end are 3 magnificent pillars in a row, the collective length of which is 190 feet- the exact measurements being 63ft. 8in- 64ft & 63 ft. Below these is an enormous stone deeper than the 3 others. Wilson gives it at 69 feet long- 13 deep & 18 broad (being a corner stone)- but I think he is wrong- he takes one very long stone & a short one next as one- & says it has been "overlooked probably because irregularly cut in the outer surface". I did not overlook this fact- in which he is right- but there is a joint so nicely made that I don't wonder at his not seeing that. He is plainly wrong. Round on the north side is a wall outside the greater Temple & in it are 9 stones averaging 31ft long, 9 feet 7 inches broad & 13 ft deep. The whole of the building has been raised in order to give height & effect. (268/1021 "Baalbek to Sahleh") We went into 2 long vaults built of magnificent masonry. Indeed I never saw stones in any building before so beautifully fitted together- an additionally difficult thing I should think when the stones are so large- as it would be hard to move them to fit them. They probably must have been cut with mathematical precision in the first place. How they ever put them into their places I can't think. At the East end we made out clearly the foot of the pillars built up between. This front must have been magnificent. A flight of many steps probably led up to the portico. I thought I could make out traces of its' having extended the whole length of the pillars- of which there were (I think) 14 with a wing at each end, with pillasters. The walls of the wings have had horizontal lines as well. The top & bottom of each stone has a lip or moulding which meet & form a line or string course. There is a single Doorway along the N: side of the flat building & 3 or 4 windows on the South side. The general plan of the building has been as follows- beginning at the East. A long flight of steps leading up to a Portico with wings at each end. From this you pass by a large central doorway with 2 smaller side doors into a large Hexagon. This has the remains of niches all around. From this you pass into a sort of Quadrangular space. The sides of which have been divided (269/1022 "Baalbek to Sahleh") the walls being partly straight & partly curved. The whole is niched with much beautiful work- not a figure remains anywhere about the buildings. Westward of

---

[43] These are the Emperor Hadrian's names and later Roman writers credited either Hadrian or his successor Antoninus Pius with the construction. Neither is known to have built at Baalbek.
[44] The date was shifted to 3 May in the 1950s.

this you pass by a recessed doorway, which was probably arched, into the Great Temple. Six grand pillars with their entablature still stand on the South side & on the North the Turks have built up the wall, building in the remains of many pillars. I am sorry I did not count them. The size of this too I could not take- the ground was so broken. I must go to the British Museum when I get back to England & see Wood & Dawkins work [Wood 1757]. Between this & the smaller Temple. South, are the remains of a vaulted way which runs diagonally- what this can have been I don't know. The Smaller Temple has all its' walls perfect & 8 pillars on one side (the S.) & 15 on the other still standing of the peristyle, with capitals, cornices & entablature, very perfect. The carving of the under part of the stones which form the roof of the peristyle, is something wonderful. The Turks built up a wall across the portico- you pass through a hole into the vestibule & enter by a grand doorway 20 ft broad & 40 or 50 high, into the Temple. The doorway is worth coming all the way to see. The Southern part of the East wall has given & the heavy stone has subsided- but (270/1023 "Baalbek to Sahleh") still hangs, pinched by the others. The architraves are broad & beautifully carved- with foliage & vine leave & grapes & heads of wheat etc. etc. The walls inside are divided into 7 compartments, each with its highly finished nich(e), by 6 fluted Corinthian pillars (not pillasters). This as far as the beginning of the Sanctuary, which is marked by the remains of 2 fluted pillars & of arches which united them & the side walls. Within the walls have <u>pillasters</u>. The cornices are very rich. To the South of the whole building & disconnected are the remains of a small round Temple- surrounded by a circular peristyle. We left Baalbek about ½ past 8. Away on our right some miles across the plain we could see a solitory column of which nothing is known & passed the remains of a building made of 7 unequal sized red granite pillars with the {kaybah} showing its Muhammadan origin. Our road lay all up the plain- the Lebanon range on our right & Ante Lebanon on our left- each covered with snow. The whole plain is under cultivation. We saw one or 2 of the usual reed Villages on the slopes of the hills. We stopped to luncheon on the banks of a small stream running across the Valley when we found some close green turf. I lay on the grass & read the Epistle of {Joshel}& Keith. We got to {Sahleh} early- about 3. I walked away (271/1024 "Baalbek to Sahleh") to the Village. I found most of the people are Christians. I came upon a Church, with a Cross carved on the door- the <u>first</u> I have seen <u>anywhere</u>. A boy came up & I signed to him to get the key- but he could not understand. He asked if I was a Frank & when I said English he patted me on the shoulder patronisingly & expressed great satisfaction. He then said something which of course I could not understand & put his head on his hands, as a person on [*** (?)] & then led the way. I followed & he brought me to a house where was evidently "good lodging for man & beast". Two ladies received me in a most flattering way & by putting their fingers in their mouths suggested that I should have dinner- but I then began to suspect that the boy had been offering to take me to a Khan- so I laughed & said my Salahma & retired. In the street I met a Caravan of I can't say how many Camels- the longest I ever saw. It was endless. My boy accompanied me- evidently suspecting I had dropped from the moon & was much relieved when we came in sight of the Tents. Since I have been writing this & waiting for dinner a bell has been ringing, I suppose for Vespers- they are Greek Roman Catholics principally. It is a blessed sound & took me away to Littlemore- I suppose the bell must have like the bell there- or the way (272/1025 "Sahleh to Khan Hussein") in which it was rung.

Tuesday. May 2nd.

A cold blowing night. I was obliged to go out for a run to get warm before going to bed. But then our Tents were pitched some 4 or 3000 ft: above the Sea. We were off early this morning- passed through the Village & so skirted along the hills beyond- but long after leaving the Village we came to a large Sarcophagus lying by the side of the road. We passed near several villages lying on the slope of the hills & the whole country generally wore a more domesticated & civilised look & the people too, only that they all carry guns. It seems that some of the Villages are Druze & some Maronite & they took different sides when Ibrahim Pacha was in the Country & there have been feuds ever since. Our road lay over the mountains; over which we toiled all the day after nine o'clock. We passed a good deal of snow. We had luncheon at a Khan, from which we first saw the Mediterranean & country below. The mountains were very fine. But I am disappointed in the Lebanon. I had formed extravagant notions of pastoral richness & beauty & instead the whole country is stones. They <u>plough stones</u> not earth. We met & passed great caravans of mules, donkeys & a few camels. We are on the great road between Damascus & Beyrout. And only fancy that in all these many years the Turkish Government has (273/1026 "Sahleh to Khan Hussein" "Khan Hussein to Beyrout") never made a road in the country. I don't wonder at their hating the Turks so. As they say, the Government takes a great deal from its people & makes no return at all. Giovanni informed me today that in Damascus Christians at this minute are annoyed. The water is stopped from their horses every night, while it runs all night through in the Muselmen's (sic) houses. We camped at the the Khan Hussein about 4. I went out with Shakespeare in my hand & lay in the shadow of a rock & read till dinner. Tommorow we have only about 4 hours to Beyrout & there I hope I may find some letters. A line or two at least I do hope. At Beyrout I shall feel that I have only one stage more. This is our last night in our Tent Home. I have enjoyed it very much.

Wednesday May 3.

Though we had but about four hours' work before us, Giovanni called us at the usual time, ½ past 4. We got off about 6 & reached Beyrout by eleven or before by the worst road I ever saw. The last descent gave us some pretty scenery- the valleys highly cultivated & rich with mulberry & other trees. Still I have seen nowhere the beautiful scenery I was led to expect. I went to my room to have a wash & dress & while doing so, Giovanni bought me an envelope containing 2 most delicious letters from Sam & Frank. They told me (274/1027 "Beyrout") everything, just what I wanted & have made me very happy. I am most thankful to be here, at my last stage, only waiting for the steamer to take me Home. The luggage came & I unpacked & then went to seek for Kennedy who has disappeared. An Austrian Lloyd's Steamer is in the harbour & I half thought he might make up his mind to go by that. I had a bit of bread & cheese & went out- sauntering through the bazaars & then came home & most important went (and) turned myself into an Englishman again by shaving off my beard. At last I feel clean & <u>cool</u> again. Read in my room till Dinner time, ½ past 5. Six Frenchmen apparently & no one else. After dinner Kennedy & I sat by the window watching the steamer getting under weigh. I sat & read W. Irving's Sketch book, which I found in the room & then to my own room & to bed. The view from the Terrace roof outside my room door is very lovely. Mountain & sea & moon. Kennedy has been trying to persuade me to go on to Constantinople. No- No. My face is set towards England now & I won't turn it away again. I think of nothing else but Home & England. To bed.

Thursday. May 4. [Beirut]

A sad day of idleness. After breakfast we had a reckoning & settling with John our Dragoman & then we strolled out with him into the Bazaars, where I bought a few pipe heads for John & Frank. Home & read till about 3- when Kennedy came (275/1028 "Beyrout") & we went out- along the sea, to the Bellevue Hotel- to see if Freeman was there. The Landlord who had been at one of the Indian Engagements & waiter at the Travellers Club, showed us over the house & prepared us to drink a bottle of wine or beer. On our way back we were amused by 3 boys batting, one of whom, a splendid little fellow, was riding a donkey into the water & persuading him to go out of his depth & swim. The scene was comical when nothing remained of the donkey above water but his head & ears pointing forward. Home to dinner at the awkward hour of 5.30. After which I sat on the roof of the house with a book. Half of the other houses I noticed were occupied by people walking in the lovely evening. I spent the rest of the evening reading in my room.

Friday. May 5. [Beirut - Dog River - Beirut]

I got up early- dressed & walked out along the shore till I came to a nice place, where I bathed. I found a place where I could take a header from the rocks into deep clear water & also get out again- which was the difficulty. Walking home I met Dr. Bryce, Wakefield & Freeman, just riding in from Damascus. After breakfast Kennedy & I started in a boat to sail across the bay to "Dog River". There was a nice little breeze when we started- but it slid away when we had gone about 6 out of the 10 miles & we had to row most of the rest of the way. We landed just (276/1029 "Beyrout") this side of river & went up the rocks to see the inscriptions. There are several sort of niches cut in the rocks, on which are the remains of a figure with one arm extended. On one I found a Latin Inscription of 8 lines, beginning:[45]

IMP. CAES. M. AVRELIVS
ANTONINUS PIVS FELIX AVGVSTVS

the other six lines were not so legible. The river is a fine stream, running down between high rocky banks- spanned by a bridge of 3 arches, built about 30 years ago. On the right as we went up to the bridge an old Turk in his mulberry ground showed me another inscription- it was long & cut on a piece of the rock that had been smoothened for the purpose- but I could not even make out what the character was. We crossed by the bridge & walked down to a sort of Café there is- where Kennedy solaced himself with a pipe & coffee. I found it was possible to get a donkey here & settled for one at 9 piastres back to Beyrout. Kennedy would sail (or not, I thought, in the calm). I mounted my donkey & started- getting over the distance in about 3 hours. They call it 15 miles- but it is not so much. When I got down to dinner I found Briggs & Rycroft & Mr. Ward & Yateman. Much talk- after dinner I adjourned to the roof- where Messrs Ward, Yateman & Rycroft smoked their cigars. I came in & wrote journal. (277/1030 "Beyrout")

---

[45] This is Caracalla, sole emperor from AD 211-217. See Chapter 11 for this and the other inscriptions at this location.

Saturday. May 6. [Beirut]

I sat reading in my room & didn't go out before breakfast. After breakfast I went to the English Consuls' (sic) to get my passport viewed. I took a turn through the Town & sat at the gate we entered by & back & then sat in my room for some time, till Kennedy came & proposed a walk. We went first to the Consuls & then along to the other Bellevue Hotel. In going in we met Mr Rudewaldt (sic) & had some talk. Mr Rycroft overtook us & joined us. I had some talk during dinner to a Frenchman, whom I sat next. Mr Ward had received letters from the East, which bring word that Japan has consented to open her ports to all nations in 1 year's time. I have nothing to say of today & so will save time & paper.

Sunday. May 7. [Beirut]

On getting up I found there were 2 steamers in the harbour. News from Constantinople that 4 English steamers & 2 French ships had taken Odessa & that the Baltic fleet had taken 5 ships. After breakfast I went to the packet office about my berth & then went to the Chapel of the American Mission, which I found was Presbyterian. I sat & stood as other people did while a hymn & extempore prayer were performed & then followed a sermon- which I thought was to last forever- but surprised me by coming to an end after {5} & then I went away. I can't call it "worship". I sat in my room & at half past 3 went out with one (278/1031 "Beyrout") of the boys of the house to visit the Maronite Church. I found a fairish building, with some bad painting & a great many very tawdry altars. But something besides all this that struck me very much. A priest was sitting in a common chair in the middle of the nave, instructing the people, who were sitting literally at his feet- men & boys, some 50 or 60 perhaps, were sitting on the floor all round him & some women behind the skreen (sic) at the west end. He was reading out of a printed book & now & then making remarks- he looked at his watch & saw it was near the hour of Vespers (I suppose) shut the book & added some remarks, in fact preached a sermon for some 10 minutes in a very animated way & then 2 other priests came in & all 3 went inside the rails reaching the 3 last altars & standing inside in front of the north one of the 3 & leaning on crutches facing the people, began a service in the usual monotonous eastern nasal drawl & I came away. The boy took me then to the Greek Catholic Church- as they call these Greek Churches which have submitted to the Pope. It was a fine old building of 3 aisles with grand clustered pillars & pointed arches. It was to all appearance exactly the same as any Greek Church not subject to the Pope. The same skreen (sic), the same pictures, the same stand with the pictures of the day, which the people were (279/1032 "Beyrout") kissing- the same high gilded pulpit. From here I went to a small chapel, where I found a priest in alb & stole sitting on a chair on the steps of the altar, preaching extempore to the people. He was very earnest & took their attention. I thought he was a Frenchman. The chapel itself was fair- the fittings most miserably tawdry. A Grecian pediment etc.were painted on the wall behind the altar & there were 2 coarse angular statues of naked angels on each side. I learnt outside, from a Frenchman who saluted me, that it belonged to the Jesuits Mission. Then the boy took me to a house & chapel of the Sisters of Charity- where I found a number of girls & young women dressed all alike in gown & bonnets with some of the sisters singing in the Chapel. The Chapel was fairish with 3 great pictures over the 3 arches. The centre one representing a number of sisters of Mary in their hideous great {waygon} bonnets kneeling before an altar, above which a priest in alb & stole seated on a clump of naked cherubs & {smoke} was giving them his blessing. A pretty piece of idolatry to set before the Turks. Home to dinner. I walked on the house top sometime with Mr Rycroft & then came to my room & packed my things & to bed.

Monday. May 8.

I had to wait a long time for my money & (280/1033 Leave Beyrout) at last finding Kennedy I had to go to Mr Heald & ask for it. It was with some difficulty that I got any. Giovanni too had disappeared & he had my plaid & the Sinai Dates. I met him just as he was leaving for the steamer & he got both for me & went on board with me in the most fatherly way. At 5 we left the anchorage & sat down to dinner a few minutes afterwards. After dinner I sat & walked on deck all the evening. Indeed Mr Ward (the American) & I walked the deck till nearly 12. We had one of the most beautiful sunset effects I ever saw. About 12 we were somewhere off Carmel- but we could see nothing. I reluctantly went down to my close netted bed.

Tuesday. May 9.

I got up early & went on deck, where I sat reading the Psalms & Lessons- till Freeman came up. About 9 we anchored off Jaffa. It looks well enough from the Sea, standing on some rising ground. In the near are a good many gardens with orange & other trees & some palms- a nasty reef of rocks runs across in front, over which a heavy surf breaks. The landing is awkward. Mr & Mrs Hardy whom we had left at Jerusalem, the latter damaged by a kick from a horse, came on board- poor thing- it was found out by accident only a few days back that the bone was actually broken & the stupid old Dr McGowan (sic)[46] had (281/1034 "Off Jaffa") been treating it for a strain only. A Frenchman who has been in Abyssinia 10 years or more & has married a Princess there & is Commander in Chief there, came on board with 3 horses & some attendants. In getting the 3rd horse on board, she lept overboard out of the boat & swam away. There was a chase & she was caught. It was doubtful whether she would not be drowned, as the halter got round her legs. She was got on board safely at last- very much exhausted. We got off again about 12 & went out to sea. A nasty swell & people sick & uncomfortable. I walked the deck till 12- alone- very cold.

May 10. Wednesday.

All the discomforts of a small, old & close vessel. A fine day however & we came in sight of Damietta & Rosetta. A good deal of talk with Lady Falkland. She had talked to me about the Jerusalem matters the day before & we began to see that we sympathized. She had seen a good deal of the Bishop & his wife & spoke very nicely of them. But she says he is a thorough Bunsenist, & insists at the German services when Bunsen's liturgy is used. She speaks of the whole set as more German than English. She asked for Bishop's house to send them out an altar cloth & suggested what it should be- a plain velvet cloth- with gold fringe & a cross marked on it. He hesitated & in the evening came a note saying that they must decline it, as they were all unanimous not to admit the Cross any where into the building. She called next day & had a long talk- but in vain & it's ended in their (282/1035 "Betw(ee)n Jaffa & Alexandria") compounding by a pulpit cushion instead. She produced Williams's Via Vita Eternal & asked if I had seen it & then she told me that Lady Fielding was her niece. She also told me about Church matters in Bombay, in which she seems to have taken much interest. I sat & read & walked on the bridge. Dr Bryce came up & walked with me. He told me that in Jerusalem one

---

[46] Although this is the only context in which TB mentions this man, he had presumably met him in Jerusalem- and been unimpressed. Dr Edward MacGowan (1795-1860) had founded what is regarded as the first modern hospital in Jerusalem for its Jewish population. He lived in the city for 18 years from 1842 and died and is buried there (Lev and Perry 2008).

day at Dr Barclay's house a little boy came in - whom he was told, was 13 years old & married & father of a child. His little wife was not 10. But said Dr Barclay I could show you a younger husband- a boy not more than 10- also married & lately or about to be a father, his wife 9. I never heard such things. After dinner I sat & read & walked. I had a long walk with Mr Ward- a lovely night. We were going ½ speed, with the Alexandria light not far off & finally just about 12 the engines were stopped & we lay off all night.

May 11. Thursday. Alexandria.

When I awoke I found they were just setting the paddles in motion again & before I was dressed we were at anchor in the harbour of Alexandria. I found everybody was going ashore for the day & night & finally determined to do so too- so I pack up some things & Rycroft & I went together. The first person I came across in the Hotel was Paulo, who said the Fentons had arrived from Cairo last night having had 2 months in the desert. I went out to get some buttons put on my boots & in my way went into the new English Church- a Saracenic (283/1036 "Alexandria") building- not by any means so bad- a parallelogram with apse & a further tower at W. end. The architect had worked in 3 Crosses in medallions very nicely over the apse arch. It wants colour much- which it won't have. Home to breakfast & much talk with the Fentons. Then out with Rycroft to Cleopatra's needle & Pompey's Pillar & the Mahmoudie Canal, where I went on board several of the Nile boats. After luncheon we went to the Pacha's Palace- where there are a few grandish rooms, with French floors & millinery & furniture. Miserably carried out. In the streets we met a procession of a boy going in state according to custom through the streets previous to the rite of Circumcision being administered tomorrow. The boy is dressed in very smart clothes & put on a young caparisoned horse & preceded by drums & pipe etc. Until this he may not enter the mosques. We went out to the Library, where I got a French & English book & a Translation of Bulwer's England & the English, in the hopes of picking up something of French. I left Rycroft reading the papers while I went out. I wandered down to the Roman Catholic Church- where I found a number of people & services going on. A monk was just finishing a sermon when I went in. Then the Litany to the Virgin was sung & there followed what I suppose was the benediction of the Sacrament or whatever the service was called. It was very blessed to be inside a nice Church with many worshippers- it was very sad to be no worshipper myself. One does especially feel the bitterness of schism. The disunion (284/1037 "Alexandria") of Christendom when in the midst of Islamism: anywhere in a strange land where one naturally goes to the House of God as Home- more especially when that strange land is the land of Turks & Infidels. We have got a Church finishing in Alexandria, but I fear it will never be the Home that a Roman Catholic Church is. It won't be open for similar daily warm glowing services like theirs. I do wish the English Church kept her Churches open & that her services were in every practicable place frequent & rich & beautiful. As I stood inside the door & listened to the Litany, I thought how much more beautiful as well as pure our Evening service would have been. I thought how the Psalms would have rung out with the rich organ in some fine church of praise & there the Holy Lessons read solemnly as the Word of God & the Canticles swelling through the fine Church after them. I do yearn for such. But if one finds our service it is mostly poor & niggardly, as if it was wrong to have a solemn service. I can only take care that it shan't be so wherever I am, please God. Home to dinner. After which much talk till ten o'clock with Mr [blank] the Shanghai Interpreter. A lovely night but I came upstairs to write instead of going for a turn. Tomorrow we start again towards Home. Fancy

Stobart & his charge passed (285/1038 "Alexandria - leaving") through here a fortnight ago on their way home- the 2 via Trieste. Schomberg Kerr by Gibraltar- and now to bed.

Friday. May 12. [Alexandria]

I got up & found the Fenton's having breakfast previous to starting & joined them. When they were gone I went out & bought a pair of boots & had my hair cut & then finding it too hot to be out, (the Humseen was blowing) I sat talking to the Chinese Interpreter. I had a long talk to him about Christianity in China. He evidently does not believe much in the Xtianising influence of the Insurgents. Then on to Missionary work, he says of Protestants & Catholics that they are more of them sufficiently versed in the language or the literature or habits of thought of the Chinese. They are a very thoughtful metaphysical race. Their language is very exact & no missionaries have been good enough Chinese Scholars to apply the fitting Chinese terms to Xtian terms. He believes if it would be done that more good might be done by writing than by preaching- but then the difficulty arises of finding a sufficiently good Chinese Scholar. He says he is surprised at the ignorance of Chinese displayed by the R(oman) Cath(olic) Miss(iona) ies. He gave me the population of Canton as 700 000. About ½ past 2 I went (286/1039 "Leave Alexandria") on board the steamer, where I found a great crowd. We were delayed by the Post Office till about ½ past 4. The navigation of the Harbour is difficult- a very nasty reef runs off the mouth. We went down to dinner & I spent the evening on deck till 12.

Saturday. May 13.

A fair wind & the fore sail set- the steamer making 8 knots all day. Mr Ward lent me the 'Home News' of 24 April, which occupied me all the morning after having had a long talk with Lady Falkland. I got into conversation with a young Scotchman, who told me he was going on by a screw Steamer from Malta to Liverpool. In the Evening after dinner he joined me on the paddleboat. He was in the same office with Percival {Franlis friend}. I am a little uncomfortable at finding, on thinking it over, that my money will run rather short. It is a horrid nuisance. I don't think I shall have enough to go home across France & certainly not by the P. & O. Steamer. I don't know what to do. The evening spent much in the usual way- i.e. in doing nothing.

Sunday. May 14. [In the steamer from Alexandria to Malta]

I sat on deck before breakfast reading & after breakfast shut myself into the hole dignified by the name of Cabin & said Morning Prayer. While sitting on deck afterwards a passenger came up & asked me if we could not have morning service. I confess it seemed to me impossible. We are a set of strangers in a foreign vessel. If in an English steamer we c(oul)d of (287/1040 "Fr. Str. Mentor") course have claimed the right to use the Cabin. I did not encourage the idea. The applicant suggested that we were so many English- forgetting that though all English we are not all Churchmen. I believe hardly the majority could be found to be churchmen. At dinner Dr Bryce got or tried to get into an argument with me about Church matters- a state Church- the Sacraments- Forms etc.- till I declined the discussion. Lady Falkland lent me a ½ yearly Vol.(ume) of the Churchman's Companion with which I was well occupied most of the day. After dinner I sat on the paddleboat with Rycroft & we read a sermon from it together. I had many longings to be at work at home. God grant I may be soon. The island of Candy [= Candia = Crete] was just seen on our starboard beam. I couldn't make it out.

Monday. May 15. [In the steamer from Alexandria to Malta]

I sat on deck before breakfast reading the Psalms & Lessons & learning some of the Articles by heart. After breakfast I read all the morning till near one- the Chunais Confession & in it a very good Easter Sermon by Scott (?) of Hoxton on Church Decoration from which I travelled to Henry & his work. It will be great pleasure looking over his portfolio & hearing all that he has been & is doing.[47] Besides I hope to go with him & see some of his works. I almost wish he was not leaving Guildford. We have got another (288/1041 "Fr. Str. Mentor") nice day- blue & fresh, with a pleasant white curl on the Sea. We hope to reach Malta about Wed. morning. What I shall do I don't know yet- I have a great mind to try & borrow enough to get home with from Rycroft. Or shall I get 10£ from Arthur Coussemaker (sic)- everyway it is an unpleasant fix. The wind got up a jot as & the sea too & by dinner time there was a great deal too much pitch & toss for many people on board. A very unpleasant Evening with drizzling rain at times & spray & very cold. I sat in the rain till I was quite cold- the ship making 4 knotts (sic) only. I went to bed early.

Tuesday. May 16

A much finer morning- the wind & the Sea both going down. I sat before breakfast reading & after breakfast read whole morning. Lady Falkland had lent me again the Via Vitae Aeternae. The sea got quite calm. At Noon we were 180 miles from Malta- hopes that we shall get there tomorrow before noon. The whole day was spent in reading. After dinner I had a talk to Rycroft & he at once said he had a letter of introduction to Mr Bell the Banker there & he had little doubt he could get me some money & would gladly go with me the 1st thing. So I have set my mind at rest. A long talk with Lady Falkland on Church matters. A walk on deck & down to the Cabin to read. A lovely night, with smooth sea. (289/1042 "Malta")

Wednesday. May 17

Before I was dressed I heard that Malta was in sight. I got up on deck as soon as I could. I found we were some 6 or 7 miles from the land. This yellow, dry brooding place & you do not see the size of Valletta till you get near. We went to the Quarantine harbour unfortunately for the thing to do is to see the Town from the great harbour. I got a boat & went ashore as soon as ever we were moored. I walked up the Town to the Hotel nearly opposite the Public Library- {Duraford's}- when I had a wash from head to foot & breakfast. Delicious to get decent Tea & bread & butter & toast an <u>innocent</u> breakfast instead of the Early Dinner they give us on board the "Mentor" . After that Rycroft went with me to Mr Bell's. He is a banker. They did not seem to make much objection to giving me some money, but asked if I knew anyone in the place. I mentioned Arthur Coussemaker & he said if he would come & back my bill, he would accept a bill on Willis. So I went to see for A. Coussemaker- I got a man to show me the way to the Barracks. I found him living at the Mess House. I sat some time with him & then told him about Bell & he offered to go at once. We walked to Bell's & got £10 or rather £9.14. Six shillings being charged for Commission etc. & then he took me to S. John's Church. I had just looked in before breakfast. It was the old Church of the Knights. It is a parallelogram with N. & S. Aisles. The whole surface of the stone (290/1043 "Malta") carved & coloured & encrusted with marbles. The ceiling is painted in subjects- now much injured. I can't imagine how- for

---

[47] A reference to his brother-in-law, the church architect, Henry Woodyer, widower of TB's sister. See Chapters 1 and 3.

surely there can be no damp! I believe the Church was despoiled of all its pictures by the French. One Chapel that at the E. end of the S. Aisle, is skreened (sic) by a silver skreen (sic). This & a large silver lamp before the high altar were saved by being painted. Near it hang the keys of Rhodes. The floor is inlaid in marbles. It was covered with matting, but we had some of it unrolled. The altar is handsome with lapis lazuli & marbles. What pleased me most I think was a piece of inlaid floor in front of the high altar- a small piece. The whole place was dusty & (to me) looking uncared for. We went thence to the Public Library. A long oblong room with books & tables at which a good many people were reading & writing. The books seemed mostly Theological. I noticed several on Architecture. Then to the Military Hospital- which was beautifully clean & in nice order & so on to the Barracks where the Buffs are at present. Here I saw the prisoner's cells & the rooms. I went through with Arthur Coussemaker, whose business it was this day to visit the different places. Back to the Inn to pay for my breakfast & down to the Stairs where I got a boat & went off to the Steamer. I met Major Foley on the way just coming back. He has found letters ordering him to join the Army in Turkey at Constantinople. It is a (291/1044 "Malta & away) nuisance for him after 5 years absence not to get a peep at England & I think he is vexed now that he delayed in Egypt & Syria- but he is too lucky in getting this appointment to grumble. We loosed from moorings & put to sea & as soon as we got outside got into a heavy rolling sea, a nasty hazy Sirocco having got up. It came on worse & worse, blowing very fresh when we went down to dinner & all the Evening. The usual mishaps occurring during dinner, Lady Falkland sitting opposite me & laughing & enjoying it she said. After dinner we all went on deck- a large brig passed us. Our wretched Steamer could not do anything but tumble about & the engine was only working at ½ power. I sat reading Shakespeare all the day in the Cabin & turned in about ½ past 10- when I had a most disturbed uncomfortable night.

Thursday. May 18.

Heavy rain came on in the night, which toward morning- put the heavy sea down some what. After breakfast we found ourselves off Sicily- but there was so thick a haze that we could see but little. All this is rather fine in the lovely, beautiful, Mediterranean! It is only in England that we have fogs & rain & wind & disagreeable weather!! All through breakfast I have been listening to Lady Falkland abuse (sic) of England- she says it so 'triste' fancy "Merrie" England being 'triste'!! (292/1045 "Fr. Str. Mentor") The sun broke out at last- but it was cold all day. We were all tired with the rough night previous & I think all were sleepy. I fell asleep on deck & awoke with a headache. After dinner it rained & I sat talking to Lady Falkland till after Tea. Before going to bed I had a walk on deck, when the rain was over & the stars were out & we were making 8 knotts (sic).

Friday. May 19.

The 1st thing I heard this morning was that Sardinia was in sight & that we should get into Marseilles tomorrow Evening (Saturday). The coast of Sardinia is very bold- the land high & precipitous. After breakfast I had a long talk with 2 Frenchmen- I had Murray's France open before me, with its' maps & they rushed to it with delight. I never saw 2 men so delighted at the sight of their country. One lived near Limouse [= Limoux] in the South, the other between Cosne [= Cosne-Cours-sur-Loire] & Bourges- & his Grandfather he said lived between Versailles & Rouen. He spoke with delight of the Country about Rouen. These men speak English perfectly & yet have never been in England. I asked how & they said they had had

English nurses. Rycroft came to fetch me on deck- we were some 10 miles from the shore perhaps- a fine coast- the wind gone down- the sea small & altogther what the Mediterranean (by all accounts) should be. About 2 we got close under the land & about 3 got among some islands, turning in towards the passage of 'Bonifacio'. The rocks, (293/1046 "Fr. Str. Mentor") limestone I think, were very rugged & picturesque. I made out one house on Sardinia with a red tiled roof. The wind was against us & a fresh sea on. Presently we came to an old fort on our right or the north & when I went down to dinner we were opposite a small town. After dinner we were in the open channel & before sunset passed out into the open sea again, westward of the Islands. The wind dead against us & a nasty pitching head sea.

May 20. Saturday.

A rough sea & a good deal of pitching all night but going down in the morning. Many complaints of want of sleep. Great questioning whether we should reach Marseilles. I had a long talk before breakfast with one of the 2 Frenchmen- on Church Matters & tried to put the English Church in a better light before him. He spoke nicely & I hope did not look upon me as quite such a heretic after it. After breakfast I sat talking with Lady Falkland, who was reading my Christian Year. She showed me some sketches. At 12 it was announced that we should get in late in the evening- so we gave up all hopes of sleeping ashore that night. Before dinner we made the land East of Marseilles & after dinner it was pleasant to sit on the paddle box & watch the shores of Europe rising again. I spent much of the lovely Evening on deck. Then went down & read for an hour & again walked the deck, indulging home thoughts, till we got into the Port at half past eleven. (294/1047 "Marseilles")

Sunday. May 21. Rogation Sunday.

I was still in bed when word was brought me that the things were wanted for the Custom House. So I made a short toilet & let my things go- taking a change of {suien} in my pocket. I went off in a boat to the "Hotel Beauvoir", where I washed & dressed & then went to the Douane. When I got there I found my Passport was not forthcoming. It seems that it never was sent from Beyrout- or else was left at Malta to which place I took my passage. While sitting at breakfast at theTable d'Hôte, Rycroft came in. After he had finished his breakfast we went together to the English Service. We found the sermon just going to begin. At the door of the Consulate as we came out I caught the Vice Consul & went to his office & got a fresh passport- & then we went home to the Hotel, where I packed my things. After that we went for a turn- round by the port & up above by a fort where we had a pretty view of the Entrance to the Harbour & saw a steamer start with troops for Constantinople- Poor fellows- they are sick enough by this time & wish themselves back I expect- though in the excitement of going "Glory" was everything I dare say & many of them perhaps will never see France again. We went to the afternoon service & thence to a Restaurant, where we dined- meanwhile our baggage went to the Station & Rycroft's servant got our Tickets. At ½ past 6 we went to the Station. The drive through the town was very pretty. (295/1048 "Marseilles to Avignon") The streets, some of which had trees in the middle, were full of people- <u>crowds</u> were out enjoying the evening. It was delicious to find oneself in a Train again. For 19 months I have gone back to the various modes of travelling in use amongst our forefathers. I can't say how pleasant it was to be in a train once more. I seem to have really got back into the world again. We sat enjoying it & the country. I expected to see nothing round Marseilles. On the contrary I thought it very pretty. We skirted along the sea for some time- among rich fields, with fine rocks & mountainous hills about. And so we flew along apparently, though I believe it was no such great pace compared to an English variety- till it got dark when we both went to sleep & awoke about 10. In ½ an hour we were at Avignon. I never saw luggage better managed. We had it in a minute. We got to the Hotel "Palais Royal" & went to bed.

Monday. May 22.

I was in bed so late last night & so tired with walking about etc. that I did not awake till late this morning. After breakfast at the Table d'Hôte at 10 I went to the bureau about my baggage & then got a Valet du place with whom I started through the Town. There is but little to see at Avignon. I went to the Cathedral. A poor & very small building. (296/1049 "Avignon") I sent back the man while I was there, to try & get me a "Guide" or some account of Avignon. While he was gone I went into the Church. It consists of a nave & 2 aisles, originally- but someone built a stone tribune, or gallery, all round the Church, cutting off the aisles more than they were originally seperated & now the aisles consist more of a number of chapels. There is really nothing see in the Church. Two or 3 of the windows have been filled with modern painted glass- not very good. The sexton showed me into what I think has been a Transept- in which is the Tomb of John XXII- (I think). It is a high long erection of open work. Once covered with statuettes- all broken & gone. In the N. Aisle of the Church is the Tomb of another Pope. This is the scene of the grand schism of the Roman Church. I went on to a rock near the Church planted out in a sort of garden- with seats etc. From this was a very fine view over the country. The Rhone at our feet. The rich country beyond, cheerful & blessed with cultivation & far away the Alps- but not the Snowy Alps. Still they were very fine indeed. Thence my guide took me to a house of sisters of the S. Charles- where poor insane people are taken care of. In the Chapel is a celebrated piece of Sculpture of Canova's. It is a Crucifixion in Ivory. I suppose the figures of Our LORD (297/1050 "Avignon") must be 3 feet high. It is very beautiful indeed. I never saw anything much more beautifully done. The face wears an expression of calm suffering. The Chapel was a long oblong building- with a great deal of White & gold etc. & some fairish pictures. Next I went to the Palace of the Popes. It is now a barrack & there is nothing to see. I went into several of the rooms, full of old [*** *** (?)] told that they once framed the Chapel. It was a vaulted building- but I could not make out much appearance of a Chapel- but I dare say it was. On one side is the high Tower- notorious for the [*** (?)] rites practised there. There is literally nothing to see. I went into 2 other small Churches on my way home. But I forgot, I went to the Museum- where there is nothing whatever to see- except a room upstairs with some pictures. The further end occupied entirely by pictures of the Vernet family, one or 2 Poussins & a few others that I thought grand. I came home & dismissed my guide & went to the Bureau & took my place in the Steamer & then went for a walk along the Rhone. I saw a steamer come in swarming with people. I amused myself watching some men fishing with some large nets which they let down from the quay- but caught (298/1051 "Avignon to Vienne") nothing. I walked along under the Walls, which run for some way parallel with the river; leaving room for a wide road between. The walls are very perfect & a beautiful grey colour. Indeed everything seems pretty here. Such a contrast to the dirt of the East- I assume it would seem more strange if I had just come from England. Home to dinner. Dupuis & the other man (whose country I can't make out) came in to dinner & then hurried away again to go by the Diligence. I came up to my room- wrote journal- read etc.

Tuesday. May 23

I got up at 3 & soon after 4 the porter came for my luggage. I found an Omnibus in the yard & we drove down to the quay- as we passed out of the gate of the Town, my heart misgave me about the rest of my luggage left at the Bureau, but the man, I understood, said it would be all right & so I left it. At the quay I found 2 boats. My bus drove to the "Express". I went on board to see for my things. They were not there. I showed my ticket to a man there & he said it was for the other boat- which probably would take 2 days in getting to Lyons- bad news for me. I posted off to the other steamer- but still could see no luggage. I asked a man who looked like an important personage & he said it was doubtless at the bureau. So I sent a man fast & it came in time. Got on board the steamer & we were off- the other boat, the Express, starting one minute before us. On (299/1052 "Avignon to Vienne") board I found the French American, Green & the tall Frenchman- who welcomed me. It was a satisfaction to find somebody else in the slow boat as well. The Rhone is very beautiful, more so I think than the Rhine. It was very pleasant all the morning. The country very rich. Fine hills near the river & magnificent mountains to the East, the French Alps & the old towns on the river very picturesque. I got a capital breakfast on board. In the afternoon it began to be suspected & whispered that we should not reach Lyons that evening & rain came on. About 5 I had some dinner & went on deck again, the rain having lulled. At 9 we came to anchor at Vienne. Where it was announced we should spend the night & on at 4 next morning. I got my Carpet bag & went ashore at once in the rain to the "Hotel Table ronde", where I got a bed & went into it at once, in order to get as much as I could out of it.

Wednesday. May 24

I was called at 3 again & got on board the boat as the clocks were striking 4 & in another minute we were off. A very pretty bit of river all the way to Lyons, which we reached at ½ past 7. I took my things at once across the Town to the other river, the "Saone", to the steamer there & took my ticket- but alas I only to Chalons. At 9 we started. There was several English on board, men & women & a nice (300/1053 "Vienne to Lyons - on to Paris") carriage with a viscount's coronet- a young fellow & a young wife inside. It came on to rain before long & rained the whole day. At a quarter past 6 we got to Chalons. The train was to start at 6.35 & I had made up my mind to go all the night- it was doubtful whether I should manage that train. Most people had taken their tickets from Lyons & so would have no trouble with luggage. I got a porter at once- who was very slow- I did most of the work myself. He got ½ my things on to an Omnibus & while he came back for the rest, the Omnibus went. I got myself & things into another- which was extra slow. When I got to the Station, the ½ of my baggage taken by the former Omnibus could not be found & while hunting for them the booking office closed & there was an end of that train. There was no help for it, but to take it kindly & wait till 8- when another train would go- but 3rd class & slow- taking till 8 next morning- 12 hours & then I should miss the 9 o'clock train to Rouen & all this owing to the dishonesty of the people at Avignon, cheating me into going by their slow boat instead of the "Express". I took a walk through the Town & found a Church open & numbers of people going in. It was very pleasant to see so many people within, saying their prayers- men & women- young & old. I knelt & said my own Evening Prayers as well. Surely it was a right thing (301/1054 "[*** (?)] France to Paris & to Rouen") to do. I determined to be a spend thrift & go first class. I had the carriage to myself- put a cushion across from seat to seat- put on a pair of dry socks & slippers. Made myself snug & went to sleep.

May 25 Thursday. Feast of Ascension

I awoke from time to time- but had a good deal of sleep till 4 o'clock. The morning broke very fine & about six 2 old gentlemen got in & spoilt my bed. We got to the Paris Terminus at 8.5. The Inspector was very civil on my saying I wished to get across to the other train & only had my portmanteau unlocked & locked again. I got into a cab & was unlucky again. The more the man whipped his 2 obstinate beasts of horses, the more they determined not to go fast & the end of it was, that I was just in time at the Station to see the Ticket Office closed again! All my chance of seeing Rouen quietly at an end. I took my carpet bag & went to the "Hotel de Dieppe" opposite the Station- where I had a regular ablution with cold water & dressed & had breakfast. I found having his breakfast a young lieutenant of the 3rd Chasseurs- he spoke English fluently & told me his Mother was an English woman. He seemed a very nice fellow. At 12 I started in the Train. It is a beautiful road the whole way. I never saw so many nice places one after another as along that line- little & big. We pass a "Les Batignolles" Montmoreau, St Denis & (302/1055 "Paris to Rouen") Neuilly. Maison- Laffitte built in 1658- where the Count d'Artois lived. Napoleon gave it to Marshal Lannes. Then we passed by S. Germain, the birth place of Louis XIV. Next Poissy- here Charles le Chauve [= Bald], in 860, held court & here in 1215 was born Louis IX 'Saint Louis'. Here too in 1561 was held the meeting between the Catholics & Calvinists. At "Meulan" we entered Normandy. Pretty churches, with high walls & roofs & apses, all along the road not far from "Mautes" we passed the "Chateau de Rosny", where lived & died the Marquis de Sully, the friend of Henry le Grand. Beyond this the Towns of La Roche-Guyon & Vernon- the later joining their name to the Vernon Harcourt. It is believed to have been a Roman Camp. It was enlarged & fortified in 1113 AD by Henry 1st of England. Next the Railway cuts through the fine avenue of the Chateau Vernon or Bizy it belonged to the Counts d'Eu & so at last to the Dowager Duchess D'Orleans. There the Chateau Gaillon, now a prison. The Town of Pont de l'Arche- named from its bridge of some 20 arches. The forest of Rouvray stretched along the hills to the left & then we came to Rouen. I got a room at the "Hotel D'Angleterre" & then went out for the ½ hour before dinner. I went of course straight to the Cathedral. The W. front is wonderful. I can't describe it. I went in. It was the Feast of our LORD's (303/1056 "Rouen") Ascension. Service was going on & the glorious Church was full from the Choir doors to within a few yards of the West door. Delicious & solemn music was gushing from the Great Organ high overhead at the West End, alternately with a smaller but still beautiful organ in the N. aisle of the Choir & a long procession of many priests was slowly pacing down the Church. I suppose 80 or 100 manly voices chanting with the organ. It was very solemn & to me as usual very sad. They were Christians worshipping God & I couldn't join- nor even knew what they were saying. They passed up the N. aisle, round behind the altar, down the S. Aisle & up the Nave again & when they passed into the Choir, the whole knelt before the altar & some most splendid music began. I have seldom heard anything so beautiful. I think it was Mozart. It was a very striking scene. I stood against the W. Door & could just make out the kneeling white figures in the far distance through the skreen (sic), (such a picture as Pugin liked to paint) & could see the lights flashing on the censers as they were tossed. This is certainly the service a Cathedral is built for. It was no use staying, though I did not like to come away. I found dinner ½ over. At the further end of the Table some Englishmen were talking very loud about Catholics (304/1057 "Rouen") & Protestants. I could not make out what they said exactly- but I heard they were laughing at the Idolatry & Ignorance of the people they had seen in Church that day & I saw that as their voices grew louder, for there seemed to be some difference of opinion, the rest of the table grew quiet at first & then nearly all were laughing. Dinner ended & I came away to my own room. It was still quite light- but I could not go out- as it had come on to rain. So I sat & read in my room & wrote up this & now to bed that I may be up early to go & see S. Ouen & S. Maclou.[48] My train goes at 11.30 & I must see what I can before Dinner at 10. This time tomorrow I hope please God I shall be on the English Channel making fast for Southampton- where this journal will come to an end. Fancy being with them again on Saturday Evening! I hope they will be all at Milton Hill.

---

[48] Two immense late medieval Gothic churches in Rouen.

Appendices

Appendix 1: The Anglo-American Group at Petra 26-29 March 1854.

Name	Origin	DPOB	DPOD	Age at Petra	Age at Death	Occupation	Death Order
Bowles, Thomas	British (England)	Milton, Berks, 5 January, 1822	12th January 1899, Paddington, London	32	77	Clergyman	11
Bryce, William	British (Scotland)	Killaig, Co. Coleraine on 28 April 1821	Morningside, Edinburgh on 21 February 1914	Almost 33	92.5	Medical doctor	14
Eames, Rev. James	American (Massachusetts)	Dedham, MA, 29 November 1814	On a ship in harbour, Hamilton, Bermuda, 17 December 1877	39.25	63	Clergyman	5
Eames, Jane Anthony	American (Rhode Island/ Massachusetts)	Wellington/ now Dighton, Mass, 21 January 1816	Boston, MA, 8 July, 1894	38	78.5	Author	10
Fenton, Samuel Greame	British (England)	Leeds, 24 December 1821	Belfast, 6 June 1892	33	70.5		9
Fenton, Rev. George Metcalfe	British (England)	Leeds, 24 September 1826	Christchurch, Hants, 18 Dec 1879	27.5	53	Clergyman	6
Fish	American (Alabama)						
Freeman	British (?)			very early 20s?			
Kennedy, Henry Hyndman	British (England)	Newland, Colford, Gloucs., bap. 18 March, 1814	Fort Augustus, Invernesshire, 20 June 1880	40	66	Merchant?	7
Rodewald	German (Bremen) and American (New York)	Bremen, 7 March 1811	Kehrsiten, Switzerland, 11 August 1884	43	73.5	Merchant?	8
Rogers, John Leverett	American (Massachusetts)	Ipswich, MA bap. 23 Oct. 1808	St Nicholas Hotel, NYC, NY 2 Sep 1869	45	61	Merchant	3
Rogers, Virginia Beverley	American (New York)	New York City, NY, 27 Mar 1827	Manhattan, NYC, 30 Apr 1900	27	73		12
Ross, Robert	British (England)	Westminster, 19 February 1813	Leghorn/ Livorno 28 November 1859	41	46.5	Lawyer; Gentleman	2
Ward, Henry Veazey	American (Maryland)	Sassafras Neck, Fredericktown, Cecil Co, Maryland, 26 Sept 1806	Chateau de Coppet, near Lake Ouchy, Geneva 15 March, 1873	44	66.5		4
Wakefield, John Edward	British (England)	Kendal, Cumbria, 8 August 1830	Great Malvern, Worcestershire, 30 July 1858	23.5	27	Medical student	1
Yeatman, Henry Clay	American (Tennessee)	22 September 1831 Nashville, TN	1 August 1910. Hamilton Place, Columbia, TN	22.5 years	78.8 years		13

Appendix 2: Timeline of Bowles' Journey in Egypt and the Levant.

Date (all 1854)	Place	Journal Page(s)
Saturday 28 January	Galle. Departs on the steamer "Bombay" in the evening.	411-414
Sunday 29 January	On board steamer "Bombay".	414-415
Monday 6 February	Aden. Day ashore	432-439
Tuesday 7 February	Depart Aden overnight. Red Sea: on board steamer "Bombay".	439-450
Sunday 12 February	Arrival in Suez	450-458
Monday 13 February	'Van' from Suez to Cairo through Desert	458-463 & 1-2
Tuesday 14 February	Cairo. Excursion to Citadel Mosque & Turkish Bazaar	2-13
Wednesday 15 February	Cairo.	13-16
Thursday 16 February	Cairo. Excursion to Shoobra Gardens	16-19
Friday 17 February	Cairo. Excursion to Tomb of the Memlooks (Mamluks), Whirling Dervishes, Nilometer on the Island of Roda & Copts Quarter	19-26
Saturday 18 February	Cairo. Excursion to the Pyramids of Giza, Sphinx & surrounding tombs	26-42
Sunday 19 February	Cairo.	42-44
Monday 20 February	Cairo. Excursion to Heliopolis	44-52
Tuesday 21 February	Cairo. Excursion to Mosque of Sultan Tayloon	52-54
Wednesday 22 February	Cairo. Excursion to Sakkara (Saqqara) & Memphis	54-56
Thursday 23 February	Cairo.	56-58
Friday 24 February	Cairo. Excursion to Sultan Hassan's Mosque	58-60
Saturday 25 February	Cairo.	60-61
Sunday 26 February	Cairo.	62
Monday 27 February	Cairo.	62-64
Tuesday 28 February	Depart Cairo for the Desert. Start of journey to Suez	64-68
Wednesday 1 March	Desert	68-70
Thursday 2 March	Desert	70-75
Friday 3 March	Suez	75-76
Saturday 4 March	Suez to Ayun Musa	76-80
Sunday 5 March	Ayun Musa to near Wadi Wardan	80-83
Monday 6 March	Wadi Gharandel via Wadi Wardan, Wadi Amarah & Marah	83-87
Tuesday 7 March	Wadi Gharandel to the Sea via Wadi Thall & Wadi Teiyibah	87-90
Wednesday 8 March	Seaside to Wadi Mucatteb (Valley of Inscriptions) via Wadi Shellal, Wadi Sedr, Wadi Maghara (excursion to see Egyptian Inscriptions)	90-96
Thursday 9 March	Wadi Mucatteb (Valley of Inscriptions) to Wadi Feiran	97-100
Friday 10 March	Wadi Feiran to Wadi Shekh	100-102
Saturday 11 March	Wadi Shekh to Mt Sinai via Wadi Ed Deir. Climbed Mt Sinai (Djebel Musa)	102-112
Sunday 12 March	Mt Sinai	112-114
Monday 13 March	Mt Sinai. Excursion to Wadi Er Rahah and "Smitten Rock"	114-117
Tuesday 14 March	Mt Sinai to Wadi Sahl? Excursion to Tomb of Sheikh El Saleh	117-121
Wednesday 15 March	Wadi Sahl?	121-122

APPENDICES

Thursday 16 March	Wadi El Ain (Valley of the Springs)	122-125
Friday 17 March	Along the Gulf of Aqaba	125-126
Saturday 18 March	Along the Gulf of Aqaba to Aqaba	126-128
Sunday 19 March	Aqaba	129-134
Thursday 23 March	Start of journey to Petra	134-137
Friday 24 March	Wadi Araba	137-138
Saturday 25 March	Wadi Gharandel and Seir	138-141
Sunday 26 March	Arrive Mt Hor and Petra	141-145
Monday 27 March	Explore Petra: El Deir, Qasr al-Bint	145-149
Tuesday 28 March	Explore Petra: Al-Khazneh, Siq	149-156
Wednesday 29 March	Depart Petra to Wadi Araba via Mt Hor	156-158
Thursday 30 March	Wadi Araba	158-160
Friday 31 March	Wadi Araba	160-162
Saturday 1 April	Wadi Kurnub and El Tellal?	162-163
Sunday 2 April	Hebron	163-166
Monday 3 April	Hebron to Jerusalem. Excursion to Pools of Solomon and Bethlehem: Church to see Holy Place	166-170
Tuesday 4 April	Jerusalem: City Walls, Pool of Siloam, Garden of Gethsemane, Church of the Holy Sepulchre	170-172
Wednesday 5 April	Jerusalem: Church of the Ascension, Bethany (Tomb of Lazarus)	173-177
Thursday 6 April	Jerusalem: Convent of the Knights of St Stephen, Tomb of the Kings, Tomb of the Judges, Jeremiah's cave	178-182
Friday 7 April	Jerusalem: Pool of Siloam, King's Gardens, Wailing Wall, Armenian Convent, Birket Es Sultan	182-186
Saturday 8 April	Jerusalem: Excursion to Village of Hanina, Nebi Samwil, Er Ram, Anata (Anathoth)	186-189
Sunday 9 April	Jerusalem: Valley of Hinnom, Hill of "Evil Counsel, Armenian & Latin Convents, Valley of Jehoshaphat	189-193
Monday 10 April	Jerusalem to Mar Sabba via Valley of Hinnom, Valley of Jehoshaphat, Valley of Kidron	193-198
Tuesday 11 April	Mar Sabba to Jericho via Dead Sea & River Jordan where Jesus was baptised	198-204
Wednesday 12 April	Jericho to Jerusalem via Spring of Elisha, Bethany, Mt of Olives, Valley of Jehoshaphat	205-206
Thursday 13 April	Jerusalem: Mt of Olives, Upper pool of Gideon	206-207
Friday 14 April (Good Friday)	Jerusalem: Mt. of Olives, Tombs in the Valley of Jehoshaphat, Church of the Holy Sepulchre	207-210
Saturday 15 April	Jerusalem: Greek Convent, Church of the Holy Sepulchre	210-214
Sunday 16 April (Easter Day)	Jerusalem: Church of the Holy Sepulchre	214-216
Monday 17 April (Easter Monday)	Jerusalem to Ain Es Yebrad via Shafat and Nebi Samwil	216-220
Tuesday 18 April	Ain Es Yebrad to Nablous via Ain el Haramiyan, Sinjil, Khan Lubban, Jacob's Well, Gerizim, Samaritan Synagogue	220-226
Wednesday 19 April	Nablous to Jenin via Sebastiyah	226-230
Thursday 20 April	Jenin to Nazareth: Greek Church, Latin Convent	230-236

Friday 21 April	Nazareth to Tiberias via Erana, Cana, Plain of Esdraelon, Es Suk	236-241
Saturday 22 April	Tiberias to Wadi Mellahah via Mijdal, Ain et Tin, Tell Husn, Khan Minyeh, Khan Jubb Yusuf	241-246
Sunday 23 April	Wadi Mellahah to Banias (Paneas, Caesarea Philippi) via Ain Mallahad, Kalat el Hunin	246-250
Monday 24 April	Banias to Bejam	250-253
Tuesday 25 April	Bejam to Jedehda	254-255
Wednesday 26 April	Jedehda to Damascus: visit to Bazaar and Khan	256-258
Thursday 27 April	Damascus: Bazaar, Banker's house, Street called Straight, Khan	259-261
Friday 28 April	Damascus: Bazaar	262
Saturday 29 April	Damascus to Ez Zebedani via Valley of the Barada, Souk Wadi Barada (Abila)	262-265
Sunday 30 April	Ez Zebedani to Baalbec: Temple of Jupiter	265-266
Monday 1 May	Baalbec to Sahleh	266-272
Tuesday 2 May	Sahleh to Khan Hussein	272-273
Wednesday 3 May	Khan Hussein to Beirut	273-274
Thursday 4 May	Beirut	274-275
Friday 5 May	Beirut: Excursion to Dog River	275-276
Saturday 6 May	Beirut	277
Sunday 7 May	Beirut: Chapel of the American Mission, Maronite Church, Chapel of the Sisters of Charity	277-279
Monday 8 May	Leave Beirut on Steamer	279-280
Tuesday 9 May	On steamer off Jaffa	280-281
Wednesday 10 May	On steamer between Jaffa and Alexandria	281-282
Thursday 11 May	Alexandria: English Church, Cleopatra's needle, Pompey's Pillar, Mahmoudie Canal, Pacha's Palace, Library, Roman Catholic Church	282-285
Friday 12 May	Leave Alexandria onboard Steamer	285-286
Saturday 13 May	Onboard Steamer "Mentor"	286
Sunday 14 May	Onboard Steamer "Mentor"	286-289
Wednesday 17 May	Valletta, Malta: St. John's Church, leave Malta onboard Steamer "Mentor"	289-291
Thursday 18 May	Onboard Steamer "Mentor"	291-292
Friday 19 May	Onboard Steamer "Mentor"	292-293
Saturday 20 May	Onboard Steamer "Mentor". Arrive Marseilles.	293
Sunday 21 May	Marseilles to Avignon	294-295
Monday 22 May	Avignon: Cathedral, Palace of the Popes, Museum	296-298
Tuesday 23 May	Avignon to Vienne	298-299
Wednesday 24 May	Vienne to Lyons and on train from Chalons to Paris	299-301
Thursday 25 May	Paris to Rouen: Cathedral	301-304
Friday 26 May	(On the Channel making for Southampton)	304
Saturday 27 May	(Home with family at Mill Hill)	304

Bibliography

Unpublished Sources

Badger, G. *Journal of Rev George Percy Badger (1815-88), chaplain and oriental scholar, at Aden 1846-62, staff chaplain and Arabic interpreter under Lt-Gen Sir James Outram in the Persian Expedition 1856-7, with additional material on his travels in Egypt.* British Library (Mss Eur B377).

Bowles, T. 1853-54. (*Daily narrative of trip from Port Resolution, Tanna Island, New Hebrides to Cairo. Voyage took place in vessels Early Bird to Hong Kong, S.S. Singapore to Galle, S.S. Bombay to Cairo*). Unpublished Diary, National Library of Australia, Canberra. (NLA MS 2264).

Bowles, T. 1854. *Travel Diary. Near East 1854.* Unpublished Diary, University of Western Australia (Special Collections and Archives).

Bryce, W. *Journal of a Visit to the Middle East by Dr William Bryce of Edinburgh.* National Library of Scotland, Edinburgh.

Macintyre, J. 1854. *Walks on Deck and Rambles on Shore During a Voyage of Circumnavigation of the Globe [1853].* Papers of James J. Macintyre, National Library of Australia, Canberra, Vol. IV (Add MS 41745). https://nla.gov.au/nla.obj-752644206/findingaid

Stobart, H. 1846-1856. *Letters from [Henry] Stobart to his mother, Mrs Thomas Chilton, 15 August 1846, October 1852 - April 1856, on board the ships Resolute. Early Bird, Sjngapore. Bengal, Calcutra. Indus and Lotus and from Sydney, Brisbane, Wollongong, Ceylon. India. the Nile. Cairo. Jerusalem, Paris and Cannes.* Unpublished Documents, National Library of Australia, Canberra. (NLA Reel M467).

Vessey, F. G. = Warriner, H. 2022. *Francis Gerald Vessey's Journal. Rome, Egypt, Sinai and Palestine.* (Published Privately).

Yeatman, H. C. *Henry Clay Yeatman Diaries.* Tennessee State Library and Archives (Yeatman-Polk Collection. Box 91)

Published Sources

Abrahams, Beth-Zion Lask 1978-80. James Finn: Her Britannic Majesty's Consul at Jerusalem between 1846 and 1863. *Transactions & Miscellanies (Jewish Historical Society of England)* 27: 40-50.

Abudanah, F. 2016. Sheikh Hussein Alawin (Ibn Injad) and his role in escorting travels to Petra, travel books as a source. *Jordan Journal of History and Archaeology* 9.2: 125-148.

Adlgasser, F. 2011. *Viktor Franz Freiherr von Andrian-Werburg. Tagebücher 1839-1858,* 3 vols, Köln: Bohlau.

Annand, A. McK. 1965. Lieutenant-Colonel William Shirriff, 7th Madras Native Cavalry, c. 1800. *Journal of the Society for Army Historical Research* 43, No. 173 (March): 1-4.

Anonymous ['Our Correspondent'] 1854a. The Late Captain Butler and the Siege of Silistra. *Illustrated London News* 29 July: 94-96.

Anonymous 1854. The overland route. From Point de Galle to Aden. *The Sydney Morning Herald*, Wednesday July 26, 1854: 3.

Anonymous 1858. Bishop Gobat and the social evil in Jerusalem. *The Ecclesiastic and Theologian* XX: 277-284.

Anonymous 1860. Death of Miss Cooper. *Jewish Intelligence* (1 Jan.): 14-16.

Anonymous 1887. (Notice under Medical Staff), *British Medical Journal*, 4 June: 1244.
Anonymous (n.d.) Cigarettes: Women. Coffin nails: the tobacco controversy in the 19th century. *Text, Cartoons and Ads from the pages of Harper's Weekly 1857-1912.* Viewed 19 July 2022. https://tobacco.harpweek.com/hubpages/CommentaryPage.asp?Commentary=Women
Anonymous 1914. William Bryce, M.D., Edinburgh. *The British Medical Journal*, 1, No. 2774 (Feb. 28): 513.
Anonymous (revised by Anita McConnell) 2004. Bryce, James (1806–1877). *Oxford Dictionary of National Biography.* Accessed online 16 March 2021.
Anonymous (revised by David Huddleston) (2006) "Bryce, James (1767–1857). *Oxford Dictionary of National Biography.* Accessed online 18 March 2021.
Ari, N. 2021. Competition in the cultural sector: Handicrafts and the rise of the trade fair in British Mandate Palestine. In K.S. Summerer and S. Zananiri (eds) 2021. *European Cultural Diplomacy and Arab Christians in Palestine, 1918-1948,* Cham: Palgrave Macmillan: 213-246.
Ashworth, P. and Kinder, J. 1998. *Westwood, Normandy, The Story of a Surrey Estate,* Guildford: Westwood Place Management.
Badger, G.P. 1838. *Description of Malta and Gozo,* Malta: M. Weiss (plus five further editions till 1879).
Badger, G.P. 1852. *The Nestorians and Their Rituals with the Narrative of a Mission to Mesopotamia and Coordistan in 1842 44, and of a Late Visit to Those Countries in 1850. Also, Researches into the Present Condition of the Syrian Jacobites, Papal Syrians, and Chaldeans and an Inquiry into the Religious Tenets of the Yezeedees.* 2 vols. London: Joseph Masters.
Badger, G.P. 1863. Introduction. In L. di Varthema 1510/1863. *The Travels of Ludovico de Varthema in Egypt, Syria, Arabia Deserta and Arabia Felix, in Persia, India, and Ethiopia, AD 1503 to 1508,* trans and Preface by J.W. Jones, London: Hakluyt Society: xvii-cxxi.
Badger, G.P. 1871. *History of the Imams and Seyyids of Oman by Salil ibn-Razik from AD 661 to 1856.* London: Hakluyt Society.
Badger, G.P. 1881. *An English-Arabic Lexicon, in which the Equivalents for English Words and Idiomatic Sentences are Rendered into Literary and Colloquial Arabic. Kitab al-Dhakhira al-'llmlya fi 1-lughatayn al-Inkillziya wa'l-'Arabiya.* London: C. Kegan Paul and Co.
Barak, O. 2015. Outsourcing: energy and empire in the age of coal, 1820-1911. *International Journal of Middle East Studies* 47.3: 425-445
Barber, J. (Captain, H.C.S.) 1845. *The Overland Guide-Book; a Complete Vade-Mecum for the Overland Traveller, to India via Egypt.* London: William H Allen & Co. (2nd. Ed. 1850)
Beamont, W. 1856. *A Diary of a Journey to the East in 1854,* 2 vols, London: Longman, Brown, Green and Longmans.
Beamont, W. 1862. *To Sinai and Syene and Back in 1860 and 1861.* Warrington: A. Mackie, Guardian Office (2nd ed. 1872)
Beamont, Rev. W.J. 1861. *Cairo to Sinai and Sinai to Cairo. Being an Account of a Journey in the Desert of Arabia, November and December, 1860.* Cambridge: Deighton, Bell and Co.
Bird, M. 1957. *Samuel Shepheard of Cairo: A Portrait.* London: Michael Joseph.
Blau, O. 1855. Inschriften aus Petra. *Zeitschrift der Deutschen Morgenländischen Gesellschaft* IX: 230-237.
Blumberg, A. 1980. *A View from Jerusalem, 1849-1858: The Consular Diary of James and Elizabeth Anne Finn,* Cranbury, NJ: Associated University Presses.
Bonar, H. 1857. *The Desert of Sinai. Notes on a Spring-Journey from Cairo to Beersheba.* New York: Robert Carter.

Bolsover, G.H. 1936. David Urquhart and the Eastern Question, 1833-37: A study in publicity and diplomacy. *The Journal of Modern History* 8.4: 444-467.

Bonello, G. (ed.) 2017. *Celebrating 200 years of Schranz,* Valetta: Allied Publications Malta for Fondazzioni Patrimonju Malti.

Boyd, D.H.A. 2005. William Henderson (1810-1872) and homeopathy in Edinburgh. *Journal of the Royal College of Physicians Edinburgh* 36:170-178.

Briggs, A. 1954. *Victorian People. A Reassessment of Persons and Themes, 1851-67.* Chicago: University of Chicago Press.

Bromley, J. and D. 2012. *Wellington's Men Remembered: A Register of Memorials to Soldiers who Fought in the Peninsular War and at Waterloo.* Vol. 1, London: Praetorian Press.

Brünnow, R.E. and von Domaszewski, A. 1904-09. *Die Provincia Arabia,* 3 vols, Strassburg: Trübner.

Buckingham, J.S. 1821. *Travels in Palestine through the Countries of Bashan and Gilead, East of the River Jordan: Including a Visit to the cities of Geraza and Gamala in the Decapolis,* 2 vols, London: Longman, Hurst, Rees, Orme, and Brown.

Burckhardt, J.L. 1819. *Travels in Nubia by the Late John Lewis Burckhardt.* London: John Murray (includes a 'Memoir on the Life and Travels of John Lewis Burckhardt' on pp. i-xlix)

Burckhardt, J.L. 1822. *Travels in Syria and the Holy Land: by the Late John Lewis Burckhardt,* London: John Murray.

Burckhardt, J.L. 1824. (Letters to E.D. Clarke from Damascus and Cairo). In W. Otter (ed.) (1824) *The Life and Remains of the Rev. Edward Daniel Clarke, LL.D., Professor of Mineralogy in the University of Cambridge.* 584-605. London: Printed for George Cowie and co.

Butler, J.A. 1854. Journal of Captain J.A. Butler at the Siege of Silistra. *National Army Museum* (NAM1968-03-45).

Butler, P. 1874. *Axel and Valborg. A Tragedy in Five Acts and Other Poems,* trans from Danish, edited by Professor (E. H.) Palmer. London: Trübner.

Canaan, T. 1930. *Studies in the Topography and Folklore of Petra.* Ch. V "The Liatneh: 'The Bedouin of Petra'". Jerusalem: Beyt Ul-Makdes Press.

Castlereagh, Viscount 1847. *A Journey to Damascus through Egypt, Nubia, Arabia Petraea, Palestine and Syria.* 2 vols. London: Henry Colburn.

Chadwick, F.E., Gould, J.H., Kelley, J.D.L., Rideing, W.H. and Seaton, A.E. 1892. *Development, Management and Appliances, Ocean Steamships. A Popular Account of their Construction,* London: Scribners.

Charles, E. R. 1866. *Wanderings over Bible Lands and Seas,* London, Edinburgh and New York: Nelson.

Chesney, L.F. and J. O'Donnell 1885. *The Life of the Late General F. R. Chesney, Colonel Commandant, Royal Artillery.* Edited by Stanley Lane-Poole. London: W. Allen and Co.

Crawford, A. 2008. Graham, James (1806-1869). Scottish itinerant photographer. In J. Hannavy (ed.) *Encyclopedia of Nineteenth-Century Photography.* 605-6. London: Routledge.

Cubley, L.M. 1860. *The Hills and Plains of Palestine.* London: Day & Son.

Cummings, W.F. 1839. *Notes of a Wanderer in Search of Health through Italy, Egypt, Greece, Turkey, up the Danube, and Down the Rhine.* 2 vols. London: Saunders and Ottley.

Cust, R.N. 1880. A tour in Palestine. *Linguistic and Oriental Essays. Written from the Years 1846 to 1878.* 252-288. London: Trübner.

Cust, R. N. 1899. *Memoirs of Past Years of a Septuagenarian,* Privately Printed.

Danvers, F.C. et al. 1894. *Memorials of Old Haileybury College,* Westminster: Constable

Darvall, J. 1845. *The Wreck on the Andamans: Being a narrative of the very remarkable preservation, and ultimate deliverance, of the Soldiers and Seamen, who formed the Ships Companies of the Runnymede and Briton Troop-Ships, both wrecked on the Morning of the 12th November, 1844, upon one of the Andaman Islands, in the Bay of Bengal. Taken from authentic Documents.* London: Pelham Richardson.

Dawe, J. 2019. William Henderson, controversial Professor of Pathology, Edinburgh. *The Grange Newsletter,* January 2019: 3

Dingley, R. 2004. Tupper, Martin Farquar. *Oxford Dictionary of National Biography* (accessed 27 October 2021)

Ditson, G.L: 1858. *The Para Papers on France, Egypt and Ethiopia.* Paris: Fowler and New York: Mason Brothers

Drew, Sir R. 1968. *Commissioned Officers in the Medical Service of the British Army, 1660-1960.* 2 vols. London: Wellcome Historical Medical Library.

Durbin, J.P. 1845. *Observations in the East chiefly in Egypt, Palestine, Syria, and Asia Minor.* 2 vols, New York: Harper & Brothers.

Eames, J.A. 1847. *A Budget of Letters.* Boston: Ticknor and Fields.

Eames, J.A. 1855. *Another Budget, or Things which I saw in the East.* Boston: Ticknor and Fields.

Eames, J.A. 1860. *The Budget Closed.* Boston: Ticknor and Fields.

Eames, J.A. 1875. *Letters from Bermuda.* Concord, RI: Republican Press Association.

Anon. [= Jane Eames?] .1878. *In Memoriam. Rev. James H. Eames, D. D., Rector of St. Paul's Church, Concord, N. H.* Concord, NH: Privately Published.

Eliav, M. 1992. The rise and fall of Consul James Finn. *Cathedra*, 65 (September 1992): 37-81 (Hebrew).

Eliav, M. 1997. *Britain and the Holy Land, 1838-1914. Selected Documents from the British Consulate in Jerusalem.* Jerusalem: Yad Izhak Ben-Zvi Press.

Ellis, W.A. (1878) *Rev. James Henry Eames, A.M., D.D.* Northfield, VT: Norwich University

Falkland, The Viscountess (Amelia FitzClarence Cary) 1857. *Chow-chow: Being Selections from a Journal Kept in India, Egypt, and Syria.* 2 vols. London: Hurst and Blackett.

Fiema, Z. and J. Frösén (eds) 2004. *Petra – The Mountain of Aaron. The Finnish Archaeological Project in Jordan. I. The Church and the Chapel.* Helsinki: Societas Scientiarum Fennica.

Finati, G. 1830. *The Narrative of the Life and Adventures of Giovanni Finati.* London: John Murray.

Finn, E.A. 1866. *A Home in the Holy Land. A Tale Illustrating Customs and Incidents in Modern Jerusalem.* London: James Nisbet and Co.

Finn, E.A. 1869. *A Third Year in Jerusalem. A Tale Illustrating Customs and Incidents in Modern Jerusalem.* London: James Nisbet and Co.

Finn, E.A. 1929. *Reminiscences of Mrs. Finn, Member of the Royal Asiatic Society.* London: Marshall, Morgan and Scott.

Finn, J. 1867. *Byeways in Palestine.* London: James Nisbet and Co.

Finn, J. 1878. *Stirring Times or Records from Jerusalem Consular Chronicles of 1853 to 1856.* London: Kegan Paul.

Fisher, H. 2001, *From a Tramp's Wallet: a Life of Douglas William Freshfield.* Wymondham: Erskine Press.

Fisher, H.A.L. 1927. *James Bryce. (Viscount Bryce of Dechmont, O. M.).* 2 vols. London: MacMillan.

Fitzcook, H. 1850-1852. *Route of the Overland Mail to India.* London: Atchley & Co.

Formby, H. 1843. *A Visit to the East Comprising Germany and the Danube, Constantinople, Asia Minor, Egypt, and Idumaea.* London: Burns.

BIBLIOGRAPHY

Forster, C. 1844. *The Historical Geography of Arabia; or the Patriarchal Evidences of Revealed Religion: A Memoir with Illustrative Maps; and an Appendix, containing Translations, with an Alphabet and Glossary, of the Hamyaritic Inscriptions Recently Discovered in Hadramaut.* London: Duncan and Malcolm.

Forster, C. 1851. *The One Primeval Language. Traced Experimentally Through Ancient Inscriptions in Alphabetic Characters of Lost Powers from Four Continents including The Voice of Israel from the Rocks of Sinai: and the Vestiges of Patriarchal Tradition from the Monuments of Egypt, Etruria, and Southern Arabia.* London: Richard Bentley.

Forster, C. 1853. *The Monuments of Egypt, and their Vestiges of Patriarchal Traditions,* London: Richard Bentley.

Forster, C. 1862. *Sinai Photographed. Contemporary Records of Israel in the Wilderness,* London: Richard Bentley.

Foster, J. 1893. *Oxford Men and their Colleges,* Volume 1. Oxford: James Parker and Co.

Freshfield, D. W. 1869. *Travels in the Central Caucasus and Bashan: Including Visits to Ararat and Tabreez and Ascents of Kazbek and Elburz.* London: Longmans Green.

Foster, C. (ed.) 2004. *Travellers in the Near East.* London: Stacey International.

Frösén, J., Arjava, A., Lehtinen, M. Kaimio, J., Buchholz, M. and Gagos, T. 2002-2018. *The Petra Papyri,* vols I-V. Amman American Center of Oriental Research Publications.

Galt, D. 2017. *Mowbray International Numismatic Auction #17, 11th March 2017,* Wellington: Mowbray Collectibles.

Garfield, B. 2007. *The Meinertzhagen Mystery. The Life and Legend of a Colossal Fraud.* Washington: Potomac Books.

Gavin, R.J. 1975. *Aden under British rule, 1839–1967.* London: Hurst and Co.

Gavin, C.E.S. and Rosovsky, N. 1995. Mendel John Diness of Jerusalem and Cincinnati. *History of Photography* 19.3: 224-228.

Gibson, S. and Chapman, R.L. 1995. The Mediterranean Hotel in nineteenth century Jerusalem. *Palestine Exploration Quarterly* 217: 93-105.

Gibson, S., T. Shapira, and R.L. Chapman. 2013. *Tourists, Travellers and Hotels in Nineteenth Century Jerusalem.* Leeds: Maney Publishing.

Gilmour, C. 2016. Alexander Henry Rhind (1833–63). *Proceedings of the Society of Antiquaries of Scotland* 145: 427-440.

Gluckman, L. 2000. *Touching on Deaths: a Medical History of Early Auckland Based on the First 384 Inquests,* Auckland: Doppelganger.

Goodwin, M. 1996. "The church of St. Andrew, Grafham".

Graham, J. 1858. *Jerusalem; its Missions, Schools, Converts etc. under Bishop Gobat.* London: David Batten.

Habegger, A. 2014. *Masked. The Life of Anna Leonowens, Schoolmistress at the Court of Siam.* Madison: University of Wisconsin Press.

Hammond, N. 2004. Tombs of Abingdon. *Oxfordshire Family Historian,* 18.3: 154-6.

Hawes, D. 2004. Cattermole, Richard (1795?-1858). *Oxford Dictionary of National Biography* accessed online 13 September 2021.

Hawkins, J.W. 2015. *Frederick Huth and Company 1809-1936: the Partial History of a City of London Merchant Bank.* Padstow: Privately Published (free download online).

Hillam, D. (ed.) 2011. *Tig's Boys. Letters to Sir from the Trenches.* Stroud: The History Press.

A Young Pilgrim (= Hindley, H. 1851. *Life in the Tent; or Travels in the Desert and Syria in 1850, by a Young Pilgrim.* London: Longman, Brown, Green, and Longmans.

Hughes, T. 1857. *Tom Brown's School Days.* Cambridge: MacMillan.

Hughes, T. 1859. *The Scouring of the White Horse.* Boston: Ticknor and Fields.
Humphreys, A.L. 1923. *East Hendred: A Berkshire Parish Historically Treated.* London: Hatchards.
Hunt, W.H. 1858. *Bishop Gobat in re Hanna Hadoub.* London: Joseph Masters.
Hunt, W.H. 1905. *Pre-Raphaelitism and the Pre-Raphaelite Brotherhood.* 2 vols. London: MacMillan.
Irby, C.L. and Mangles, J. 1823. *Travels in Egypt and Nubia, Syria and Asia Minor During the Years 1817 and 1818.* London: Private Printing
Irving, R. and Maitland, M. 2015. An innovative Antiquarian: Alexander Henry Rhind's excavations in Egypt and his collection in the National Museums Scotland. In N. Cooke and V. Daubney (eds) *Every Traveller Needs a Compass. Travel and Collecting in Egypt and the Near East.* 87-100. Oxford: Oxbow.
Keith, A. 1848. *Evidence of the Truth of the Christian Religion derived from the Literal Fulfilment of Prophecy; particularly as illustrated by the History of the Jews, and by the Discoveries of Recent Travellers.* 36th ed. Edinburgh: Waugh and Innes.
Keith, A. 1948. Thomas Hastie Bryce, 1862-1946. *Royal Society* 5.16: 658-665.
Kennedy, Captain H.A. 1860. Some reminiscences of the life of Augustus Fitzsnob, Esq. *The Chess Monthly* 4 (January): 1-11.
Kennedy, Captain H.A. 1876. *Waifs and Strays. Chiefly from the Chess-board.* 2nd ed. London: W.W. Morgan.
Kennedy, D.L. 2018 Travellers to 1857 to Petra. In Z.M. Al-Salameen and M.B. Tarawneh (eds) *Proceedings of the First Conference on the Archaeology and Tourism of the Ma'an Governorate.* 187-207. Ma'an, Jordan: Al-Hussein Bin Talal University.
Kennedy, D.L. 2004. *The Roman Army in Jordan.* 2nd edition. London: Council for British Research in the Levant.
Kennedy, D.L. and R.H. Bewley 2004. *Ancient Jordan from the Air.* Oxford: Council for British Research in the Levant.
Laborde, L. de 1830. *Voyage de l'Arabie Pétrée par Leon de Laborde et L. M. A. Linant de Bellefonds,* Paris: Giard.
Laborde, L. de 1836. *Journey through Arabia Petraea, to Mount Sinai and the Excavated City of Petra, the Edom of the Prophecies.* London: Murray.
Laborde, L. de and Linant de Bellefonds, L.-M.-A. 1994. *Petra Retrouvée, Voyage de l'Arabie Pétrée, 1828, augmenté d'extraits du carnet de voyage inédit de L.-M.-A. Linant de Bellefonds.* Edited with notes by C. Augé and Pascale Linant de Bellefonds. Paris: Pygmalion.
Lawrenson, R. 2004. Medical Practice in New Zealand 1769-1860. *Vesalius* X: 1,4-9.
Lazard, B. 1990. The photographs of James Graham in the Middle East. *The Photo Historian Supplement,* Summer 89: 1-18.
Lear, E. (ed.) F(ranklin) L(ushington) 1896-7. A leaf from the journals of a landscape painter. *Macmillan's Magazine,* 75: 410-430.
Legh, T. 1819. A Continuation of the Route to Jerusalem. In W.A. MacMichael *Journey from Moscow to Constantinople.* Chapter 4. London: John Murray.
Lev, E. and Perry, Y. 2008. Dr Edward Macgowan (1795-1860), a long-term pioneer physician in mid-nineteenth century Jerusalem: founder and director of the first modern hospital in the Holy Land. *Journal of Medical Biography* 16(1): 52-6.
Lewis, N. N. 2003. Petra: the first comers. In G. Markoe (ed.) *Petra Rediscovered: Lost City of the Nabataeans.* 112-116; 267. New York: Harry N. Abrams Inc. in association with the Cincinnati Art Museum.
Lewis, N.N. 2004. Travellers, tribesmen and troubles: journeys to Petra, 1812-1914. In C. Foster (ed.) *Travellers in the Near East.* 135-154. London: Stacey International.

Lewis, N.N. (ed.) 2007. A transcript of the Journals written or dictated by W.J. Bankes during his Journeys to and from Petra in 1818 with a Foreword, Introduction and Notes, a brief paper entitled The End of the Journey' and a commentary on 'Letter V' of Irby and Mangles' *Travels in Egypt and Nubia, Syria, and Asia Minor, Unpublished Mss in Dorset Record Office*. See notice by Lewis/ Macdonald 2008.

Lewis, N.N. 2007. The rediscovery of Petra, 1807-1818. In K.D. Politis (ed.) *The World of the Nabataeans*. 9-24. Stuttgart: Franz Steiner Verlag.

Lewis, N. N. (and M. A. Macdonald) 2008. A description of the first exploration of Petra Norman N. Lewis. *Syria* 85: 377-8.

Lindsay, A.C. (Lord) 1838 *Letters on Egypt, Edom, and the Holy Land*. 2 vols. London: Henry Colburn.

Llewellyn, B. 2017. Across cultures: Joseph Schranz and his associates in mid-19th century Constaninople. In G. Bonello (ed.) 2017. *Celebrating 200 years of Schranz*. 185-250. Valetta: Allied Publications Malta for Fondazzioni Patrimonju Malti.

Llorca-Jaña, M. 2012. The economic activities of a global merchant-banker in Chile: Huth & co. of London, 1820s-1850s. *Historia* (Santiago) 45.2: 399-432.

Lottin de Laval, P.V. 1855-1859. *Voyage dans la péninsule arabique du Sinaï et l'Egypte Moyenne. Histoire, géographie, épigraphie*. Paris: Gide et Cie.

Lushington, Mrs Charles [= Sarah Gascoyne Lushington] 1829. *Narrative of a Journey from Calcutta to Europe by Way of Egypt in the Years 1827 and 1828*. London: John Murray.

Macintyre, J. 1854. *Walks on Deck and Rambles on Shore During a Voyage of Circumnavigation of the Globe [1853]*. Papers of James J. Macintyre, National Library of Australia, Vol. IV (Add MS 41745). https://nla.gov.au/nla.obj-752644206/findingaid

Manginis, G. 2016. *Mount Sinai. A History of Travellers and Pilgrims*. London: Haus Publishing

Martineau, H. 1848. *Eastern Life, Present and Past*. 3 vols. Philadelphia and London: Moxon.

Melman, B. 1995. 'Domestic life in Palestine': Evangelical ethnography — faith and prejudice. In: *Women's Orients: English Women and the Middle East, 1718-1918*, 2nd ed., London: Palgrave Macmillan.

McCavitt, J. and George, C.T. 2016. *The Man Who Captured Washington: Major General Robert Ross and the War of 1812*. Norman: University of Oklahoma Press.

McKenzie, J.M. 1991. The Beduin at Petra: the historical sources. *Levant* 23: 139-145.

McKenzie, J.M. 1995. *The Architecture of Petra*. Oxford: Oxford University Press.

Mendel, Z. and Protasov, A. 2019. The entomofauna on Eucalyptus in Israel: A review. *European Journal of Entomology* 116: 450-460.

Miettunen, P. 2008. Jabal Harun: history, past explorations, monuments, and pilgrimages. In Z. Fiema and J. Frösén (eds) *Petra - The Mountain of Aaron. The Finnish Archaeological Project in Jordan. I. The Church and the Chapel*. Ch. 2: 27-49. Helsinki: Societas Scientiarum Fennica.

Milner-Gibson-Cullum, G. and Macaulay, F.C. 1906. *The Inscriptions in the Old British Cemetery of Leghorn*. Leghorn: Raffaello Giusti.

Mitchell, D. 1991. The 'New Woman' as Prometheus: Women artists depict women smoking. *Woman's Art Journal* 12. 1: 3-9.

Monk, C.J. 1851. *The Golden Horn; and Sketches in Asia Minor, Egypt, Syria, and the Hauran*. 2 vols. London: Bentley.

Murchison, R.I. 1868. Rev. Pierce Butler. *The Journal of the Royal Geographical Society of London*, 38: cxxxiii-cxviii.

Nance, S. 2009. The Ottoman Empire and the American flag: patriotic travel before the age of package tours, 1830–1870. *Journal of Tourism History* 1:1: 7-26.

Olin, S. 1843. *Travels in Egypt, Arabia Petraea and the Holy Land*. 2 vols. New York: Harper.

Oliver, A. 2014. *American Travellers on the Nile. Early U.S. Visitors to Egypt, 1774-1839.* Cairo/ New York: American University of Cairo Press.

Palmer, E.H. 1871a. The Desert of the Tih and the Country of Moab. *Palestine Exploration Fund Quarterly Statement* 3-76.

Palmer, E.H. 1871b. *Desert of the Exodus: Journeys on Foot in the Wilderness of the Forty Years' Wanderings.* 2 vols. Cambridge: Deighton, Bell & Co.

Palmer, E.H. 1874. Biographical sketch. In P. Butler *Axel and Valborg. A Tragedy in Five Acts and Other Poems,* trans from Danish, edited by Professor (E. H.) Palmer. v-ix. London: Trübner.

Palmer, W. 1861. *Egyptian Chronicles.* 2 vols. London: Longman Green.

Palmer, R. 1896. *Memorials. Part I. Family and Personal 1766-1865.* 2 vols. London: Macmillan.

Palmer, S.M. 1883. An adventure at Petra. *Macmillan's Magazine* 47 (January): 187-197.

Patterson, J. 1867. *Egypt and the Nile Considered as a Winter Resort for Pulmonary and other Invalids.* London: Churchill.

Peake, F.G. 1958. *A History of Jordan and its Tribes.* Miami: University of Miami Press.

Penner, P. 1987 *Robert Needham Cust, 1821-1909: A Personal Biography.* Lewiston, New York and Queenston, Ontario: Mellen Books..

P and O n.d. *P&O Heritage Fact Sheet: 'Bombay 1852'.* Pdf online.

Pococke, R. 1743. *A Description of the East, etc.* London: Printed by W. Boyer for J. and P. Knapton and Others.

Polk, W.H. 1912. *Polk Family and Kinsmen,* Louisville, Ky (Bradley & Gilbert Co)

Polk, W.M. 1915. *Leonidas Polk, Bishop and General.* 2 vols. New Edition. New York: Longmans, Green, and Co.

Polk, W.R. 2000. *Polk's Folly. An American Family History.* New York: Doubleday.

Porter, J. L. 1855. *Five Years in Damascus: Including an Account of the History, Topography, and Antiquities of that City; with Travels and Researches in Palmyra, Lebanon, and the Hauran.* 2 vols. London: Murray.

Powerscourt, Viscount 1903. *A Description and History of Powerscourt.* London: Mitchell and Hughes.

Quiney, A. 1995. 'Altogether a Capital Fellow and a Serious Fellow Too': A Brief Account of the Life and Work of Henry Woodyer, 1816-1896. *Architectural History* 38: 192-219.

Rhind, A.H. 1856. *Egypt, its Climate, Character and Resources as a Winter Resort.* Edinburgh: Constable.

Ridding, Lady Laura 1919. *Sophia Matilda Palmer: Comtesse De Franqueville, 1852-1915. A Memoir.* London: Murray.

Ridgaway, H.B. 1876. *The Lord's Land: a Narrative of Travels in Sinai, Arabia Petræa, and Palestine, from the Red Sea to the Entering in of Hamath.* New York: Nelson and Phillips.

Robinson, G. 1837. *Travels in Palestine and Syria,* 2 vols. London: Henry Colburn.

Roper, G.J. 1984. George Percy Badger (1815–1888). *British Society for Middle Eastern Studies Bulletin* 11: 140-55.

Roper, G.J. 2008. Badger, George Percy, (1815–1888). *Oxford Dictionary of National Biography* accessed online 25 February 2021.

Ross, D. (ed.) 1836. *Opinions of the European Press on the Eastern Question. Translated and Extracted from Turkish, German, French and English Papers and Reviews.* London: James Ridgway.

Ross, J. (ed.) 1902. *Letters from the East by Henry James Ross, 1837-1857.* London: Dent.

Saini, A. 2020. The harrowing tale of two British ships wrecked on the Andaman Islands in 1844. *The Hindu* (13 January). Accessed online 4 February 2021. https://www.thehindu.

com/society/history-and-culture/the-harrowing-tale-of-two-british-ships-wrecked-onan-andamans-island-in-1844/article30533448.ece

Sartre-Fauriat, A. 2021. *Aventuriers, voyageurs et savants. A la découverte archéologique de la Syrie (XVIIe-XXIe siècle)*. Paris: CNRS

Schranz, J.C and J.J. 2017. Four generations of Schranz artists. In G. Bonello (ed.) 2017. *Celebrating 200 years of Schranz*. 1-17. Valetta: Allied Publications Malta for Fondazzioni Patrimonju Malti.

Schranz, J.J. (2016-18) (A series of articles about the Schranz artists beginning on 26 May 2016 through till 22 July 2018), *The Times of Malta*. Accessed online 3 December 2021. https://timesofmalta.com/articles/view/Antonio-Schranz-and-the-Travelling-Artists-Grecian-Ideal.649282

Schranz, J.J. 2017. Antonio Schranz's wanderlust - Looking for his tracks in the sands of many deserts. In G. Bonello (ed.) 2017. *Celebrating 200 years of Schranz*. 135-184. Valetta: Allied Publications Malta for Fondazzioni Patrimonju Malti.

Sekunda, N. (ed.) 2007. *Corolla Cosmo Rodewald*. Gdansk: Gdansk University.

Seton-Karr, W.S. (n.d.) *Autobiography*. 3 vols. (unpublished manuscripts held in British Library Africa and Asia collection).

Smith, W.B. (1986) *America's Diplomats and Consuls of 1776-1865: A Geographic and Biographic Directory of the Foreign Service from the Declaration of Independence to the End of the Civil War*. Washington, DC: US Government Printing Office.

Sperling, D.C. 2004. Cattermole, George (1800-1868). *Oxford Dictionary of National Biography*. Accessed online 13 September 2021.

Stanley, A.P. 1844. *The Life and Correspondence of Thomas Arnold*. 2 vols. London: B. Fellowes.

An American [Stephens, J.L.] 1837. *Incidents of Travel in Egypt, Arabia Petraea and the Holy Land*. 2 vols. New York: Harper and Brothers

Stephens, H.M. 2004. Butler, James Armar (1827-1854). *Oxford Dictionary of National Biography*. Accessed online 17 March 2020.

Stewart, R.W. 1857. *Tent and the Khan. A Journey to Sinai and Palestine*. Edinburgh: W. Oliphant and Sons.

Stobart, H. 1855. *Egyptian Antiquities collected on a voyage made in Upper Egypt in the years 1854, 1855, and published by the Rev. H. Stobart, M A., Queen's College, Oxford*. Berlin: Värsch and Happe lithogr. fac. sim. under the direction of Dr. H. Brugsch.

Stockdale, N.L. 2006. Danger and the missionary enterprise: The murder of Miss Matilda Creasy. In H. Murre-van den Berg (ed.) *New Faith in Ancient Lands: Western Missions in the Middle East in the Nineteenth and Early Twentieth Centuries*. 113-132. Leiden: Brill.

Taylor, D. 2009. 'I must write'. *Carlyle Studies Annual* 25: 190-196.Top of Form

Taylor, D. 2015. *Top of Form'Under the Cedar' The Lushingtons of Pyports. A Victorian Family in Cobham - and elsewhere in Surrey*. Tolworth: Grosvenor House Publishing.

Taylor, D. 2020. *The Remarkable Lushington Family: Reformers, Pre-Raphaelites, Positivists, and the Bloomsbury Group*. London: Lexington Books

Taylor, J. 2001. *Petra and the Lost Kingdom of the Nabataeans*. London: I. B. Tauris.

Taylor, M.N. 1966. *The Journal of Ensign Best. 1837-1843*. Wellington, New Zealand: R. E. Owen, Government Printer.

Tupper, M.F. (ed.) 1856. *Out and "Home", With a Few other Memorials of the Late Rev. William George Tupper, M. A., Trinity College Oxford*. London: Bosworth and Harrison.

Turner, G. 2001. Sennacherib's Palace at Nineveh: The Drawings of H. A. Churchill and the Discoveries of H. J. Ross", *Iraq* 63: 107-138

Urquhart, D. 1838. *Spirit of the East: Illustrated in a Journal of Travels Through Roumeli in an Eventful Period.* 2 vols. London: Henry Colburn.

Van der Steen, E. 2013. *Near Eastern Tribal Societies during the Nineteenth Century. Economy, Society and Politics between Tent and Town.* Sheffield: Equinox.

Vincent, J.E. 1906 *Highways and Byways in Berkshire.* London: MacMillan and Co.

Wakeling, G. 1895. *The Oxford Church Movement. Sketches and Recollections.* London: Swan and Sonnenschein.

Ward, L. 2015. *The London County Council. Bomb Damage Maps. 1939-1956.* London: Thames and Hudson.

Warriner, H. 2022. *Francis Gerald Vessey's Journal. Rome, Egypt, Sinai and Palestine.* (Published Privately).

Waterfield, G. 1963. *Layard of Nineveh.* London: John Murray.

Waterson, C.D. and C. Macmillan Shearer (eds) 2006. *Former Fellows of the Royal Society of Edinburgh, 1783-2002. Biographical Index.* 2 vols. Edinburgh: Royal Society of Edinburgh.

Watson, C. 1909. *Life of Major-General Sir Charles Wilson, K.C.B., K.C.M.G.* London: John Murray.

Weeks, E.M. 2014 *Cultures Crossed: John Frederick Lewis and the Art of Orientalism,* The Paul Mellon Centre for Studies in British Art. London and New Haven: Yale University Press.

Wenning, R. 1990. Two forgotten Nabataean inscription. *ARAM* 2: 143-150.

Wheeler, R. 2006. *Palmer's Pilgrimage: The Life of William Palmer of Magdalen.* Pieterlen: Peter Lang AG.

Whitehead, D. 1998. David Ross of Bladensburg a nineteenth-century Ulsterman in the Mediterranean. *Hermathena* 164: 89-99.

Whitehead, D. 1999. From Smyrna to Stewartstown: A numismatist's epigraphic notebook. *Proceedings of the Royal Irish Academy* 99C.3: 73-113.

Whitehead, D. 2014. The epigraphical transcripts and travels of David Ross of Bladensburg. *Zeitschrift für Papyrologie und Epigraphik* 189: 159-174.

Wilkinson, G. 1843. *Modern Egypt and Thebes: Being a Description of Egypt; Including the Information Required for Travellers in that Country.* 2 vols. London: Murray.

Wilkinson, G. 1847. *Handbook for Travellers in Egypt.* London: Murray

Wilkinson, T. 1976. *Two Monsoons.* London: Duckworth.

Wilson, J. 1847. *The Lands of the Bible Visited and Described, in an Extensive Journey Undertaken with Special Reference to the Promotion of Biblical Research and the Advancement of the Cause of Philanthropy.* 2 vols. Edinburgh: Whyte.

Wood, R. 1757. *The Ruins of Balbec, Otherwise Heliopolis in Coelo-Syria.* London: Private Publication.

Wosford, A. 2020. A lone figure in the distance: James Graham and 19th century photography in Palestine. *Blog: Palestine Exploration Fund.* Accessed 25 July 2022. https://www.pef.org.uk/a-lone-figure-in-the-distance-james-graham-and-19th-century-photography-in-palestine/

Wright-St Clair, R.E. 2003. *Historia Nunc Vivat: Medical Practitioners in New Zealand 1840–1930,* [NZ]. Christchurch: University of Otago.

Yeatman, T.P. 1984. Side arm of a Confederate staff officer. *North South Trader* Vol. XI, No. 2 (Jan-Feb): 25-6.

Yeatman, T.P. 2000. *Frank and Jesse James.The Story Behind the Legend.* Nashville: Cumberland House.

Zayadine, F. 1974. Excavations at Petra (1973-1974). *Annual of the Department of Antiquities of Jordan* 19: 135-150.